Systems Engineering for Critical Infrastructure in a Cyber World

RIVER PUBLISHERS SERIES IN DIGITAL SECURITY AND FORENSICS

The "River Publishers Series in Security and Digital Forensics" is a series of comprehensive academic and professional books which focus on the theory and applications of Cyber Security, including Data Security, Mobile and Network Security, Cryptography and Digital Forensics. Topics in Prevention and Threat Management are also included in the scope of the book series, as are general business Standards in this domain.

Books published in the series include research monographs, edited volumes, handbooks and textbooks. The books provide professionals, researchers, educators, and advanced students in the field with an invaluable insight into the latest research and developments.

Topics covered in the series include:-

- Blockchain for secure transactions
- Cryptography
- Cyber Security
- Data and App Security
- Digital Forensics
- Hardware Security
- IoT Security
- Mobile Security
- Network Security
- Privacy
- Software Security
- Standardization
- Threat Management

For a list of other books in this series, visit www.riverpublishers.com

Systems Engineering for Critical Infrastructure in a Cyber World

Aleksandra Scalco
Department of Defense (DOD),
Defense Acquisition University (DAU),
USA

Steven Simske
Colorado State University,
USA

River Publishers

Routledge
Taylor & Francis Group
NEW YORK AND LONDON

Published 2025 by River Publishers
River Publishers
Broagervej 10, 9260 Gistrup, Denmark
www.riverpublishers.com

Distributed exclusively by Routledge
605 Third Avenue, New York, NY 10158, USA
4 Park Square, Milton Park, Abingdon, Oxon OX14 4RN

Systems Engineering for Critical Infrastructure in a Cyber World / Aleksandra Scalco and Steven Simske.

Routledge is an imprint of the Taylor & Francis Group, an informa business

ISBN 978-87-4380-911-1 (hardback)
ISBN 978-87-4380-916-6 (paperback)
ISBN 978-87-4380-913-5 (online)
ISBN 978-8-743-80912-8 (ebook master)
DOI: 10.1201/9788743809128

The views expressed in this book do not represent the views of Defense Acquisition University, the Department of Defense (DOD), or its components, or the United States.

The authors and publisher made every effort to ensure that the information was correct at time of publication. In some instances, change is underway. For example, as of September 2025, the U.S. DOD was designated as the Department of War through an executive order. The DOD reference is used in this book since that was the designation for most of the references used. Special care was taken to ensure information presented was obtained from publicly available sources and referenced in citations. The authors and publisher do not assume and hereby disclaim any liability to any party for any perceived loss, damage, or disruption that may be caused by errors or omissions, whether such errors or omissions are the result of oversight, negligence, accident, or any other cause. Any perceived slight of any individual or organization is purely unintentional. The resources provided in this book are based on research and are shared for informational purposes, and should not be used to replace any professional judgement, nor specialized training.

Contents

Preface

"A rising tide lifts all boats."
– John F. Kennedy (1917–1963),
president of the United States from 1961 to 1963

Abstract

This book offers a practical, inter-systems, systems engineering, and stakeholder alignment approach that system owners, stakeholders, and workforces can use to overcome cyber vulnerabilities arising within control systems in critical infrastructure. This book shares ideas and references to best practices, using composed solutions and approaches that can be immediately helpful to teams in successfully managing operations with a cyber interface. The goal is to ensure that by sharing knowledge, a rising tide will lift all boats to help achieve the desired outcome of overcoming cyber vulnerabilities in critical infrastructure environments using a disciplined systems engineering approach for fielding cybersecurity mitigations for industrial control systems (ICS) and supervisory control and data acquisition (SCADA) systems found in critical infrastructure. The objective is to define and demonstrate how to redefine enduring security concepts and processes using systems engineering practices and novel methods at an important level that facilitates necessary stakeholder alignment. Anyone who utilizes the concepts in these pages will find valuable tools to engage the entire team, from technicians to system designers to resource decision-makers. These tools help create alignment between existing resources and organizational and mission objectives. The contents of this book share practical applications using a disciplined, consistent, and tested systems engineering approach that can be useful to any reader, regardless of their varying perspectives, experience, level of expertise, and knowledge about the systems they must apply their talents to protect. Technical leaders can use the book to articulate to their team and key stakeholders what needs to be done to achieve success.

Systems engineering, SE, cyber, heating, ventilation, and air conditioning, HVAC, remote terminal unit, RTU, distributed control systems, DCS, industrial control systems, ICS, supervisory control and data acquisition, SCADA, programmable logic controller, PLC, human–machine interfaces, HMI, knowledge, skills, abilities, and tasks, KSATs

Introduction

If you have ever wondered where to start to help unlock a team's full potential to overcome cyber vulnerability in critical infrastructure, then this book may be a helpful resource. Technical and operational professionals who perform essential work activities in water and wastewater treatment facilities or building controls such as heating, ventilation, and air conditioning (HVAC) may otherwise rarely interact with offensive or defensive cyber security professionals on a day-to-day basis but may also find out that they are suddenly expected to be high-performing teammates leveraging the cyber medium[1]. Now, the water/wastewater treatment system operator with on-the-job training to monitor meters and gauges and specialized knowledge of the right amount of ammonia or chlorine chemicals to add for water disinfection and environmental co-treatment must use new software, systems, and applications that have been conceptualized, designed, and developed by computer scientists and engineers. At the same time, computer scientists and engineers are designing and developing applications and solutions to improve system efficiencies, monitor operating conditions, and, in the case of heating, ventilation, and air conditioning (HVAC) systems, control temperature, humidity, and air quality in buildings, often without firsthand mechanical expertise on the systems. These upgrades to systems interface critical infrastructure components and sensors with other technologies and systems to connect reliable, near-real-time data and to bring new analytics into the decision-making process in the cyber world. The systems used for monitoring and controlling industrial processes and devices consist of both hardware and software components. These components create an interface with the physics-based mechanical pumps and control valves. Remote terminal unit (RTU) components interface devices to distributed control systems (DCS),

[1] The authors use the term "cyber medium" to refer to the materials, such as hardware or software, used in the transfer of data from one source to another. The US Department of Defense (DOD) uses the term "cyber domain" to describe the DOD warfighting domain referring to a specific operational area where military forces engage in combat or strategic warfighting actions that uses cyber as a medium.

industrial control systems (ICS), or supervisory control and data acquisition (SCADA) systems, aggregating data using the protocol supported by the field device, collecting field data, and allowing the operator to perform and control system tasks. Programmable logic controllers (PLCs) are used as economical standard hardware solutions for SCADA RTU. The commonality of critical infrastructure systems in the cyber world lies in the interface between physics-based physical systems, electronic communication systems, the data component, and the collaborative integration of human–machine interfaces (HMI), which allow human input and expertise to direct the functions. Field I/O (input/output) devices, such as sensors and actuators, transfer data between physical control tasks and computers.

Decisions must be made by various stakeholders regarding critical infrastructure systems within municipal water and wastewater treatment facilities, as well as building facilities. These systems comprise devices that are controlled or instrumented to gather data, enabling the operator to remotely control the system. Challenges include interoperability of components, maintenance, integration, reliability, environmental conditions, and security. In a cyber world, critical infrastructure not only adds to the complexity but also introduces new cyber vulnerabilities and urgency to the challenges. It is possible to unlock everyone's understanding to develop impactful teams for critical infrastructure in a cyber world.

The objective of this book is to present enduring concepts (and meta-conceptual approaches akin to design patterns) that someone can apply to secure data monitoring and control processes through systems engineering practice and novel methods for necessary stakeholder alignment. Anyone utilizing the ideas in these pages will find valuable tools to engage the entire team, from technicians to system designers, software engineers to control system engineers, computer scientists to construction project managers, and resource decision-makers to quality and safety managers. This engagement, in turn, will create alignment from existing resources with organizational and mission objectives. We try not to focus solely on a United States-centric view, even though that is our home base. Approaches to asset ownership of critical infrastructure vary by jurisdiction, as does the industry's decomposition into sectors, policies, and governance. However, many examples are based on efforts from the United States. The systems engineering approach is generally similar across countries in terms of engineering principles and practices.

It is necessary to address critical infrastructure in an increasingly cyber world because mission functions depend on the underlying infrastructure. A data center needs power distribution to function. The business applications running in a data center will only function with power distribution to the servers located in the facility, regardless of the source of power. However, physical process controls can operate without a cyber interface. Mechanical processes have been in use for a long time, even before the internet emerged, prior to the 1990s. The systems must function as expected during any distributed power disruptions, using some backup power generation, regardless of the cause of the power failure. The bottom line is that the system needs power. Cyber vulnerabilities can lead to disruptions in the power supply, potentially resulting in system failure. Critical infrastructure in a cyber world is vulnerable to cyber-induced failure, just like any other type of cause. Industrial internet adoption (or revolution) has been quick to adopt innovative technologies, yet slow to change. What does that mean? It means that as system stakeholders added new technology capabilities to critical infrastructure industries, such as remote access monitoring and control, the associated workforce, management, and operations processes did not change at the same pace. The lack of upgrades to older equipment in operations that have been in place for 30 years or more has not kept pace with the efficiencies gained in business operations in the new cyber world. Where that leaves people working in the critical infrastructure sector is a boatload of challenges, including retrofitting older equipment with the latest technologies, a push to overcome skills and labor gaps, the need to rapidly invest in new construction, and the need to create successful teams that can successfully overcome all these challenges. In many ways, the critical infrastructure sector is the "front line" for cyber-physical security, and so addressing its exigencies – the purpose of this book – is readily repurposed to other formerly "offline" systems that will increasingly move online to cyber-physical systems.

Key Concepts

The perspective of this book extends beyond the scope of national policies, guidelines, instructions, and directives. Readers may follow the best option to meet unhampered organizational objectives and any constraints readers may have (such as statutes or laws as appropriate by the country in which they operate) without direct, explicit instructions. The authors presume readers will follow the prescribed models based on their country of residence. While references are provided to offer useful insights into best and emerging practices, the content is generic enough to be generally helpful to anyone looking to improve their team's performance when working with systems that have cyber interfaces.

How can we simplify the approach to addressing the numerous variations and complexities of cyber vulnerabilities found in critical infrastructure industrial control systems (ICS) and supervisory control and data acquisition (SCADA)? This book uses three key conceptual frameworks: inter-system, systems engineering, and stakeholder alignment.

An inter-system framework is one in which applications are hosted across varying locations connected to use or share the same data and information across resources. Inter-system scenarios exist between two or more systems, such as when multiple subsystems are integrated to function for a coordinated, desired output. The nature of inter-system scenarios given cyber interfaces is that "a rising tide lifts all boats," an adage often attributed to the United States' 35th President, John F. Kennedy, due to his 1963 speech about the shared defense of Europe. The idea was that the shared defense of Europe would be good for all – the success of one benefits all those around. Similarly, in today's cyber-connected world, an improved cybersecurity posture, defensive measures, and shared situational awareness of cyber threats and vulnerabilities lead to a body of knowledge that someone can directly apply to improve cybersecurity globally. We are interconnected. Systems are inter-systems.

Inter-system

Most systems in today's critical infrastructure involve integrating various systems, sub-systems, and system components. The first conceptual framework is inter-system, essentially referring to the communication protocols between two devices. This communication can represent the data flow of a process or a process control system used in operations to manage a physical process control device, such as a sensor, controller, or actuator. There is an important distinction between data flow architectures and process control systems. The data flow architectures may maintain the data within a single, intra-system application, or the data flows may traverse between connected, inter-system scenarios. When data is used across coexisting resources, as found in critical infrastructure, it is inter-system and requires inter-system testing or end-to-end system testing for interoperability. Regardless of how the data strategy is implemented, it relies on technology, people, and processes. The interface functions in communication systems and software have a significant impact on the system and are critical to the design and development process, which are essential elements to consider in any system. The Zero Trust security model is a data, access control, and authentication strategy that defines the services, methods, architectures, technologies, processes, people, and rules that support an organization's decision-making process and data management. The Zero Trust security model is discussed in more detail in Chapter 16, "Modernization Strategies" (John Kindervag, 2012). Control systems found in critical infrastructure underpin all aspects of an information enterprise. These systems and the operational assets they manage must be resilient to cyber vulnerabilities. The process control system involves controls (analog in/out, digital in/out), fieldwork management interactions, and troubleshooting equipment and processes. Process control systems can be highly automated. These systems should be used in a manner that ensures the health and safety of humans. The process architecture is an abstraction of the field system processes used to move the physical world by producing something, such as a movement function or an output function. Process control systems are systems that gather and process near real-time data transmitted from sensors and control processes that regulate values, pumps, and actuators, which produce force, torque, or displacement, thereby moving or operating something to perform a specific function and establishing a higher level of control within inter-systems. Organizations can apply a Zero Trust strategy to both traditional and non-traditional IT systems. We will use the term "Operational Technology (OT)," as coined by Gartner, to refer to these systems (Perkins et al., 2016). Gartner defines Operational Technology (OT) as "hardware and software that detects or causes a change through direct monitoring and control of industrial equipment, assets, processes, and events" (Gartner, 2024). Cyber vulnerabilities found in hardware are discussed in Chapter 6, "Cyber Vulnerabilities Start at the Hardware Level," and software approaches to addressing cyber vulnerabilities in critical infrastructure systems are explored in Chapter 18, "Software Approaches."

Achieving a state of shared situational awareness about potential cyberattacks before a cyber event requires a comprehensive inter-system approach. The integration functions between systems using common data sources must be sharable, consistent, accurate, and have an understandable mental model of what is happening. It is essential to understand the baseline architecture of each system, including its functional, physical, and allocated components, as well as its interfaces and interconnections. There are many resources available to better understand these models, which leads to another key conceptual framework: systems engineering.

Systems engineering

The second conceptual framework involves the application of systems engineering principles and practices. Systems engineering is an interdisciplinary field focused on the development and oversight of complex system concepts, architecture, and design, as well as the subsequent lifecycle management of products and systems that support the realization of solutions to complex problems and challenges. The evolution of systems engineering practices can be traced back to the Bell Telephone Laboratories as early as the 1900s, with significant applications of the systems approach to systems engineering described in use during World War II (Buede, 2009). The US Department of Defense (DOD) later adopted systems engineering practice to develop missiles and missile defense systems. Systems engineering relies heavily on models throughout the system development and lifecycle process, from initially defining the actual design problem to articulating the design problem into specific high-level mission statements of what needs to be accomplished to well-defined, derived, detailed requirements about what the system will do across every relevant phase of the system's lifecycle. The systematic approach of systems engineering provides a method for defining, verifying, validating, and testing the system or system-of-systems against the requirements to achieve stakeholder acceptance. Due to the heavy reliance on models, digital engineering is a key component of system engineering.

Stakeholder alignment

The third conceptual framework is stakeholder alignment. Stakeholder alignment is the process of gathering valuable input and achieving consensus among individuals with a vested interest in system decisions and outcomes. Stakeholder misalignment may be greater than innate system vulnerabilities (Scalco, 2022). In cybersecurity, stakeholder misalignment can lead to a system that is less secure than any of the individual stakeholder systems, for example, installing two anti-malware systems that treat each other as malware.

Central Theme and Navigation Aids

The central theme of this book is that it is possible to create successful teams that know how to work together to overcome the complexity of vulnerability presented in the cyber medium. We can think of "cyberspace" much like Lewis Carroll's "Alice's Adventures in Wonderland" – a strange, confusing world of contradictory qualities despite valid reasoning – keep data in a system secure (behind a locked door) but also ensure the data is available and accessible (so Alice can get to it) (Carroll, 1865). In the physics-based world, we want to regulate the granularity of material at the appropriate level and secure access to physical process control areas. Problems occur whenever there is change, such as in "The Adventures of Alice in Wonderland" (when Alice grows too large or becomes too small) (Carroll, 1865).

Navigational aids are found throughout the content and are broken down into the knowledge, skills, abilities, and tasks (KSATs) the team needs to be successful when using cyber as a medium. Two work roles of interest, as found in the DOD Cyber Exchange, are used to illustrate these Knowledge, Skills, Abilities, and Tasks (KSATs): Control Systems Security Specialist (Work Role ID: 462) and Cyber Defense Infrastructure Support Specialist (Work Role ID: 521/NIST: PR-INF-001). The DOD Cyber Exchange is a publicly available resource that provides a lexicon for individual performance, helping guide better teams (DOD, 2023). The resource can be found at the following link: https://public.cyber.mil/wid/dcwf/. These navigational aids enable readers to track the content of this book in relation to knowledge, skills, abilities, and tasks (KSAT) elements. The authors refer to these KSATs as a navigational aid to help readers navigate their way to an impactful team. These KSATs, or "breadcrumbs," enable the reader to better understand what is required to be an effective cybersecurity team member.

1. The Control Systems Security Specialist (Work Role ID: 462) is "[r]esponsible for device, equipment, and system-level cybersecurity configuration and day-to-day security operations of control systems, including security monitoring and maintenance along with stakeholder coordination to ensure the system and its interconnections are secure in support of mission operations" (DOD, 2023).
2. The Cyber Defense Infrastructure Support Specialist (Work Role ID: 521/NIST: PR-INF-001) "[t]ests, implements, deploys, maintains, and administers the infrastructure hardware and software" (DOD, 2023).

The book is organized with systems engineering practice and principles in mind. Most chapters open with a forward, or introductory analogy, based on either team sports or literature to illustrate the ideas, such as found in Antoine de Saint-Exupéry's "The Little Prince." As another example, the game of ice hockey is used to illustrate the systems engineering practice and principles used to address the challenges of cyber interfaces. In ice hockey, the shooting game requires accuracy, speed, and a quick release to put the puck past the goalie and into the net, evading defensive players. Players need skating skills, balance, and control of skate edges on the ice. The best players make stick handling look easy, keeping control of the puck close to their bodies. The players need a solid physical presence, intensity, cardiovascular endurance, strength, and a sense of the game to anticipate where the puck will be next. In the cyber-connected world, these skills are similar. What do these two types of topics have in common? They allow the reader to relate cybersecurity and control system cyber-physical security to the real world, an association known to be helpful in addressing the wide range of vulnerabilities in systems.

Organization of the Book

This book is organized into three parts, following the book's key concepts described above – inter-systems, systems engineering, and stakeholder alignment – to guide practitioners through systems engineering and cybersecurity mitigation considerations for critical infrastructure industrial control systems. Each chapter gives insight for practical application.

Part I – Inter-system communications and networking

Part I – The "Inter-system communications and networking" approach is a dialectical meta-framework to facilitate the exploration of cybersecurity principles. Part I provides an overview of the operational need for control systems, the problem that needs to be addressed, and a summary of some of the modernization strategies that can be applied to critical infrastructure in a cyber world.

Part II – Systems engineering methods and practice

Part II – "Systems engineering methods and practice" describes how systems engineering principles and practices are used to solve the problem of integrating cybersecurity into control systems. The section describes the foundations of systems engineering and security engineering. It includes a case study on fielding a cybersecurity defensive capability for control systems used in commercial or government-owned systems. This section provides an example of capability implementation through the systems engineering lifecycle, from concept exploration and development to requirements engineering, systems architecture, system design and development, integration phases, and finally, engineering and post-development transition to operations and maintenance.

Part III – Stakeholder alignment

Part III – "Stakeholder alignment" presents a thorough consideration of determining misalignment, disagreement, vulnerability, and likely foci of exploitation, which is more comprehensive than previously published. Cyber introduces new capabilities to control systems. System vulnerability creates uncertainty and a lack of agreement among professionals about the use of a cyber interface. Such disagreement leads to misalignment, which in turn leads to vulnerability. The exposure induced by disagreement and misalignment may be greater than the innate system design vulnerability. Previously published research by the authors presents an approach to measuring variance within segments of the cybersecurity profession, aiming to identify vulnerabilities. This section describes how to align stakeholders with the threats outlined in Part I, "Inter-system communications and networking," following the systems engineering methodology presented in Part II, "Systems engineering methods and practice." It provides more than 100 questions that can serve as a launching point for a questionnaire survey tool, as outlined in Appendix A: "Stakeholder Alignment Survey Questions." Each previous chapter addresses questions that align with the stakeholder alignment questionnaire tool presented in Appendix A: "Stakeholder Alignment Survey Questions." Hence, everyone can understand the problem-solving approach and logic behind each question, including how the questions relate to critical infrastructure in a cyber world.

Part IV – Conclusion and questionnaire resource

Part IV – "Conclusion and questionnaire resource" summarizes how the practical, inter-system, and systems engineering framework found in this book may support owners, stakeholders, and workforce of an organization looking to integrate solutions into operations to overcome cyber vulnerabilities in critical infrastructure control systems, regardless of the strategy, whether by New Product Introduction (NPI), which includes new technology and integration solutions, or training and "upskilling" the workforce. Appendix A: Questionnaire Resource is offered at the end of this book. This book shares practical applications using a disciplined, consistent, and tested systems engineering approach that can be useful to any reader, regardless of their varying perspectives, experience, level of expertise, and knowledge about a given system.

Who is the Book For?

This book is for anyone addressing critical infrastructure or other systems moving increasingly to a cyber-physical reality. This book is also for anyone who wants to build a better team in a cyber-connected world. This book is for people who find a small key to a door in a seemingly mad cyber world where the physics of time and space are relative (think of quantum computing), and we are either too small or too large to fit through a passageway to get to a place that seems like where we might want to be. It is the cyber paradox. It seems illogical to have contradictory qualities despite valid reasoning. Keep data in a secure system (behind a locked door) while ensuring the data is available and accessible (so "Alice," a professional accessing data using the cyber medium can get to it) and ensure variables such as the measures and output are regulated to optimal performance (just the right size for fit and function). The question is often asked: Where do I start? Where do I start as a system owner to assemble the right team to protect my system from cyber vulnerabilities? As a manager, whom do I hire with the right skills to defend my networks, keep my data safe, and harden my systems against cyberattacks? As a professional, what do I need to stay relevant? As a student, how do I get into the game? We connect technical experts and system operators with cybersecurity practitioners and systems engineers, providing everyone with access to systems engineering expertise to gain a deeper understanding of cybersecurity. This book is also for anyone who just wants to know where to start and what to do next, given the complexity cyber presents.

This book provides an inter-systems and systems engineering framework for system owners and control system stakeholders, using a disciplined approach for fielding and operating cybersecurity mitigations in critical infrastructure. The objective is to present enduring security concepts and processes using systems engineering practice and novel methods for necessary stakeholder alignment. For an organization to be successful, everyone from a cross-functional team needs to be involved throughout the lifecycle of cybersecurity initiatives to be successful, from the Chief Information Security Officer (CISO) setting the course of the strategy to the Chief Information Officer (CIO) guiding the team responsible for making it so 24 hours, seven days a week (24/7) — the infrastructure and operations teams, enterprise-level architecture teams, applications teams, and the security and risk management team. This book is for professionals who are students of the game. As in ice hockey, you need to know how to skate first to get into the game. Let us lace up and get on the rink.

References

Buede, D. M. 2009. *The engineering design of systems: models and methods,* Hoboken, New Jersey, John Wiley & Sons, Inc.

Carroll, L. 1865. *Alice's Adventures in Wonderland,* London, England, Macmillan Publishers.

DOD. 2023. *DoD Cyber Workforce Framework* [Online]. DOD Cyber Exchange Public: Department of Defense (DOD). Available: https://

public.cyber.mil/dcwf-work-role/control-systems-security-specialist/ [Accessed 2023].

Gartner. 2024. *Operational Technology* [Online]. www.gartner.com: Gartner. Available: https://www.gartner.com/reviews/market/operational-technology-security [Accessed 2024].

John Kindervag, S.b., Kelley Mak 2012. Build Security Into Your Network's DNA: The Zero Trust Network Architecture. Forrester: Forrester.

Perkins, E., Rueb, D., Moran, M., Steenstrup, K. & Zimmerman, T. 2016. Predicts 2017: IT and OT Convergence Will Create New Challenges and Opportunities. Gartner.

Saint-Exupéry, A. D. 1943. *The Little Prince,* New York, Reynal & Hitchcock.

Scalco, A. 2022. Measuring Disagreement in Segments of the Cybersecurity Profession as a Means of Identifying Vulnerabilities, Colorado State University.

List of Figures

List of Tables

List of Abbreviations

ABAC	Attribute-based access control
ACI TTP	Advanced cyber industrial control system tactics, techniques, and procedures
AI	Artificial intelligence
AMI	Advanced metering infrastructure
AoA	Analysis of alternative
API	Application programming interface
APPI	Act on the Protection of Personal Information
APT	Advanced persistent threat
ASIL	Automotive safety integrity level
ATM	Automated teller machine
ATO	Authority to operate
AWS	Amazon Web Services
BAN	Building area network
BIOS	Basic input/output system
CA	Certificate authority
CaaS	Cloud as a service
CAGR	Compounded annual growth rate
CAPTCHA	Completely automated public turing test to tell computers and humans apart
CAS	Chinese Academy of Sciences
CCB	Control configuration boards
CCP	Chinese Communist Party
CDD	Capability development document
CDP	Capability development plan
CDPF	Cyber Defence Policy Framework
CISA	Cyber Infrastructure Security Agency
CISO	Chief information security officer
CM	Configuration management
CND	Computer network defense
CNE	Computer network exploitation
COA	Courses of action

COTS	Commercial-off-the-shelf
CPD	Capability production document
CPS	Cyber-physical system
CPU	Central processing unit
CRAM	Cyber Resilient Assessment Model
CRISP	Cybersecurity Risk Information Sharing Program
CRM	Continuous risk management
CRM	Customer relationship management
CS	Control system
CS-CIDC	Context-sensitive critical infrastructure dynamic classes
CSDP	Common security and defense policy
CSPE	Control Systems Platform Enclave
CSS	Central security service
CSU	Colorado State University
CTE	Critical technology elements
CWS	Community water system
DAA	Designated approval authority
DBS	Dynamic biometrics system
DCIP	Defense Critical Infrastructure Program
DCS	Distributed control system
DDoS	distributed denial-of-service
DE	Digital engineering
DFA	Deterministic Finite Automata
DfM	Design for maintainability
DfM	Design for manufacturing
DfR	Design for reliability
DHS	Department of Homeland Security
DIB	Defense Industrial Base
DiD	Defense-in-depth
DLP	Data loss prevention
DMZ	Demilitarized zone
DOD	Department of Defense
DoS	Denial of service
DOTMLPF-P	Doctrine, Organization, Training, Materiel, Leadership, Education, Personnel, Facilities, and Policy
DR	Disaster recovery
DRAM	Dynamic random access memory
DRM	Data rights management
ECG	Electrocardiogram
ECP	Engineering change proposal

ECRA	European Cyber Resilience Act
EFS	Encrypting file systems
EMS	Electronic manufacturing services
EPA	Environmental Protection Agency
ERP	Enterprise resource planning
EU	European Union
EU-GDPR	European General Data Protection Regulation
EV	Electronic vehicle
FAN	Field area network
FBI	Federal Bureau of Investigation
FDE	Full disk encryption
FFBD	Functional flow block diagrams
FIA	Fédération Internationale de l'Automobile
FMC	Fully mission-capable
FMECA	Failure modes, effects, and criticality analysis
FMS	Foreign military sales
FRCS	Facility-related control systems
FTA	Fault tree analysis
GAO	Government Accountability Office
GOTS	Government off-the-shelf
GPU	Graphics processing unit
GSA	General services administration
GTE	Grand Touring Endurance
GUI	Graphical user interface
GWAC	Government-wide acquisition contract
HAN	Home area network
HART	Highway addressable remote transducer
HHS	Health and Human Services
HICV	Human-introduced cyber vulnerabilities
HIPAA	Health Insurance Portability and Accountability Act
HMI	Human–machine interface
HP	Hewlett-Packard
HR	Human resources
HVAC	Heating, ventilation, and air conditioning
I/O	Input or output
IA	Information assurance
IACD	Integrated adaptive cyber defense
IACS	Industrial Automation and Control Systems
IC	Integrated circuit
ICD	Initial capabilities document

ICS	Industrial control systems
IDD	Interface design descriptions
IDS	Intrusion detection system
IEC	International Electrotechnical Commission
IEEE	Institute of Electrical and Electronic Engineers
IIoT	Industrial Internet of Things
IMP	Integrated master plan
IMS	Integrated master schedule
IMT	Integrated management team
INL	Idaho National Laboratory
IOA	Indicators of attack
IOC	Indicators of compromise
IoT	Internet of Things
IPDS	Intrusion detection/protection system
IPO	Initial public offering
IPT	Integrated product team
IR	Infrared
IRB	Institutional Review Board
IRR	Internal rate of return
ISA	International Society of Automation
ISAC	Information Sharing and Analysis Center
IT	Information Technology
IV&V	Independent verification and validation
J-BASICS	Joint base architecture for secure industrial control systems
JCTD	Joint Capability Technology Demonstration
JHU APL	Johns Hopkins University Applied Physics Laboratory
JIT	Just-in-time
JTANICS	Joint Threat Assessment and Negation for Installation Infrastructure Control Systems
JWAG	Joint Warfighter Advisory Group
KPP	Key performance parameters
KSA	Key system attribute
KSAT	Knowledge, skills, abilities, and task
LAN	Local area network
LCA	Life cycle analysis
LD	Ladder diagram
LE	Likert Entropy
LIDAR	Light detection and ranging
LMP	Le Mans prototypes
LOE	Line of effort
MAC	Mandatory access control

MAC	Media access control
MBSAP	Model-based systems architecture practice
MBSE	Model-based systems engineering
MES	Manufacturing execution systems
MFA	Multi-factor authentication
MIP	Mixed integer programming
ML	Machine learning
MNA	Multi-network authentication
MOE	Measure of effectiveness
MOSAICS	More situational awareness for industrial control systems
MTBF	Mean-time-between-failure
MTBM	Mean-time-between-maintenance
MTTD	Mean time to detect
MTTR	Mean time to repair
MTTR	Mean time to respond
MUA	Military utility assessment
NASA	National Aeronautics and Space Administration
NATO	North Atlantic Treaty Organization
NCCC	National Cyber Coordination Center
NCSC	National Cyber Security Center
NFA	Nondeterministic Finite Automata
NIST	National Institute of Standards and Technology
NPI	New product introduction
NRLM	Network-resident latent malware
NSA	National Security Agency
NSM	National Security Memorandum
NTNCWS	Non-transient non-community water system
OEM	Original equipment manufacturer
OJT	On-the-job training
OMB	Office of Management and Budget
OO	Object-oriented
OOP	Object-oriented programming
ORD	Operational requirements document
OS	Operating system
OSD	Office of the Secretary of Defense
OSI	Open Systems Interconnection
OT	Operational Technology
OT-SDN	Operational technology software-defined networking
OV	Operational view
PC	Personal computer
PCBA	Printed circuit board assembly

PCI DSS	Payment Card Industry Data Security Standard
PCU	Power consumption unit
PERA	Purdue Enterprise Reference Architecture
PII	Personally identifiable information
PIT	Platform information technology
PKI	Public key infrastructure
PLA	People's Liberation Army
PLA	Programmable logic array
PLC	Programmable logic controller
PM	Project manager
PNNL	Pacific Northwest National Laboratory
POC	Points of contact
PPR	Person-to-population ratio
PRA	Probabilistic risk analysis
PRC	People's Republic of China
QRT	Quick reaction test
RAID	Redundant array of independent disk
RAM	Random access memory
RDP	Remote desktop protocol
RDT&E	Research, development, test, and evaluation
RMF	Risk Management Framework
RPI	Rensselaer Polytechnic Institute
RTU	Remote terminal unit
SAS	Substation automation system
SCDA	Supervisory control and data acquisition
SDN	Software-defined networking
SDP	Software-defined perimeter
SEMP	Systems engineering management plan
SES	Senior Executive Service
SIEM	Security information and event management
SME	Subject matter expert
SMT	Surface mount technology
SNL	Sandia National Laboratory
SNR	Signal-to-noise ratio
SOA	Service-oriented architecture
SOAR	Security orchestration, automation, and response
SOC	Security operations center
SoS	System of systems
SOW	Statement of work
SPIDERS	Smart Power Infrastructure Demonstration for Energy Reliability and Security

SPOT	Smart personal object technology
SSD	Solid-state drive
SSL	Secure sockets layer
STIG	Security technical implementation guide
STIPL	Science and technology integrated priority list
SysML	Systems modeling language
TCP	Transmission control protocol
TDS	Technology development strategy
TEMP	Test and evaluation master plan
TMOT	Test management oversight team
TNCWS	Transient non-community water system
TPM	Technical performance measure
TPM	Trusted platform module
TRL	Technology readiness level
TSA	Transportation Security Administration
TTP	Tactic, technique, and procedure
TWH	Technical warrant holder
UEBA	User and entity behavior analytics
UEFI	Unified extensible firmware interface
UFC	Unified Facilities Criteria
UMC	University of Missouri in Columbia
UML	Unified modeling language
UNECE	United Nations Economic Commission for Europe
USB	Universal serial bus
USSR	Union of Soviet Socialist Republics
VCG	Vectorcardiogram
VLAN	Virtual local area network
VM	Virtual machine
VoIP	Voice over internet protocol
VPN	Virtual private network
VRF	Virtual routing and forwarding
VTS	Vessel traffic system
WAN	Wide area network
WBS	Work breakdown structure
WEF	World Economic Forum
WLAN	Wireless local area network
WPA	Wi-Fi-protected access
WWW	World wide web

1

Understanding the Game

"A good hockey player plays where the puck is. A great hockey player plays where the puck is going to be."
– Wayne Gretzky (1961–), Canadian ice hockey player and coach

Abstract

A metaphor for the art of cybersecurity can be drawn from the game of ice hockey. The most important phase of responding to an incident is preparing for it before it happens. The same is true for ice hockey. A great player always has a sense of where the puck will be, not just where it is at the moment. Tabletop exercises are run for both offensive and defensive practice of cyber knowledge, skills, and abilities, just as drills are run for practice in ice hockey. Essential skills are required, and effective organizational strategies must be understood and applied. Understanding the game is essential for comprehending the systems of systems inter-system meta-framework from both an engineering and a stakeholder perspective. All cross-functional roles need to be concerned with potential system vulnerabilities, such as social engineering, worm infections, insider threats, phishing, and ransomware. Cross-functional teams in critical infrastructure sectors also need to understand control system field devices, practical approaches for securing physical network components using the cyber medium, critical thinking, and Zero Trust security model strategies and principles. All of these skills come together in network defense, control system field device security, and network support and security. It just takes one successful malicious actor to score against the most effective cyber-defensive operations. Every goal in ice hockey is worth the same value of one point. There are no extra points to gain after scoring a shot – just one single point. Cyber threat management, like ice hockey, is a team sport. One single breach is all it takes for the threat agent to score.

DOI: 10.1201/9788743809128-1

1.1 Introduction

The blue lines comprise the most strategic zone in ice hockey. The lines divide the ice rink into three zones. The defensive zone is where the goal is defended; the neutral zone is between the defensive and offensive zones, and the offensive zone is where the opposing team's goal is located. Possession inside the offensive zone is a dangerous situation. Getting inside the opponent's blue line zone gains the attacker a scoring chance. As sports spectators, we can understand what needs to be defended because we can see the goal and simultaneously recognize when there is an attack on it. There is often a lack of clarity at the same level about what to protect from cyberattacks when considering a critical infrastructure entity. However, just as in ice hockey, it just takes one successful strike. Every goal in ice hockey is worth the same value of one point – no additional points for longer-distance shots on goal. There are no extra points to gain after scoring a shot – just one single point. The objective is to score more points than the opposing team. There are three 20-minute periods in each game. Players can take a shot at the goal from anywhere on the ice, as long as they are not ahead of the hockey puck when taking the shot (i.e., they are not offside). The goal is scored if the entire puck crosses the goal line. The player uses their stick to hit the puck only when it is in contact with the ice. Stickhandling is key to controlling the puck. Doing it well is a strategic advantage.

A point in ice hockey can be scored if it is deflected off a player, including the goalie – we can refer to this as an "insider threat." In the cyber-connected world, a Zero Trust security model means that no one is trusted inside or outside a network by default, limiting an attacker's chances of scoring either by direct shots on goal or by deflection off your teammate, such as an unintentional insider who leverages susceptibility to exploitation. John Kindervag and his colleagues at Forrester Research first introduced the idea of a Zero Trust security model in 2012 in the paper "Build Security Into Your Network's DNA: The Zero Trust Network Architecture" (John Kindervag, 2012). The DOD Chief Information Office for Cybersecurity (DOD/CS) Senior Advisor and Director of the Zero Trust Portfolio Management Office, Randy Resnick, championed the Zero Trust strategy and the first DOD Zero Trust Reference Architecture as a paradigm shift in the philosophy of how infrastructure, networks, and data are secured within the DOD (DOD, 2024). The key tenet of Zero Trust is that everything and anything attempting to establish access to a system, whether from within or outside the system perimeter, must be verified. The Zero Trust approach involves continually verifying everything and anything that touches the system, which lends itself to addressing both physical and cyber threats (DISA et al., 2021). The Zero Trust security model is revisited in Chapter 9, "Road to Commoditization," and in Chapter 16, "Modernization Strategies."

1.2 Advanced Persistent Threats (APT)

Cyber threats generated by phishing attacks, coupled with an advanced persistent threat (APT) payload, can expose overlooked attack vectors. These adversarial tactics make detection considerably more difficult. Using APTs on critical infrastructure systems to achieve long-term objectives gives adversaries time and, consequently, an advantage in exploiting conditions that remain undetected for extended periods. Industrial control system (ICS) operators unfamiliar with the effects of a cyber event on their systems – and how an attack on one component may present another risk – can overlook areas where the system is susceptible to exploitation (Terry Merz, 2019). Traditional security network models rely on a single access control layer at the perimeter, described by Kindervag as a "crunch outside, chewy center" perimeter. Every organizational segment has a blue line game in a Zero Trust environment. The adversary aims to score a point by taking shots on goal – the "Crown Jewels." The Crown Jewels are the most critical systems and most sensitive data. It just takes one single point to win the game.

A power play in ice hockey occurs when one or more players are sent to the penalty box, leaving the penalized team shorthanded for a specific amount of time until the player(s) are allowed back into the game, which gives the opposing team a power play – a distinct scoring advantage. Using cyber, the opponent always has a power play, and the game never stops. There are no periods. There are no 15-minute intermissions between periods. Cyber is all overtime, shootouts, and sometimes injuries. No timeouts. The key to this game is to know your organization's system's crown jewels," to mind your blue lines, and to keep the shots out of the goal.

Remember, it only takes one point of entry, one successful exploit of a vulnerability, for the threat agent to compromise a system. It is easier to know your system's "crown jewels" with access to the physical network topography. Still, access to the physical network topography also makes it easier for an adversary to know how to access the "crown jewels" that need to be defended. You also want to ensure that "critical assets related to control systems in the organization have been identified (*yes, no, I do not know*)" and are "well-defined (*yes, no, I do not know*)," and that the "single points of failure are known (*yes, no, I do not know*)" (Scalco, 2022). The blue line is the most strategic zone, and allowing the opponent to enter the blue line zone gives the attacker a scoring chance. Deny by default. It takes practice. It involves ensuring that "defensive countermeasures are in place to detect threat activity (*yes, no, I do not know*)" (Scalco, 2022). It takes the ability to handle a stick and a puck. And the ability to skate. Ice hockey is played on ice. You have to skate – and usually rather well – before you can play. The need for the ability to skate makes ice hockey a uniquely challenging sport because you cannot even begin to master the skills of the game if you have not mastered skating on the ice. Mastery of critical infrastructure connected via a cyber medium requires knowledge about how vendors can connect to the system, an understanding of the system's network topology, and awareness of when the systems can be accessed and by whom. Once you can skate within the infrastructure, you can then move to protect it.

Questions about vendors' authorization to connect to the network using a Zero Trust strategy might include: "Vendors remotely support the network (*Yes, No, I do not know*)"; "Vendors directly connect their equipment to the network (*Yes, No, I do not know*)," or "Vendors can directly update their software over the network (*Yes, No, I do not know*)" (Scalco and Simske, 2022). Examples of suggested questions to help guide prioritization efforts and better understand the facilities and infrastructure used in operations are shared in the reference section of this book, specifically in Appendix A, "Stakeholder Alignment Survey Questions" (Scalco and Simske, 2022).

1.3 Programming Knowledge

Similarly, you cannot even begin to defend a cyber system without basic assembly programming knowledge, such as Ladder Logic. Unique knowledge of critical infrastructure components, such as programmable logic controllers (PLCs), embedded systems, and sensors, is as important as understanding IT firewalls and network segmentation. You need to develop programming skills – or someone on the team must possess these skills. Most programmable logic controllers (PLCs) can be programmed using ladder logic, a simple visual language. Still, it may not be very intuitive for programmers familiar with languages like C, JavaScript, or Python. Most control-level networks interface with peer-to-peer networks, providing an interface between controllers such as programmable logic controllers (PLCs), distributed control systems (DCS), human–machine interfaces (HMI), historical archiving, and supervisory control and data acquisition (SCADA) systems to coordinate and synchronize control between nodes. Programmable logic controllers (PLCs) are computer-based, solid-state devices that control industrial equipment and processes (Stouffer et al., 2023), as illustrated in Figure 1.1. As an example,

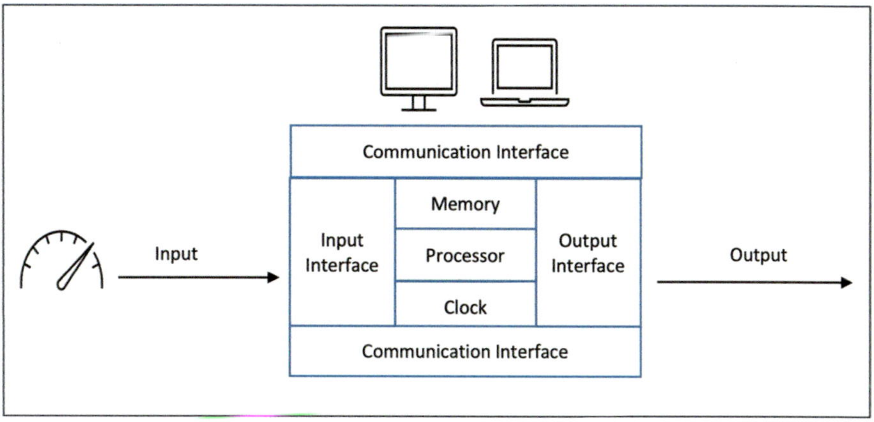

Figure 1.1 Programmable logic controller.

the DOD Cyber Workforce Framework knowledge, skills, abilities, and tasks (KSAT) for a control systems security specialist calls out knowledge of control systems technologies such as programmable logic controllers (PLCs) as needed for devices, equipment, and system security configuration and operations of control systems:

Control Systems Security Specialist (Work Role ID: 462) *KSAT* 6929 – Knowledge of control system technologies, such as programmable logic controllers (PLCs), supervisory control and data acquisition (SCADA) software, distributed control systems (DCS), and operational technology (OT) (DOD, 2023).

Most great ice hockey players began skating relatively close to the time they learned to walk. They are comfortable in a sport played on ice at incredible speeds. Most great cyber experts are comfortable using cyber as a medium. Unlike ice skating, which is challenging to learn later in life, JavaScript is not. However, learning to code using JavaScript is likely easier for "digital natives" with cyber mindsets. The distinction was coined in 2001 by Marc Prensky in "Digital Natives, Digital Immigrants." Today's critical infrastructure has not just changed incrementally. It has changed radically. Marc Prensky referred to it as a "singularity" of new technology. Suppose the first wave involved emails, messaging, and cell phones. In that case, the next wave is the distinction between how professionals working in critical infrastructure sectors adapt to new technologies, methodologies introduced into legacy systems, and future systems utilizing software-intensive capabilities. It is worth noting that integrated systems themselves – aggregating direct and indirect (e.g., procured) human design and development – should learn how

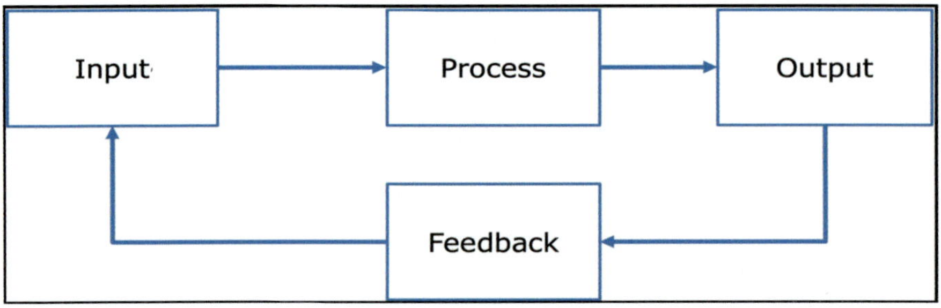

Figure 1.2 Embedded system.

to be cyber-secure in their infancy; that is, when they are learning to walk, or as the system is being designed.

Programmable logic controllers (PLCs) require programming. Ladder diagram (LD) ladder logic is the most commonly used language in programmable logic controller (PLC) programming. Java code can be used to communicate with the PLC to implement authentication. JavaScript is a relatively easy-to-learn class-based programming language. JavaScript is not an object-oriented programming language, but it supports both procedural and object-oriented programming (OOP). The syntax is straightforward, making it easy to use and integral to large-scale applications. There are other skills to develop, but learning some basic programmable logic controller (PLC) programming, such as ladder diagram (LD), function block diagram, sequential flowcharts, structured text, or instruction list, is recommended to gain a deeper understanding of the game. Code is, after all, the embodiment of the logic of the programmer. Learn logic first, then programmable logic controller (PLC) programming.

1.4 Embedded Systems

Embedded systems, found in critical infrastructure such as manufacturing, perform a specific task within a given period, synchronizing with software and hardware, as illustrated in Figure 1.2. Characteristics of embedded systems include that they generally do not multitask as well as laptop personal computers. Embedded systems are single-function devices that have lower manufacturing costs than laptops, hold less memory, operate in near real-time, and are designed to meet a fixed runtime requirement. They are also integrated into the physical world, often through the use of sensors. Examples of types of embedded systems include standalone systems, such as those with less complex analog inputs and digital outputs; real-time systems,

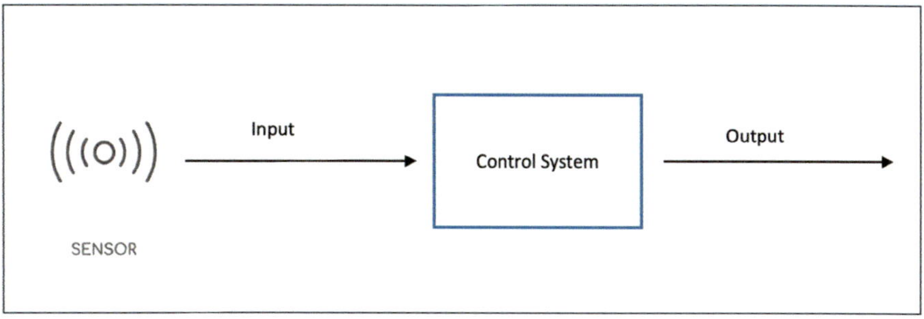

Figure 1.3 Sensor.

like a thermometer; or networked embedded systems, like a monitoring system. The DOD Cyber Workforce Framework knowledge, skills, abilities, and tasks (KSAT) for a control systems security specialist calls out knowledge of embedded systems as needed for device, equipment, and system-level security configuration and operations of control systems:

Control Systems Security Specialist (Work Role ID: 462) KSAT 43a – Knowledge of embedded systems.

At the most fundamental level, a control system used in critical infrastructure receives data from a sensor. A control system is a set of mechanical or electrical devices that control the desired response of other devices by controlling output by varying input, as shown in Figure 1.3. A control system makes course-of-action (COA) decisions about how discrete, continuous, or hybrid processes function, generally ensuring that processes operate within appropriate parameters, safely, at an appropriate rate, and within the required quality. Control systems help factories and facilities produce quality goods safely and efficiently.

When integrated, these elements enable the operations of industrial control systems (ICS) found in critical infrastructure. The ICS contains control loops that utilize various network protocols, operating with cycle times ranging from minutes to seconds to milliseconds. The set point can be based on nested or cascading, whereby the variable is determined by another control loop, as shown in Figure 1.4 (Stouffer et al., 2015).

Key components of the ICS operation include the control loop, which consists of sensors, controller hardware such as programmable logic controllers (PLCs) and actuators such as the control valves, switches, and motors, the human–machine interface (HMI) that display the process status information to the operators and engineers, and remote diagnostics and maintenance used to prevent, identify and recover from failures, as shown in Figure 1.5 (Stouffer et al., 2023).

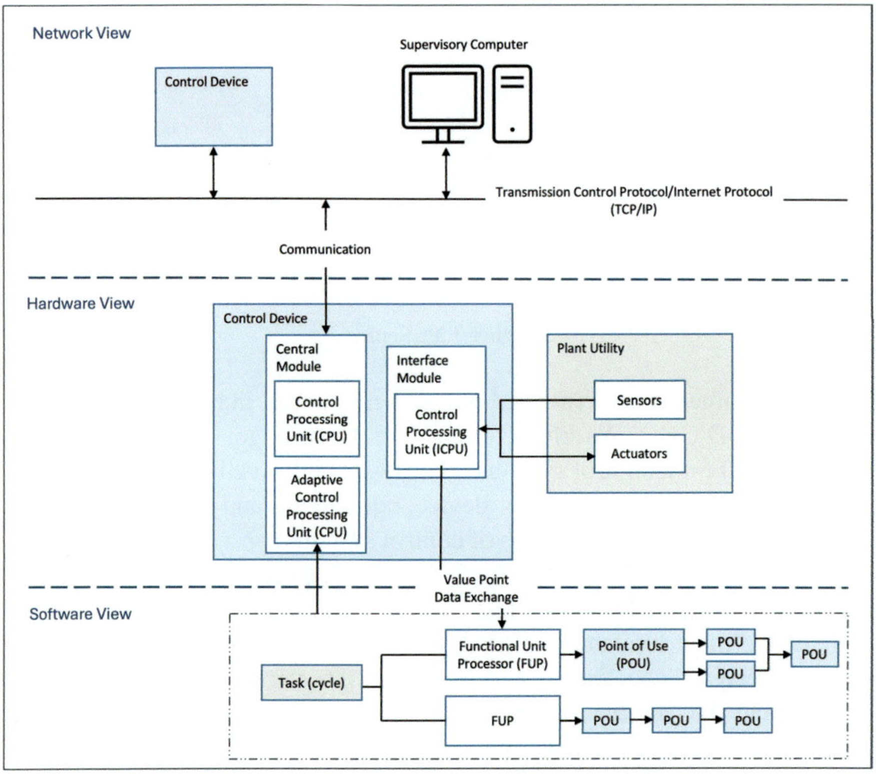

Figure 1.4 Industrial control system operation.

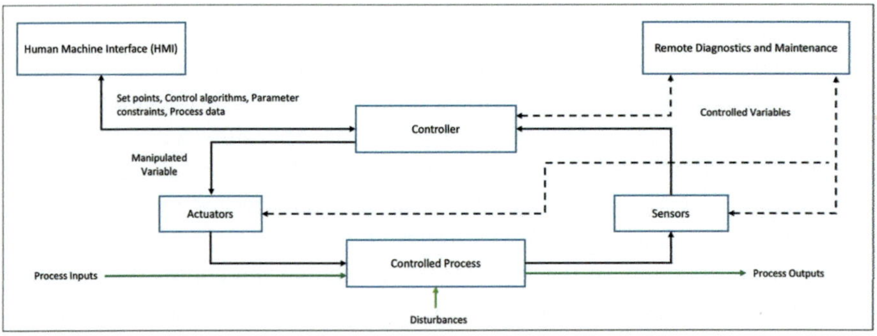

Figure 1.5 Industrial control system operation security.

1.5 Cyber Domain Knowledge

Ice hockey players can reach speeds on the ice of upwards of 20 miles per hour using skating technique, leg power, and their upper bodies to generate momentum. Cyber professionals need the same momentum, knowledge of techniques, unique skills, and love of the game. Professional hockey players constantly move in bursts of high energy, with quick acceleration and agility. Players need stamina and speed. When you get on the ice, you have to know the game. It takes years of practice. Players can swap positions. For example, a defense player can also be moved to play forward. A defensive player playing forward is similar to offensive cyber professionals who work in red team roles, testing a network's security effectiveness to strengthen network security. Playing forward is comparable to defensive cybersecurity professionals who work in network hardening and security operations roles, focusing on prevention, detection, and response to attacks. Both offensive and defensive cybersecurity professionals can benefit from swapping positions and gaining a deeper understanding by playing both roles.

For critical infrastructure, additional key players have a more specialized role in knowing the actual physical system. For example, the goalie has a more specialized job on the team. The best players have a hockey sense that lets them anticipate where the puck will be to get ahead of where it will be. Whether you have played, watched, or know nothing about the sport, ice hockey is a game that can be likened to working with a cyber-physical system. The attack on goal can be lightning-fast. No attack may work for a long period of time, and then suddenly there is a breach and a goal. Speed is a huge factor in the game. Everyone is moving fast. It would be best to assemble a team that can skate and knows the game.

Questions about the team assembled include: "I have cyber awareness training (*Yes, No, I do not know*)"; "Cyber awareness training is required for my position (*Yes, No, I do not know*)," or "Cyber awareness training is fully funded and available for my position (*Yes, No, I do not know*)" (Scalco and Simske, 2022). Examples of suggested questions to help guide the assembly of a winning team are found in Chapter 24, "Creating a Model-based Organization."

1.6 Summary

Essential cyber skills include adaptability, collaboration, and problem-solving abilities. In ice hockey, the shooting game requires accuracy, speed, and a quick release to put the puck past the goalie and into the net, evading defensive players. Players need skating skills, balance, and control of their skate edges on the ice, including knowing when to pivot, change momentum, and react to changing game conditions. The best players make stickhandling look easy, keeping control of the puck close to their bodies. The players need a solid physical presence, intensity, cardiovascular endurance, strength, and a sense of the game to play where the puck is going to be, and not just where it is at the moment. How does a practitioner stay sharp, how does a professional maximize the team, and how does a student get started? It may be easier for a native skater to learn, but it is not impossible for anyone to appreciate the game. The bottom line is that, from the coach to the trainer to all skaters on an ice hockey team, all stakeholders must have a clear understanding of the game. Understanding the game is essential for comprehending the systems of systems inter-system meta-framework from both an engineering and a stakeholder perspective. Once you know how to skate, you are not likely to forget it. Once you adopt an attitude of "cybersecurity from the ground up in design," you are also unlikely to forget it.

References

(DISA), J. D. I. S. A., (NSA), N. S. A. & TEAM, Z. T. E. 2021. Department of Defense (DOD) Zero Trust Reference Architecture. *In:* DEFENSE, D. O. (ed.). Washington, DC.

(DOD), D. O. D. 2024. *Bio Senior Advisor, Zero Trust Portfolio Management Office, Department of Defense Chief Information Officer Randy Resnick* [Online]. https://dodcio.defense.gov/About-DoD-CIO/bios/Resnick.aspx: DOD CIO. Available: https://dodcio.defense.gov/About-DoD-CIO/bios/Resnick.aspx [Accessed 2024].

DOD. 2023. *DoD Cyber Workforce Framework* [Online]. DOD Cyber Exchange Public: Department of Defense (DOD). Available: https://public.cyber.mil/dcwf-work-role/control-systems-security-specialist/ [Accessed 2023].

John Kindervag, S. B., Kelley Mak 2012. Build Security Into Your Network's DNA: The Zero Trust Network Architecture. Forrester: Forrester.

Scalco, A. 2022. *Measuring Disagreement in Segments of the Cybersecurity Profession as a Means of Identifying Vulnerabilities,* Colorado State University.

Scalco, A. & Simske, S. 2022. Measuring Stakeholder Alignment to Overcome Control System Vulnerability. Department of Homeland Security (DHS) CISA Industrial Control Systems Working Group.

Stouffer, K., Pease, M., Tang, C., Zimmerman, T., Pillitteri, V., Lightman, S., Hahn, A., Saravia, S., Sherule, A. & Thompson, M. 2023. NIST Special

Publication NIST SP 800-82r3. *Guide to Operational Technology (OT) Security.* NIST: NIST.

Stouffer, K., Pillitteri, V., Lightman, S., Abrams, M. & Hahn, A. 2015. NIST Special Publication 800-82 Revision 2: Guide to Industrial Control Systems (ICS) Security. *In:* COMMERCE, D. O. (ed.). Gaithersburg, MD 20899-8930: National Institute of Standards and Technology (NIST).

Terry Merz, C. F., Aleksandra Scalco 2019. A Context-Centered Research Approach to Phishing and Operational Technology in Industrial Control Systems. *The Journal of Information Warfare.*

2

History of the Term "Cyber"

"Give me a word, any word, and I show you that the root of that word is Greek."

– Gus Portokalos,
"My Big Fat Greek Wedding,"
2002. (Commission, 2022)

Abstract

Cyber is a widely misunderstood term. Understanding the origin of the term and its adoption in the vernacular helps us comprehend how it can inform organizational and governmental operational strategies. Furthermore, understanding how the term came into use by governmental agencies allows us to understand the term "cyber" in today's public policy. This chapter gives a general history of the word "cyber." Even today, most professionals do not understand the term "cyber" well. The goal of this chapter is to demystify the term and its trajectory into everyday use globally. When we understand the term "cyber" in its contextual meaning and the cognitive shortcuts made about the word's meaning, we can extrapolate the concept as it is applied in various instances and gain a better understanding of the term in relation to the world around us. Cyber is actually more closely related to organizational theory than it is to information technology (IT) theory. The relationship to organizational theory is an important distinction because, somehow, the term "cyber" has become predominantly associated with information technology (IT). This cognitive shortcut to information technology (IT) overlooks the broader impact of the term itself, which encompasses the application of IT in organizational theory. Cyber is cross-disciplinary. The importance of governance and all disciplines working together to achieve a shared vision becomes clear when "cyber" is viewed from an organizational perspective. "Cyber," in its origins, is the study of control systems and communications and the interface between humans and machines.

DOI: 10.1201/9788743809128-2

2.1 Introduction

"Cyber" is an interesting term. It is a word loaded with different meanings for different people. Using the word in the general lexicon became mainstream in the 1990s. The first use of the word can arguably be traced to communication and control theory by the US mathematician Norbert Wiener shortly after the Second World War. In 1948, Wiener seemed to have started the trend by using a Latinized form of a term derived from Greek origins to describe systems and organizational theory. Wiener added Latinized "-ics" to create the term "cybernetics," derived from the Greek root word "*kubernētēs*," meaning to pilot or steer. Related to kubernētēs is "*kubernēsis*," which may be translated as the gift of administration or governance. "Cyber" refers to the concept that a system and organizational leader guides a group toward a common goal, much like a safe harbor when it comes to the cybersecurity of information. Systems engineering practice can be considered the technical leadership that guides a group in cyberspace or toward a common goal, addressing cyberspace events and outcomes, such as mitigating or overcoming cyber vulnerabilities. Cybernetics evolved as a field of study of control systems, communications, and the interface between humans and machines. Popular culture began incorporating "cyber" into film and literature as part of the psychedelic, futuristic creations inspired by the 1960s achievement of human space travel. "Cyberpunk" as a science fiction sub-genre had been riding the wave of the 1960s and 1970s futuristic movements by the time George Lucas' *Star Wars* was released in 1977. In the film "Star Wars," George Lucas describes a galactic communications grid known as the "*HoloNet,*" which is utilized for holographic communication and information exchange (George Walton Lucas, 1977).

As early as 2018, researchers recognized the complexity of "cyber" in the modern military interconnected and technology-reliant operational environment (Span et al., 2018). Martin "Trae" Span, Logan O. Mailloux, and Michael R. Grimaila surveyed cybersecurity architecture analysis approaches to improve cybersecurity. They identified the definition of the term "cybersecurity" as one of the least understood within the Department of Defense (DOD). Furthermore, they identified cybersecurity approaches as ad hoc. "Despite being often cited; this definition tends to confuse because it is packed with domain-specific IT jargon: availability ensures the system is used as anticipated; integrity is the protection from unauthorized modification; confidentiality is keeping data private; authentication is a validation of the claimed identity; and, nonrepudiation is the ability to prove that an action has taken place" (Span et al., 2018). They observed that the definition is hindered by legacy terminology rather than a more straightforward definition, which would better protect critical systems from cyber threats. Further, they surveyed architectural approaches for applicability to complex system cybersecurity based on literature focused on weapon systems. Most predecessor surveys focused exclusively on computer networks and IT system security controls, such as compliance-based information assurance (IA) (Ibid.). The predecessor surveys are considered inadequate for the complexity of cybersecurity in control systems.

As the term "cyber" gained widespread use in all forms of media, relevant regulations and policies emerged to address cyber vulnerabilities. These policies and regulations are discussed in Chapter 3, "Policy and Regulations." Cyber was declared and recognized as a warfighting domain, and for the rest of us, it was something to acknowledge and about which to figure out the details. Cyber as a warfighting domain is discussed in Chapter 4, "Cyber Is a Medium (Not a Domain)."

Back to the origins of ideas and terms, the term "cyber space," originally written as two words, can be attributed to science fiction writer William Gibson in 1982 as "the online world of computer networks and especially the Internet, the environment in which communication over computer networks occurs" (Gibson, 1982). In "Burning Chrome," interestingly, the main characters are two hackers: a hardware specialist and a software expert. Gibson later used the theme in his award-winning trilogy novels in the 1980s: "Neuromancer," "Count Zero," and "Mona Lisa Overdrive." By the 1990s, anything cool was making its way into the cyber realm. Cyberspace became a compound word, giving a new meaning to the notion of the gift of administration or governance, or "kubernēsis," as a system and organizational domain.

2.2 Summary

When we ask about how the context of the term "cyber" is used, we start to sense an organization and commonality around the term at higher levels. It helps to clarify thoughts about an abstraction. We use the word differently today because of the unanticipated diversification of computing and networking resources and processes. However, knowing the word's origin provides some understanding of what was initially meant. The ideas of "cyber" and "cyber space" may be associated with IT and computer sciences, even though the term "cyber" is more closely related to organizational theory. The cognitive shortcut from cyber to IT or computer science evolves into something more integral to everyday operations, rather than remaining exclusive to the expertise of "cyber" experts or specialists. Finally, the importance of governance and all disciplines working together to achieve a shared vision makes sense when "cyber" is viewed from an organizational perspective. Control system engineers are responsible for the control subsystem. This role is physics, engineering, and mathematics-based. Control system engineers come from mechanical, electrical, and general engineering backgrounds. Systems engineers focus on managing complexity throughout the design, development, integration, and operational lifecycle of a system. Generally speaking, cybersecurity professionals monitor network traffic, look for security events and incidents, and design and operate software programs.

An IT professional can be a great cybersecurity professional. However, technical and engineering knowledge, as well as experience in critical infrastructure systems, is required to understand, design, and develop physics-based systems. Effectively understanding the peculiar idiosyncrasies of any system or network can lead to good cybersecurity insights, and thus, inform policies and practices. "Cyber" is a misunderstood term; therefore, it is accompanied by policies with wildly differing perspectives on the term. When we understand the term "cyber" in its contextual meaning and the cognitive shortcuts made about the word's meaning, we can extrapolate the concept as it is applied in various instances and gain a better understanding of the term in relation to the world around us. It takes all disciplines for an organization to work together, each from its own "cyber" perspective, to achieve an organizational vision. It is also necessary for stakeholders to have a common understanding to effectively execute cyber policy, regulations, and guidelines, which leads us to our next chapter, Chapter 3, "Cyber Policy, Regulations, and Guidelines."

References

Commission, E. 2022. The Articles of the Cyber Resilience Act (CRA). *In:* UNION, C. O. T. E. (ed.). Brussels, Belgium: General Secretariat of the Council.

Star Wars, 1977. Directed by GEORGE WALTON LUCAS, J.

Gibson, W. 1982. Burning Chrome. Omni.

Span, M. T., Mailloux, L. O. & Grimaila, M. R. 2018. Cybersecurity Architectural Analysis for Complex Cyber-Physical Systems. *The Cyber Defense Review,* 3.

3

Cyber Policy, Regulations, and Guidelines

"This is my special gift for you. Don't ever open it."
Zeus, Greek Mythology about Pandora's Box

Abstract

As "cyber" became a real-world complex challenge, a range of responses emerged to resolve and regulate industry issues and promote the public interest. Policy, regulations, and guidelines emerged to deal with the newly coined term, "cyber." Well-intended, narrowly conceived compliance rules for risk-based approaches to system security and data protection aimed to provide a structured approach to dealing with abstraction but resulted in dysfunctional post-design and development checklists. However, intentions to offer structured processes for risk management and security often fail to consider security design in the early phases of conceptualization, instead attempting to bolt on cybersecurity after the developmental test has been completed. Policy, regulation, and guidelines doubled down. Government sector policymakers' attempts to enforce accountability using standard linear approaches lack answers to questions about how to accomplish what is asked (i.e., implementation directions, decision-making guidance, and specification thresholds). Among the challenges is that if these policies, regulations, and guidelines fail to address complex cybersecurity issues fundamentally, then support for an overall "culture of cybersecurity" also fails. Reductionist approaches to regulation require an equal measure of holistic approaches to inter-systems and an understanding of how systems interact. Systems engineering modeling approaches can help inform more effective decision-making and policy-making. Evolving initiatives to draft policies, regulations, and guidelines are slowly forming what will eventually become international law governing cyber conduct. Issues related to cyber have already become the subject of diplomatic debate as nations discuss legal responsibilities and cyber conduct. Exploring these approaches provides insight into emerging norms and standards that apply across countries and international entities, such as global corporations.

DOI: 10.1201/9788743809128-3

3.1 Introduction

Cyber vulnerabilities to critical infrastructure, such as industrial control systems that underpin the delivery of essential services like power and water, can have an isolating effect. Cyber equipment relies on critical infrastructure, which is currently supported by components in electronic systems, including resistors, capacitors, transistors, and integrated circuits. Compliance actions in the information technology (IT) enterprise put a spotlight on the need to improve due diligence and risk management, extending to the lowest-level field devices and sensor networks, particularly at the business level of networks that connect outward to external network applications. The United States took a more standard linear approach based on compliance rules. The United States issued orders for government networks and guidelines for commercial entities to follow, in an effort to establish boundaries around the boundaryless abstractions of cyberspace.

Meanwhile, the European Union (EU) and its member countries, including Germany, have taken a more holistic approach to cybersecurity. The EU, wary of external interference with privacy rights, the freedom of movement of goods and services, and commonality, also formed regulations. Germany, in particular, is focused on avoiding a monoculture dependency on single IT vendors and is pursuing technical sovereignty. The European Union Agency for Cybersecurity (ENISA) developed guidance on the cybersecurity and resilience of critical infrastructure sectors, publishing the "Handbook for Cyber Stress Tests" in May 2025 (ENISA, 2025). The objective is to provide organizations with a practical and straightforward five-step guide to help them withstand and recover from cyber events. The steps begin with stakeholder engagement, which is discussed in more detail in Chapter 25, "Stakeholder Alignment," of this book. In Chapter 25, we present a mathematical model and methodology to help you identify uncertainty in environments where it is likely to exist among stakeholders and where vulnerability is also likely to be present (Scalco, 2022). Valuable questions to ask include whether the organization is aware of available government support resources and how to access them, whether the organization has a formalized relationship with government agencies, and whether cyber incident responses, policies, procedures, and logistics are well-documented within the organization, as things will inevitably go wrong. Nations are gearing up for international cyberspace law governing the Internet, electronic commerce, privacy, and, more recently, laws governing cyber warfare aimed at disrupting critical infrastructure. Initially, cyber policy focused on addressing copyright infringement, IT-related fraud, and web-based crimes, as well as the boundaries between government authorities and private sector interests. Notably, by working together, sharing lessons learned, and leveraging expertise through collaboration, many of the challenges facing the owners and operators of industrial control systems in a cyber world can be overcome, thereby protecting people, property, and system functions.

Cyber policy, regulations, and guidelines for critical infrastructure must address network convergence at the lowest level of the operational system. Integrating data from operational technology (OT) can provide business operations with valuable information to drive operational efficiency and productivity, as well as enhance business competitiveness, which is discussed in more detail in Chapter 9, "Road to Commoditization." Examples include sensors that provide reliable temperature measurements essential for operations. Sensors include motion, heat, temperature, pressure, and even biometric sensors, which are discussed in further detail in Chapter 19, "Biometrics." Industrial temperature measurement sensors include devices such as rugged and shock-immune Thermocouples that measure temperatures in ranges of –400 to 420 °F (–240 to 216 °C). In addition to the wide temperature ranges, the sensors are simple, offer quick response time, and do not require external power. Disadvantages include vulnerability to corrosion, reduced stability, and the potential for cyber tampering. Other types of sensors can provide remote read-back availability. Understanding the relationship between business objectives and technical physics-based system components is crucial for incorporating these considerations into cyber policy, regulations, and guidelines, particularly as applications become increasingly pervasive in smart cities and manufacturing, and technologies such as cloud computing and artificial intelligence (AI) emerge. Unproven elements that use innovative technologies can be attractive options. Recall from Chapter 1, "Understanding the Game," that cyber is a team sport. Each player on the team is essential, particularly when the network provides not only basic connectivity but also plays a pivotal role in functional operations.

In an industrial control system (ICS), the network controller is the main component for managing and orchestrating all aspects of the network. The network controller collects and analyzes network traffic by maintaining an inventory of devices on the network and their status. Network controllers can help to reduce operating costs by providing agility and responsiveness to changing business needs. For example, an organization can implement changes to all network devices universally without the resource-consuming need to configure each device individually. A universal approach can accelerate changes on the network, detecting potential issues and taking automated corrective action before they turn into problems that require real maintenance. Generally, plant sensors, actuators, device buses, and pumps found at the lowest level of the architecture are referred to as field devices. The job of these devices is to transfer data in a time-critical task. Timing is critical in automated processes. All elements of the control system must be thoroughly understood and comprehended. In addition to knowledge of cyber defense policies, procedures, and regulations, the concepts of computer networking and cyber defense in relation to threats and vulnerabilities must be thoroughly understood. Digital engineering (DE) can be a valuable tool.

Digital engineering (DE) and modeling approaches are better suited to help deliver solutions at a pace that is responsive to the changing environment and operational needs while still countering cyber vulnerabilities. DE can help stakeholders better understand and develop requirements, facilitating collaboration and informed decision-making. The advantages of DE and modeling approaches leveraged by the model-based systems engineering (MBSE) practice include the visualization of technical complexity into a single source of truth, which enables the ability to expedite data-centric decisions. A challenge is the need for all stakeholders to commit to integrating shared data into the decision-making process. However, when applied to complexity, the process becomes more consistent and understandable, making informed decision-making easier and traceable to unambiguous requirements.

3.2 US Laws, Regulations, and Guidelines

While the personal computer revolution of the late 1970s and the dawn of the World Wide Web (WWW) of the late 1980s and early 1990s took their foothold, the computer systems and networks found in critical infrastructure industrial control systems had yet to take hold, and catch up in adapting to the broad connectivity of the digital world. While some aspects of cyber laws and regulations associated with clearly insurable factors (i.e., money stolen from a bank account directly attributable to a cyber hack) have caught up, valuing the risk has not. Metrics need to be analytically measurable and quantifiable to effectively translate into enforceable legislation. Subjective assessments and evaluations lack the objectivity of quantifiable metrics based on real data. Today, considerations for industrial control systems are mainly based on a qualitative approach and opinions of subject matter experts (SMEs). Still, they need to be more mature and testable against data-driven metrics, just as regulatory laws are for reasonable and appropriate safeguards for protecting privacy and intellectual property. For example, cybercrime laws in the United States prohibit online identity theft, intellectual property theft, and crimes such as gambling. Compliance depends on the industry and data type. The Health Insurance Portability and Accountability Act (HIPAA) protects patient health information. At the same time, the Payment Card Industry Data Security Standard (PCI DSS) establishes security standards and requirements for the transmission of credit card data. Driving security standards for operational technology (OT) down to the lowest level component found in critical infrastructure is complex but not insurmountable. In the broader context, this encompasses the integrity of the supply chain for these components. There is value in communicating the benefits and value of security standard initiatives that protect people, property, and the environment.

Along with data privacy laws and regulations, compliance-based frameworks and executive orders emerged to modernize cybersecurity for the federal government. Compliance-based frameworks are effective in ensuring the confidentiality, integrity, and security of information. Addressing administrative, physical, technical, and nontechnical safeguards to maintain compliance can be measured against threats in a risk assessment. These approaches do not work so well for physics-based, physical control process systems. First, one must be able to discern if a control system has been hacked. Suppose a system operator is not monitoring for control system cybersecurity incidents or conducting control system cyber forensics. In that case, the likelihood of detecting a cyber event is low, and it may not even be identified as cyber-related.

In contrast to the Healthcare Information Portability and Accountability Act (HIPAA) and Payment Card Industry Data Security Standard (PCI DSS) standards in the United States, the US cyber laws, regulations, and guidelines for industrial control systems (ICS) have not yet been fully implemented and need to be effectively updated. Organizations closely monitor and measure their enterprise IT cybersecurity to ensure data protection compliance and prevent incidents from occurring. At the same time, most critical infrastructure risk assessments have been based on physical failures, attacks, or natural disasters. The potential of cyber malware or advanced persistent threats (APTs) attacks on critical infrastructure was first presented in 2010, with subsequent updates in 2014, in the original release of the ICS-CERT advisory on Stuxnet Malware Mitigation (CISA, 2014). Stuxnet rocked control system owners' and operators' understanding of cyber and physical processes within ICS and the automation industry writ large by the success of complex infection attack mechanisms in critical infrastructure. The sophisticated method of Stuxnet infection used USB drives to propagate in systems. System owners and operators rushed to disable the USB drives and put policies into place to keep them that way. Pandora's box was opened.

Cybersecurity compliance-based frameworks have emerged to help businesses understand, manage, and mitigate cybersecurity risks to their business IT networks and data. The frameworks guide best practices but fall short of offering quantitative metrics for assessment models in industrial control systems (ICS). For example, the National Institute of Standards and Technology (NIST), within the US Department of Commerce, published the NIST Cybersecurity Framework Version 1.0 in 2014 in response to the US White House Executive Order 13636 (E.O. 13636), which was released on February 12, 2013 (NIST, 2014). E.O. 13636 tasked NIST with developing a cybersecurity framework to share cyber threat information and build successful approaches to reduce cyber risks to critical infrastructure (Obama, 2013). Since then, NIST has released additional guidelines that regulate how government agencies approach cybersecurity and provide best practices for protecting the systems of both public and private entities. The NIST framework identifies cybersecurity functions: identify, protect, detect, respond, and recover. Subsequently, the NIST Cybersecurity Framework Version 2.0 included supply chain and governance. While NIST provides frameworks to make critical infrastructure more resilient, a method for quantifying metrics is not given. In the absence of a method to quantify the cyber resilience of ICS components such as programmable logic controllers (PLCs), supervisory control and data acquisition (SCADA) systems, distributed control systems (DCS), and remote terminal units (RTUs), there is limited basis for compliance-based regulatory approaches for ICS. A methodology proposed by Md Ariful Haque et al. for control systems is the ICS Cyber Resilient Assessment Model (CRAM). This methodology still has the limitations of a linear compliance approach, which overlooks measures for the complexity of behavioral attack methods; however, it does provide a weighted value approach for quantifying resilience. The approach is to decompose ICS resilience metrics of robustness (e.g., access control and authentication and segmentation), redundancy (e.g., contingency), resourcefulness (e.g., monitoring and detection and response and recovery), and rapidity (e.g., latency and restoration) into physical, organizational, and technical submatrices to assess, evaluate, and measure (Md Ariful Haque, 2018). Ways to enforce compliance, such as in enterprise IT resilience, still have not caught up in OT ICS, which makes incentivizing cybersecurity investment a challenge, and drafting policy even more so. A better understanding of the problem space has enabled standards and best practices to evolve.

Between 2012 and 2016, several broader cybersecurity efforts related to critical infrastructure emerged, paving the way for a greater understanding of the convergence of business enterprise IT systems and those IT components found in operational technology (OT). Gartner was the first to define operational technology (OT) as "hardware and software that detects or causes a change through direct monitoring and/or control of industrial equipment, assets, processes, and events" (Gartner, 2024). Operational technology (OT) security encompasses not only cybersecurity but also physical security, encompassing all the practices and technologies used to protect components from threats. This holistic approach to intersystem thinking is described in the Foreword of this book as critical to successful operations in a cyber world. Precursors included the "Handbook for Self-assessing Security Vulnerabilities & Risks of Industrial Control Systems on DOD Installations," published on December 19, 2012. Standards bodies followed with the NIST publishing the "NIST Framework and Roadmap for Smart Grid Interoperability Standards," Release 2.0 in February 2012 (NIST, 2012). Cybersecurity efforts, such as DoDI 8500 Cybersecurity and DoDI 8510 Risk Management Framework, were followed by DoDI 8140 Workforce. The Unified Facilities Criteria (UFC) cybersecurity of facility-related control systems (FRCS) emerged in 2016, followed by an updated version in 2017. Subsequent National Institute of Standards and Technology (NIST) updates followed, such as the NIST Guide to Industrial Control System (ICS) Security updated in 2023, as did other cybersecurity handbooks, guides, frameworks, and standards, such as the US Department of Homeland Security (DHS) Cybersecurity and Infrastructure Agency (CISA) guidelines for industrial control system (ICS), the UK NCSC guidelines, the International Society of Automation (ISA) and International Electro-technical Commission (IEC) ISA/IEC 62443 Standards for Security of Industrial Automation and Control Systems, and the European Union Agency for Cybersecurity (ENISA) guidelines for best practices for IIoT operators, manufacturers, and users.

3.2.1 "Handbook for Self-assessing Security Vulnerabilities & Risks of Industrial Control Systems on DOD Installations," 2012

The "Handbook for Self-assessing Security Vulnerabilities and Risks of Industrial Control Systems on DOD Installations," published on December 19, 2012, was a response to a recognized joint operational problem. It was a collaborative effort between the Joint Threat Assessment and Negation for Installation Infrastructure Control Systems (JTANIICS) Quick Reaction Test (QRT) and the Joint Test and Evaluation (JT&E) Program under the Director, Operational Test and Evaluation (OT&E), Office of the Secretary of Defense (OSD), and the combined work products of the 346th Test Squadron,

262nd Network Warfare Squadron, and the Idaho National Laboratory (INL) under the aegis of the Air Force Joint Test Program Office with the advice of Joint Warfighter Advisory Group (JWAG) (JTANICS, 2012). The primary objective of the output was to ensure mission assurance, with a focus on risk management. "Although not always obvious, many missions depend on the unfailing functioning of [Industrial Control Systems (ICS)] and therefore on the security of those systems" (JTANICS, 2012). The handbook identified how cyber vulnerabilities could pose a direct threat to the mission through the Industrial Control System (ICS) or an indirect threat through network connections. The handbook shared an eight-step process for self-assessment that remains valid today and will be used later in this book to map to systems engineering activities to overcome cyber vulnerability found in Part II – Systems Engineering Methods and Practice:

- Analyze missions (prioritize missions)
- Identify assets (inventory, map, visually inspect)
- Determine industrial control systems (ICS) connectivity (all connection points, including potential connection points)
- Determine ICS dependencies (link missions, assets, infrastructure, ICS)
- Assess risk to the mission (threats, vulnerabilities, consequences)
- Prioritize risk management approaches (based on mission priorities and resources)
- Implement actions (commit resources, assign responsibilities)
- Monitor and reassess (continuous process, not "one and done")

3.2.2 National Institute of Standards and Technology (NIST) "NIST Framework and Roadmap for Smart Grid Interoperability Standards," 2012

The National Institute of Standards and Technology (NIST) published the "NIST Framework and Roadmap for Smart Grid Interoperability Standards," Release 2.0, in February 2012 (NIST, 2012). NIST published this framework and roadmap in response to an urgent need to establish interoperability standards and protocols for the Smart Grid. NIST's plan included three phases:

1. Accelerate the identification and consensus on Smart Grid standards.

2. Establish a robust smart grid interoperability panel (SGIP) that sustains the development of the many additional standards that will be needed.
3. Create conformity testing and certification infrastructure.

3.2.3 "NIST Guide to Industrial Control System (ICS) Security," NIST Special Publication 800-82, Revision 2, updated 2023

NIST published the "NIST Guide to Industrial Control System (ICS) Security," Special Publication 800-82, Revision 2, updated in 2023. This guide provides a framework for effectively securing Industrial Control Systems (ICS) and their components, including supervisory control and data acquisition (SCADA), distributed control systems (DCS), and programmable logic controllers (PLCs). The guidance takes into account the unique functionality of industrial control systems (ICS) and operational technology (OT) requirements within a security program. A downside to NIST 800-82, Rev. 2, is its focus on compliance requirements, rather than a risk-based or more holistic approach to understanding how systems interact. This approach can have an unintentional consequence of creating well-intentioned, yet dysfunctional, post-design and development checklists.

3.3 European Union Laws, Regulations, and Guidelines

The United States was one of many countries on the path to creating conformity. However, while the United States employed a more standard linear approach that required answering the question of how to accomplish what was asked, other countries took a more holistic approach to understanding inter-systems. For example, in Europe, the European Union (EU) tried to add context to regulate industry issues and promote public interest. In Europe, cybersecurity is addressed in EU laws and regulations, including the European General Data Protection Regulation (EU-GDPR), the Security of Network and Information Systems Directive (NIS Directive), and the European Union Agency for Cybersecurity (ENISA). As mentioned earlier in this chapter, EU organizations, such as ENISA, are publishing guidance to help organizations in their member countries become resilient in a cyber world. ENISA published "Good Practices for Security of Internet of Things in the Context of Smart Manufacturing" in 2018. The guidance encompasses three main areas: policy, organizational, and technical practices for designing secure systems that operate in cyberspace, as well as for ensuring security throughout the system's lifecycle. In June 2019, the European Cybersecurity

Act was published, identifying ENISA as a permanently mandated agency to lead cyber coordination in the European Union (EU).

The European Council adopted the EU Cyber Defence Policy Framework (CDPF) in 2014, also identifying cyberspace as the fifth domain of operations as an update to the EU Common Security and Defense Policy (CSDP). Subsequent updates identify cyber defense as a key element in the EU Capability Development Plan (CDP). In 2021, European Commission President Ursula von der Leyen proposed a European Cyber Defence Policy that would encompass legislation for common standards under a new European Cyber Resilience Act (ECRA). The new legislation bolsters cybersecurity requirements for products with digital elements, including both hardware and software, throughout a product's lifecycle. Unlike US legislation, the EU CRA establishes specific objectives to create conditions that reduce the inherent vulnerabilities of hardware and software component selections. For example, four specific objectives of the EU CRA are established:

1. "Ensure that manufacturers improve the security of products with digital elements [from] the design and development phase and throughout the whole life cycle.
2. Ensure a coherent cybersecurity framework that facilitates compliance for hardware and software producers.
3. Enhance the transparency of security properties of products with digital elements.
4. Enable businesses and consumers to use products with digital elements securely (Commission, 2022).

These four objectives of the EU's CRA are impactful when considering the transition of innovation into operations, component vulnerabilities, and identifying the entity responsible for both.

3.3.1 Germany's technical sovereignty

In Germany, the Federal Office for Information Security (in German, "Bundesamt für Sicherheit in der Informationstechnik" (BSI)) was established in 1991 as the federal authority responsible for cybersecurity. Germany's BSI Act, released in 2009, is the foundational German cybersecurity act (Law, 2009). Other countries have established similar agencies to coordinate cybersecurity and provide frameworks for entities operating in their countries.

For example, the National Cyber Coordination Center (NCCC) is the leading cybersecurity agency in India. India established frameworks, such as the National Cybersecurity Policy in 2013, to help the public and private sectors protect their infrastructure from cyberattacks. Subsequently, numerous public sector and commercially developed frameworks have been designed to help simplify the complexity associated with cyber threats to critical infrastructure, aiming to achieve resiliency, security, safety, and, importantly, operational mission success. These frameworks simplify the numerous variations in interfaces between systems and applications hosted in various locations, utilizing diverse standards and protocols across centralized, decentralized, or distributed network architectures. These architectures are supported by the World Wide Web and a combination of infrastructures, people, processes, and technologies.

Science and research initiatives resulted in new ways to overcome cyber vulnerabilities. However, the experimentation did little to overcome the growing public relations issues and public opinion regarding the handling of what is increasingly considered "dangerous" technologies. Discussion in Germany focused on avoiding a "monoculture" dependency on single, large vendors, and the government sought policy goals of more open-source alternatives (Schallbruch and Skierka, 2018). Linux is an example of an open-source alternative family of operating systems based on the Linux kernel at its core, combined with tools and applications. The Linux community-developed operating system (OS) is resilient, customizable, and free to use. Germany's government pursued a comprehensive cybersecurity agenda. It strengthened the role of the Federal Office for Information Security (BSI) after Snowden's 2013 revelations and went as far as to call for German "technological sovereignty." By 2015, Germany was a pioneer in cybersecurity regulation, with the German IT cybersecurity law, "Gesetz zur Erhöhung der Sicherheit informationstechnischer Systeme (IT-Sicherheitsgesetz)," coming into full force. The corresponding European Network Information Security directive followed in 2016 (Schallbruch and Skierka, 2018).

3.3.2 DOD cybersecurity workforce framework KSAT

Even for the US government and the Department of Defense (DOD), which operate more within a structured framework for dealing with abstractions, understanding concepts such as monoculture and technical sovereignty is essential for comprehending critical infrastructure in a cyber world. The DOD Cyber Workforce framework knowledge, skills, abilities, and tasks (KSAT) for a control systems security specialist calls out knowledge of national and international laws, regulations, policies, and ethics as they relate to cybersecurity:

> Control Systems Security Specialist (Work Role ID: 462) and Cyber Defense Infrastructure Support Specialist (Work Role ID: 521/NIST: PR-INF-001) *KSAT* 1157 – Knowledge of national and international laws, regulations, policies, and ethics as they relate to cybersecurity.

3.3.3 United Kingdom, Japan, and Australia

The United Kingdom passed the Data Protection Act (DPA) in 2018, which focused on the handling of data. Japan established a basic policy for cybersecurity in the Basic Act on Cybersecurity, enacted in 2014 as an amendment to the Act on the Protection of Personal Information (APPI), the country's primary data protection legislation. The United Kingdom updated the Basic Act on Cybersecurity in 2019. Similar to Germany's vision, in 2023, Australia launched the 2023–2030 Cyber Security Strategy, setting forth the Australian Government's vision to be a world leader in cybersecurity (Government, 2023). Australia's plan consists of six elements, known as "cyber shields." Each provides a layer of cyber defense to the country's citizens and businesses (Government, 2023):

1. Strong business and citizens
2. Safe technology
3. World-class threat sharing and blocking
4. Protected critical infrastructure
5. Sovereign capabilities
6. Resilient region and global leadership

What is game-changing about both Germany's and Australia's approach is that the governments are shifting from technical topics to whole-of-nation endeavors to overcome cybersecurity abstraction in tangible ways supported by public–private partnerships. The public–private dialogue in both Germany and Australia emphasizes that strengthening security foundations requires collaboration to build national cyber resilience. Implementation will be challenging. Systems engineering modeling approaches can supplement the strategy with methods to achieve the goals of addressing complex cyber issues and ensuring the necessary support for success. Reductionist approaches to regulation require an equal measure of holistic approaches to inter-systems and an understanding of how systems interact, which is found in threat modeling and analysis tools.

3.3.4 International Society of Automation (ISA)/International Electrotechnical Commission (IEC) 62443 Standards for Security of Industrial Automation and Control Systems, 2020

Another more holistic security perspective is found in the International Society of Automation (ISA)/International Electrotechnical Commission (IEC) 62443 Standards for Security of Industrial Automation and Control Systems, updated in 2020, which are a series of commercial standards that address cybersecurity for operational technology and define the requirements and processes for implementing and maintaining secure Industrial Automation and Control Systems (IACS) from design, development, and operation throughout the system's lifecycle. ISA/IEC 62443 is the most comprehensive industry standard for Industrial Control Systems (ICS), covering all cybersecurity aspects of system integration, testing, and installation. Additionally, the ISA112 Supervisory Control and Data Acquisition (SCADA) Systems standards committee is actively developing consensus-based technical standards. Industry sector-specific standards for operational technology are based on the ISA/IEC 62443 to avoid conflicting cybersecurity requirements.

3.4 Modeling Approaches for More Effective Decision-making

Martin "Trae" Span et al., in their 2018 research, "Cybersecurity Architectural Analysis for Complex Cyber-physical Systems," found that threat modeling and analysis used to overcome cyber vulnerabilities should focus on more than just identifying problems (e.g., assessments and document-based engineering). Architectural analysis approaches for highly complex systems are often ad hoc, rather than structured, in a way that both technological and non-technological aspects are well understood (Span et al., 2018). Dynamic adversary tactics necessitate dynamic tools and countermeasures that integrate modeling approaches such as model-based systems engineering (MBSE). MBSE can help provide traceability mapped to the component level and fit-for-purpose views, enabling more effective decision-making (Scalco, 2022). The use of dynamic tools to counter dynamic adversarial tactics is further discussed in Chapter 15, "Design Tradeoffs," and shared in the More

Situational Awareness for Industrial Control Systems (MOSAICS) integrated solution example. Model-based systems engineering (MBSE) is further discussed in Chapter 24, "Creating a Model-based Organization." Martin "Trae" Span concluded that professionals lack a thorough understanding of architectural cybersecurity analysis. "Moreover, given cybersecurity's widespread interest, it was surprising to find a general lack of understanding or consistency regarding what it means to conduct architectural analysis for cybersecurity" (Span et al., 2018).

3.5 Critical Infrastructure Sectors

Policies and collaboration require some level of standardization. Notably, various countries categorize critical infrastructure sectors differently from one another. For example, when the United States issued "Presidential Policy Directive 21/PPD-21: Critical Infrastructure Security and Resilience," 16 critical infrastructure sectors were identified (Obama, 2013). Other nations identified different numbers of categories. The US PPD-21 identifies chemical, communications, dams, emergency services, financial services, government facilities, information technology (IT), transportation systems, commercial faculties, critical manufacturing, the defense industrial base (DIB), energy, food and agriculture, healthcare and public health, nuclear reactors, materials, and waste, water, and wastewater systems. The directive defines the term "all hazards" as a threat or incident, natural or man-made, that warrants action to protect life, property, the environment, and public health or safety, and to minimize disruptions to government, social, or economic activities. "All hazards," as further defined in the directive, include natural disasters, cyber incidents, industrial accidents, pandemics, acts of terrorism, sabotage, and destructive criminal activity targeting critical infrastructure. The directive defines "resilience" as the ability to prepare for and adapt to changing conditions, withstand and recover rapidly from disruptions, and withstand and recover from deliberate attacks, accidents, or naturally occurring threats or incidents (Obama, 2013).

In contrast, the United Kingdom identifies 13, Germany identifies 10 (Law, 2009), Australia identifies 11, and Canada identifies 10 critical national infrastructure sectors (Government, 2009). Adding complexity to collaboration is that there are various entities designated as the primary risk management agency for each. For example, in the United States, the Health and Human Services (HHS) is the primary agency responsible for risk management in hospitals. At the same time, the Transportation Security Administration (TSA) is the primary risk management agency for railways.

Asset ownership also varies by country, by the system asset owner, maintenance service provider, integration service provider, and product suppliers, and whether an asset is government-owned and operated, commercially owned, in some form of partnership, or foreign-owned and operated. A key question is whether long-term system monitoring capabilities are practiced, whether standard operations and activities are well-characterized and whether publicly available control system information is managed for security (i.e., on public-facing websites and in job postings).

Classification types can be confusing. For example, there are more than 148,000 public water systems in the United States. The Environmental Protection Agency (EPA) defines three types of water systems: community water system (CWS), non-transient non-community water system (NTNCWS), and transient non-community water system (TNCWS). This is important because, as connectivity is added to systems and policies and regulations are written, the type of water system must be understood at the lowest component level for the correct type. The systems engineer must have domain knowledge of water systems in the United States to realize that a community water system (CWS) is a public system that supplies water to a population year-round, that is, where people reside. A non-transient non-community water system (NTNCWS) is "A public water system that regularly supplies water to at least 25 of the same people at least six months per year. Some examples are schools, factories, office buildings, and hospitals, which have their own water systems" (EPA, 2024). At the same time, a transient non-community water system (TNCWS) is "A public water system that provides water in a place such as a gas station or campground where people do not remain for long periods of time" (EPA, 2024). Working with the system requires knowing who is served as well as the source of the water. The sector is a knowledge domain that requires specialized expertise and a thorough understanding of its terms. Further complicating the matter is that even within critical infrastructure sectors, terms such as supervisory control and data acquisition (SCADA) can have different meanings. A question to ask is whether defensive countermeasures are in place to protect operations.

3.6 Summary

Slowly but surely, nations are coming to terms with the use of cybersecurity as a tool in the warfighting domain. Globally, technical leaders influence policy to guide people and organizations toward common goals in cyberspace. The United Nations has recognized the growing challenges of cybercrimes and is working towards establishing treaties on crimes committed using the internet and computer networks. The term "cyber" is recognized by the Department of Defense (DOD) and the North Atlantic Treaty Organization (NATO) as an operational domain equal to air, land, and sea. However, the term "cyber" and its implications as a warfighting domain have yet to be fully defined (Crowther, 2017). The Department of Defense (DOD) also recognizes "space" as a domain. The ambiguity of the term's use as a domain makes resource allocation challenging. It pushes the boundaries of government authorities and private sector interests, which is an interesting challenge to define. Generally, when used by authors, the term "cyber" refers to any place where computer systems and networks are utilized. Utilizing modeling approaches, such as model-based systems engineering (MBSE), can facilitate more informed decision-making and policymaking. Systems engineering modeling approaches can help inform more effective decision-making and policymaking, which will eventually become international law governing cyber conduct. All these national-level efforts give hope. The ISA/IEC 62443 provides hope by presenting a harmonized standard that can serve as a basis for modeling approaches to achieve commonality and consensus conformity in control systems.

In ancient Greek mythology, Zeus was regarded as the ruler of all things. It was Zeus who gave Pandora a gift of a box with the instruction never to open it. When she did, Pandora released all evils upon earth. However, Zeus does not know that Prometheus secretly adds hope to the box that Pandora opens. Prometheus, according to legend, is said to have given fire and spirit to humans and also gave them hope. Similarly, despite the considerable concern about cybersecurity (rightly so), there is also hope to be found, just as in Pandora's box. Having the ability to visualize the abstraction of fire, spirit, and hope requires some form of digital engineering (DE) (i.e., digital twins, modeling, simulation, and other means of both a priori, or predictive, and a posteriori, or analytical, assessments of systems for their overall cybersecurity design, development, and resiliency) when it comes to developing laws that govern the conduct in cyber. The need for the ability to visualize abstractions about an information system used to control programmable logic controllers (PLCs) leads to Chapter 4, "Cyber is a Medium (Not a Domain)."

References

(CISA), C. I. S. A. 2014. ICS Advisory (ICSA-10-238-01B) Stuxnet Malware Mitigation (Update B). *In:* ICS-CERT (ed.) *ICS Advisory.* https://us-cert.cisa.gov/ics/advisories/ICSA-10-238-01B: Department of Defense.

(NIST), N. I. O. S. A. T. 2012. NIST Framework and Roadmap for Smart Grid Interoperability Standards, Release 2.0. *In:* Laboratory, O. O. T. N. C. F. S. G. I. E. (ed) *NIST Special Publication 1108R2*. U.S. Department of Commerce.

Commission, E. 2022. The Articles of the Cyber Resilience Act (CRA). *In:* UNION, C. O. T. E. (ed.). Brussels, Belgium: General Secretariat of the Council.

Crowther, G. A. 2017. The Cyber Domain. *The Cyber Defense Review,* 2, 68–78.

(ENISA), E. U. A. F. C. 2025. Handbook for Cyber Stress Tests. European Union Agency for Cybersecurity (ENISA).

Gartner. 2024. *Operational Technology* [Online]. www.gartner.com: Gartner. Available: https://www.gartner.com/reviews/market/operational-technology-security [Accessed 2024].

Government, A. 2023. 2023-2030 Australian Cyber Security Strategy.

Jtanics 2012. Handbook for Self-assessing Security Vulnerabilities & Risks of Industrial Control Systems on DOD Installations. *In:* Department of Defense (DOD), O. O. T. S. O. D., Operational Test and Evaluation (ed.). Director, Operational Test and Evaluation, Office of the Secretary of Defense.

Law, G. F. 2009. Act on the Federal Office for Information Security (BSI Act -- BSIG). *In:* Gazette, F. L. (ed.). Bonn: German Legislature.

MD Ariful Haque, S. S., Bheshaj Krishnappa 2018. Cyber Resilience Framework for Industrial Control Systems: Concepts, Metrics, and Insights.

NIST 2014. Framework for Improving Critical Infrastructure Cybersecurity. NIST.

Obama, B. 2013. Executive Order 13636 - Improving Critical Infrastructure Cybersecurity. *In:* HOUSE, W. (ed.). White House: White House.

Scalco, A. 2022. *Measuring Disagreement in Segments of the Cybersecurity Profession as a Means of Identifying Vulnerabilities,* Colorado State University.

(ENISA), E. U. A. F. C. 2025. Handbook for Cyber Stress Tests. European Union Agency for Cybersecurity (ENISA).

EPA 2024. Information about Public Water Systems. *In:* Agency, U. S. E. P. (ed.). https://www.epa.gov/dwreginfo/information-about-public-water-systems: United States Environmental Protection Agency.

Government, C. 2009. National Strategy for Critical Infrastructure. Canada: Canadian Government.

Law, G. F. 2009. Act on the Federal Office for Information Security (BSI Act -- BSIG). *In:* Gazette, F. L. (ed.). Bonn: German Legislature.

Obama, P. B. 2013. Presidential Policy Directive/PPD-21, Critical Infrastructure Security and Resilience. *In:* Secretary, O. O. T. P. (ed.). The White House: The White House.

Scalco, A. 2022. *Measuring Disagreement in Segments of the Cybersecurity Profession as a Means of Identifying Vulnerabilities,* Colorado State University.

Schallbruch, M. & Skierka, I. 2018. *Cybersecurity in Germany,* Berlin, Germany, Springer.

Span, M. T., Mailloux, L. O. & Grimaila, M. R. 2018. Cybersecurity Architectural Analysis for Complex Cyber-Physical Systems. *The Cyber Defense Review,* 3.

4

Cyber is a Medium (Not a Domain)

"The medium is the message."

Marshall McLuhan (1911–1980),
Canadian academic and philosopher
(McLuhan, 1967)

"Sailboat racing becomes a game of chance only when you are not prepared."

Buddy Melges (1930–2023),
American Competitive Sailing Champion,
1992 America's Cup Winner

Abstract

Domain knowledge is needed to operate successfully in any industry. The fact that domain-specific knowledge is a crucial consideration when deploying artificial intelligence (AI) technologies is essential to remember. Understanding relevant industry terminology, challenges, risks, and operational dynamics is an essential key to the successful deployment of technology. The term "domain" implies defined boundaries over which some entity has some authority and control. In computer science, the term "domain" refers to a specific name assigned to an internet protocol (IP) web address associated with a space for organizing services. Cyber is more a medium than a domain. A medium conveys a message much more than the content (McLuhan, 1967). The medium is the commonality of all the creation and consumption across it that has an impact. The "cyber domain" implies a unified meta-system encompassing everything under a defined authority and control. The term "cyber domain" is great when used as an abstraction, but not so great for drafting policies and regulations, given the boundless potential of cyber interfaces. In this chapter, we attempt to explain how "cyber" is a medium that shapes and controls human association and action rather than a domain that implies boundaries. Say "cyber medium" and see how it changes perceptions and understanding about cyber. Cyber as a medium holds greater

DOI: 10.1201/9788743809128-4

value than cyber as a domain – it implies transformation and interactive influence rather than stasis. This greater value of cyber as a medium has a profound effect on one's stance on cybersecurity. Thinking of cyber as a medium creates a stance of preparedness for change rather than complacency. Systems engineering relies heavily on models to express boundaries and interfaces throughout system development and the lifecycle. Various tools and types of modeling languages can be used to help stakeholders understand a system being modeled in the cyber medium to achieve the necessary stance of change preparedness in cyber operations. Competitive sailing is used in this chapter as an analogy to illustrate the distinction between cyber as a domain and cyber as a medium.

4.1 Introduction

The history of the term "cyber" suggests that it encompasses more than just an IP address or physical space, but rather the organizational intention behind a message. Recall from Chapter 2, "History of the Term 'Cyber,'" that its origins are derived from the Greek root word "*kubernētēs*." Cyber, based on its origin as a Greek word meaning "to steer," is more of a contextual vehicle to express organizational abstraction than it is a "domain" assigned to specific site addresses that access particular web pages or services on the Internet. "Cyber" includes both well-defined processes for preventing threats beforehand and ad hoc capabilities for responding to threats in near real-time. A network domain is a system of interconnected systems, resources, and groupings of computers that share a commonality of account information and centralized services. The content of the cyber medium is always another medium. Cyber is a ubiquitous meta-system. You cannot expect to spatially organize your services to a singularly defined cyber location on the internet. That is what makes using the term "cyber domain" such a misnomer. Other than to provide a unique website address on the internet, "cyber domain" does not make sense unless it is used to describe specific operations. For example, the Department of Defense (DOD) uses the term "cyber warfighting domain" to describe a particular operational area where military forces conduct both offensive and defensive operations or engage in strategic warfighting actions. Cyber operates at near real-time speed. Cyber operations encompass both offensive and defensive measures. Data transferred from one source to another through cyber operations in the DOD's "cyber warfighting domain" must be both safe and secure in one location while enabling kinetic warfighting in another. If cyber is a medium, then based on the origin of the cyber terms, "to steer" is the medium by which data is protected.

The term "cyber" conveys the concept of administration or governance. An operational requirement is to provide situational awareness of system operations and relevant events of the critical infrastructure target's environment. A question to ask is how awareness and active, valued communication between workforce groups about ongoing vulnerabilities and newly discovered threats is maintained? Another operational requirement is to provide analytics and decision support based on operational priorities. A key question to ask is how partner-targeted activities and threat intelligence information from external organizations are coordinated for vetting and validation (Scalco and Simske, 2022). The term "cyber domain" literally implies authority and administration over everything connected to the Internet, including all devices with an IP address. No wonder establishing metrics for control is challenging – *and misunderstood.* It is tough to rule over the entirety of the cyber medium administratively.

4.2 Multi-industry Skill Sets

Understanding multiple industries outside of one's familiar skill set helps gain a broader understanding. Professional cultural differentiation can create uncertainty in design decisions or process changes in areas outside traditional areas of competency. For example, an aerospace engineering team may be highly sensitive to the safety requirements of aircraft or spacecraft. In contrast, a computer science team may be highly sensitive to the requirements of operating systems. The intersections between these fields are becoming increasingly broader. Traditional IT personnel focus on confidentiality, integrity, and data availability, known as the principle of "C.I.A." Traditional OT personnel prioritize availability, as well as integrity and confidentiality. The systems engineer may be a valuable champion for the cultural change needed. The systems engineering role differs from the IT and functional engineering roles because of the numerous interdisciplinary connections made as part of a significant system development project. The systems engineer addresses system problems by leveraging the combined knowledge and technical expertise of engineers and specialists from various disciplines and cross-functional teams, including subject matter experts (SMEs) from both physical and cyber domains. Systems engineering can help tie the technologies and training to the business costs. The cyber medium requires a range of skills from various professions, as well as professionals with knowledge about the target critical infrastructure environment (Scalco and Simske, 2021).

The sport of ice hockey, introduced earlier in Chapter 1, "Understanding the Game," is an excellent example of the knowledge, skills, abilities, and tasks (KSATs) required to comprehend cyber operations. Shifting slightly to another sport, competitive sailing competitions known as "regattas" are a great example of the knowledge, skills, abilities, and tasks (KSATs) necessary to understand cyber operations. Like ice hockey players, competitive sailors dedicate many hours to building their skills to steer a sailing craft. Both sports require cardiovascular endurance, core strength, balance and coordination, and flexibility. Both hockey and sailing require a level of controlling and steering. In hockey, controlling and steering a puck. The medium of ice in the game of ice hockey can also change and significantly impact the game and puck control. Changing conditions, such as rain, can affect ice conditions, warming the ice and creating ruts that make it more difficult "to steer" the puck. Resurfacing the ice can smooth the ice surface, allowing for quicker handling or steering of the puck. Similarly, in sailing, changing conditions affect the control and steering of the sailboat.

However, unique to competitive sailing is the need to be good in a dynamic environment by adjusting to a variety of currents, understanding wind and wave patterns, predicting weather conditions, and maneuvering around obstacles and other vessels. Race sailboats are designed for performance, speed, seaworthiness, and safety against the elements. The boat design and crew skills need to be prepared to meet the unexpected water currents and conditions. The water medium can change conditions. Similarly, the conditions of cyber mediums are subject to change. Rapid changes to the cyber medium make it challenging for organizations to effectively steer around the evolving use of AI to enhance attack methods. Steering becomes the medium by which data is protected. Just sailing involves guiding a sailboat through the water.

One of the more prestigious and oldest yacht racing competitions is the America's Cup, but the more difficult might be the Vendée Globe. The America's Cup sailing competition is a series of regatta races between two club teams. The defending team is the team that holds the trophy, and the challenger teams are the ones that battle each other for the privilege of challenging the defending team. Each team can enter one boat. The course is set for sailing around the Isle of Wight, which is determined by the race committee. The first race was held in 1851 and was not known as the America's Cup until an American team from New York City won it. Challengers are subject to complying with protocols that govern the rules of conduct, which must be agreed upon by both the challenger and the defender (unlike cyber, where rules of conduct are less defined). There is no fixed race schedule. The race dates are set when agreed to by the challenger and defender teams.

The competition represents the latest technology incorporated into the newest boat designs, utilizing complex control systems and backed by a strong sense of competition. It features experienced, skilled sailors defying wind and water, with speeds exceeding 50 knots (1 knot equals one nautical mile). As a point of reference, a US Navy carrier is designed to sail at about 30 knots. The more difficult Vendée Globe is a solo sail around the globe, non-stop, without assistance. "One woman or one man, the world and a boat. It is very clear" (Globe, 2024). For either competition, the sailor must know the medium and the domain of the sport. The medium is the environment of the seas and winds. The domain is the competition itself. Winning represents deep, intimate knowledge and understanding of the medium and its domain. The mast, sails, navigation, keel, gauges, personal skills, personal limitations, safety protocols, and wind and ocean conditions are just a few of the foundations.

Essential skills for cybersecurity require a blend of computer network and operating system knowledge (the environment), proficiency in programming languages such as Python and Java, and scripting languages like PowerShell, for penetration testing and security incident handling. The use of AI-driven threat detection systems, predictive modeling, and large learning datasets to train and test models requires the technical skills necessary to stay ahead of exploitation and mitigation techniques. Understanding firewalls, network intrusion detection systems (IDS), virtual private networks (VPNs), and network segmentation and microsegmentation are essential technical skills for professionals. Microsegmentation is illustrated as a key concept of the Zero Trust security model in Chapter 8, "Use Cases and Overview Illustrations." While these technical skills are crucial, cyber operations also necessitate problem-solving skills and a comprehensive understanding of digital environments and emerging technologies.

4.3 Cyber is a Medium

If cyber is a medium, then critical infrastructure sectors can be considered as domains. In internet vernacular, a domain is the equivalent of a physical address for delivering and accessing services, such as water giving access to a sailboat. Cyber is the environment where cyber methodologies are applied, including the functions of software applications, communication, security operations, and access management. Successful governance involves steering the data or process controls.

Critical infrastructure is a general term used to describe elements integrated to perform both physical and virtual actions. It encompasses both

Table 4.1 General cyber hygiene measures.

General cyber hygiene measures
Update to the latest version of the operating system (e.g., Windows 10).
Use multiple-factor authentication.
Use strong passwords to protect remote desktop protocol (RDP) credentials.
Ensure anti-virus, spam filters, and firewalls are up to date, properly configured, and secure.
Audit network configurations and isolate computer systems that cannot be updated.
Audit network for systems using RDP, close unused RDP ports, and apply multi-factor authentication wherever possible, and log RDP login attempts.
Audit logs for all remote connection protocols.
Train users to identify and report attempts at social engineering.
Identify and suspend access of users exhibiting unusual activity.
Reference: FBI, C., EPA, MS-ISAC 2021. Compromise of U.S. Water Treatment Facility. In: Advisory, J. C. (ed.).

the physical and cyber realms, as well as the human actor performing these actions. A critical infrastructure sector, such as a water treatment facility, can be considered a distinct domain. A professional needs industry-specific knowledge to govern a domain effectively. For example, a professional working in the power industry needs an industry-specific understanding of power systems. Connecting to the required knowledge and skills is where systems engineers come into play. Systems engineers work within a specific industry using processes and practices specific to the domain. Systems engineering is interdisciplinary. Systems engineers provide inter-system access to all engineering domains and professionals from various other domains, such as computer science, for integrated systems of systems. Sound system engineering practices help ensure that systems in a water treatment facility are cyber-resilient. For example, limiting who can release caustic chemicals and restricting the amounts that can be released is a safety control – another necessary cyber hygiene mitigation, as well as a security measure for remote access control to the system. Remote party access is commonly employed as a means for legitimate remote control of computer systems. The remote capability can also allow unauthorized entities to gain access to control systems. General mitigation recommendations published in a Joint Cybersecurity Advisory about the compromise of the US Water Treatment Facility (FBI, 2021) are shown in Table 4.1. These recommendations would apply to any critical infrastructure system to mitigate similar risks.

4.4 Systems Modeling

Systems engineering relies heavily on models throughout the system development and lifecycle process to support stakeholder alignment. General-purpose modeling languages provide a standard way to visualize and construct documents of abstractions, concepts, and system relationships in the cyber medium. Various practical tools and types of modeling languages can help stakeholders understand the system being modeled in software-intensive systems. Models support communication about system components. Modeling is a method for normalizing all data collected from critical infrastructure domain systems in the cyber medium in a standardized manner.

For example, in the systems modeling language (SysML), an actor represents an external interface, such as a person or an object, that interacts with the system within the context of the system boundary. The keel is a flat blade that sticks into the water, forming an external interface. The keel interacts with the water. Water is the medium for the sailing domain. A boat can be sailed with the keel up, but it is the keel that helps keep the boat on a straight course and right-side up. In the unified modeling language (UML), an actor specifies a specific role that can be an operator, user, or any other entity that interacts with the system. The America's Cup regatta sailor (an actor) interacts with the physical sailboat (system) to control maneuverability and performance. Control surfaces can only be controlled by direct contact from the crew or the crew using a control system that is clearly defined in the boat class rules (Cup, 2022). Sailing skills are required for competing in the America's Cup regatta, just as cyber skills are needed for the critical infrastructure system domain, such as those in a water facility or power utility.

Many of the critical infrastructure systems of interest were not developed using model-based systems engineering (MBSE). Yet, the thorough understanding of domain knowledge, coupled with the complexity of system aspects, warrants a standard way to visualize the architecture and interfaces in the cyber medium using models to ensure that components function together as they should and that new components connected to the internet do not pose unexpected and unknown system risks. The approach is like the sailing skill of adjusting to water currents. A sailor's experience level ranges from entry-level managing a one-person keelboat on a lake to that of a highly skilled expert, a qualified sailor capable of competing in world-class regattas in challenging sea conditions, or sailing solo around the globe, competing in the Vendée Globe.

What further complicates the connection of physical systems using the cyber medium is that systems can comprise either human or non-human elements that manage, control, influence, and shape the processes and activities of the system. The sailor is not necessarily needed. Tactics, techniques, and procedures used to navigate can be automated. Automated execution of industrial processes automation found in industrial control systems such as supervisory control and data acquisition (SCADA), distributed control systems (DCS), and programmable logic controllers (PLCs) can reduce human intervention in processes by using predetermined courses of action (COA) at the lowest levels in the system hierarchy. Machine automation provides near real-time control over system power and safety controls. Data is passed between these processes and information systems to achieve efficiencies and is used for decision-making about the system and the products and services it provides. Initially, information assurance (IA) focused on data security. Information assurance (IA) evolved from communications security (COMSEC), which focused on securing information systems and networks. Processes and practices add the element of mission assurance. Systems engineering principles and practices, such as digital engineering (DE) and model-based systems engineering (MBSE), can help interdisciplinary teams immediately overcome the vulnerabilities of the cyber medium for specific critical infrastructure.

4.5 Summary

Systems engineering facilitates the successful design, development, integration, testing, transition, operations, sustainment, maintenance, and disposal of a system within a specific domain. Systems engineering can help stakeholders understand inter-system complexities by using models, various tools, and different types of modeling languages to visualize interfaces. Similar to competitive sailing, multiple interacting interdisciplinary elements, crew skills, and sailboat subsystems function together in changing conditions such as the seas and weather. An effective team needs to understand how to integrate and manage systems across their lifecycle, whether they are designed with the cyber medium in mind or built without prior knowledge of cyber, as many systems are.

Marshall McLuhan's theory that "the medium is the message," meaning that communication technology shapes content more than the message conveyed, applies to cyber technologies (McLuhan, 1967). The cyber medium influences the domain in focus, such as the warfighting domain, but cyber itself is not the domain. Content in cyberspace is predetermined by the medium, which determines how information is perceived: the cyber medium shapes and influences system interaction and behavior. Indeed, industry-specific domain knowledge is essential for operating successfully. The term "cyber domain" is great as an abstraction, but not for drafting policies and regulations. Identifying cyber as a medium changes perceptions and understanding of cyber and is perhaps better suited for expressing organizational abstraction to specify broadly applicable policies and guidelines. The first task for anyone steering should be to ensure the sailboat or system is set up correctly, pointed in the right direction, and that the navigation rules of the waterways are understood. Good advice is never assuming a clear passage to steer. Never assume the system is uncompromised when you have the responsibility to steer data or any assets in the cyber medium.

References

Cup, A. 2022. AC75 Class Rule v2.0.

Globe, V. 2024. *Vendée Globe* [Online]. www.VendeeGlobe.org. Available: https://www.vendeeglobe.org/en/what-vendee-globe#:~:text=10th%20November%202024.-,The%20course,Globe%2C%20in%202016%2D2017. [Accessed 2024].

McLuhan, M. 1967. *The Medium Is the Message*, Penguin Books.

FBI, C., EPA, MS-ISAC 2021. Compromise of U.S. Water Treatment Facility. *In:* ADVISORY, J. C. (ed.).

Scalco, A. & Simske, S. 2021. Model for Multi-concern Assurance in the Digital Transformation of Engineering of Critical Infrastructure Control Systems.

Scalco, A. & Simske, S. 2022. Measuring Stakeholder Alignment to Overcome Control System Vulnerability. Department of Homeland Security (DHS) CISA Industrial Control Systems Working Group

5

Cyber Paradox

"It takes all the running you can do to keep in the same place. If you want to get somewhere else, you must run at least twice as fast as that!"
– Mad Hatter in Lewis Carroll,

"Alice's Adventures in Wonderland"
(Carroll, 1865)

"Don't compromise yourself. You are all you've got."
– Janis Joplin (1943–1970), American singer

Abstract

The "cyber paradox" represents complexity. Professionals may wonder where to begin to deal with the increased complexity of connected devices (i.e., increased monitoring at the lowest component level), uninterrupted system availability (i.e., improved control of functionality at the system level), and ever-increasing statutes and compliance regulations across sectors (i.e., at the meta-system level). Cyber connects non-traditional IT, known as operational technology (OT) (Perkins et al., 2016), and IT professionals alike, including technicians, engineers, business enterprise planners, and IT network operators. It requires a unique understanding of both the physical system and IT systems to operate effectively. A paradox of cyber is that while it may seem the traditional enterprise IT professional is more adept at dealing with complex software-intensive systems, the IT professional might introduce cyber vulnerabilities when trying to defend them. Engineering and other technical professionals may hold the most valuable domain-specific system knowledge about actual physical components (e.g., valves and pumps), local controllers, and intelligent devices (e.g., process sensors and actuators) that interface in near real-time (e.g., microseconds) with networking and communication components than the person with only computer skills. The paradox is that cyber professionals must possess both IT and physical system skills,

DOI: 10.1201/9788743809128-5

as well as domain-specific knowledge. Few professionals have the insight and experience in both realms to effectively integrate innovative technologies and capabilities into control system operations and management practices, thereby overcoming cyber vulnerabilities. Systems engineering practices can help focus the entire team on meeting system requirements, achieve effective cybersecurity, and develop principled technical leaders who are comfortable working with cross-functional teams to prevail in a specific system domain.

5.1 Introduction

Systems engineering helps address uncertainty and complexity. There needs to be a thorough understanding of a system's functional and non-technical requirements, as well as an understanding of how all the various aspects of the system work together to meet these requirements. The approach we recommend incorporates all the individual elements and ensures they operate together as intended, despite their interdependencies and complexity. The systems engineer has a general understanding of everyone's role and contribution, aligning with the organization's goals. Requirements may be decomposed from existing requirements into detailed, lower-level parts or derived as part of a new requirement analysis process. Systems engineering methods and processes refine requirements as part of the design activity into a configuration that everyone on the cross-functional team can understand. High-level requirements are decomposed into functional, operational, and technical requirements, which are then allocated across the system. The organization and configuration management (CM) of the requirement baseline help the team to overcome system complexity. Configuration management (CM) and other tools, such as the integrated master schedule (IMS), are mapped to the work breakdown structure (WBS), and organizational structures, such as integrated product teams (IPT), are complementary to the systems engineering practice.

5.2 System Requirements

When asked for advice on what to do next in complex situations presented by cyber, we might think of an exchange between two characters in "Alice's Adventures in Wonderland" by Lewis Carroll (Carroll, 1865). In the story, the character Alice falls asleep in a meadow and dreams she follows a white rabbit down a rabbit hole. As Alice tries to find her way out of a strange, confusing world throughout the story, the Cheshire Cat in a Tree provides vague details in response to Alice's questions: "*Would you tell me, please, which way I ought to go from here?*" The Cheshire Cat, in response, asks, "*Where do you want to go?*" "*I don't much care where,*" Alice says. "*Then,*" said the cat, "*it doesn't much matter which way you go.*"

Using systems engineering practices can help guide the team by effectively focusing on defined system functionality and requirements, such as cybersecurity, throughout the entire software development lifecycle. Cyber does not belong to just one discipline. System protection from cyber vulnerabilities requires hardening the system by reducing the vulnerability surface from potential attack vectors. This act of system hardening requires a thorough understanding of the system. Without knowing the system, it is difficult to determine what normal looks like and what a compromise entails. Part of the "pre-flight" and preparation for any cyber design is to define, understand, and prevent future compromise. Recall from Chapter 3, "Cyber Policy, Regulations, and Guidelines," that the eight-step process for self-assessment for vulnerabilities and risks of industrial control systems (ICS) is a method to ensure that operations that depend on critical infrastructure systems are secure. You cannot afford to let your network be compromised. In a world where many channels of communication were once independent (phone, newspaper, mail, television, radio, etc.), they are now dependent on your cybersecurity architecture. To paraphrase Janis Joplin, it is all you have got. Without system requirements, it is difficult to know when design and integration activities are complete or to verify if the system meets the needs. Incentivizing IT professionals alone will not create a more cyber-secure system. Involvement by all member disciplines is required. Requirements communicate necessary functions and system features between components touched by various disciplines throughout the system's lifecycle. Systems engineers can help manage requirements to overcome complexity in engineered systems and communicate effectively with the various disciplines involved, from design to the management of integrated components. Multi-layered system engineering practices, such as configuration management (CM), establish the system baseline and manage it when changes are made to the system throughout the system lifecycle, including upgrades and the integration of new capabilities. The systems engineering practice of configuration management (CM) maintains traceability to requirements, system inventory, and document changes within the system.

System inventory is crucial in detecting, identifying, and responding to cybersecurity events. Without an accurate system inventory, distinguishing between an event (i.e., an observable system occurrence) and an incident (i.e., a disruption of normal system operations) is like skating on thin ice. There is a high chance of something going wrong. Typically, the role of the chief information security officer (CISO) or the security operations center (SOC) is to document a cyber incident report. The caveat is that the cyber incident must be known before it can be reported. If the system operator fails to recognize that an event occurrence may be a cyber incident, the incident will likely remain unreported. The ability to recognize an event matters because an organization has to know what to monitor and how to monitor it in order to plan incident management and response. The likelihood of taking an action that could lead to negative consequences is high. Incident management is the approach to detecting, responding to, and managing events that could cause the loss or disruption of functions, operations, or services. The entire team needs to work together as a cohesive unit.

5.3 Effective Cybersecurity

Effective cybersecurity concepts include indicators of compromise (IOC) and indicators of attack (IOA). IOCs and IOAs are used to detect, identify, and respond to cybersecurity incidents. IOCs are static data or artifacts used as evidence that indicate potential compromise or malicious activity. IOAs focus on behavioral anomalies, such as unusual lateral movements, to detect attackers. Unusual behavioral activity aligned to known tactics, techniques, and procedures (TTPs) can be used by cyber defensive operations for early detection of compromise in a system. IOAs are dynamic, behavior-based concepts used to anticipate and prevent potential attacks before they can cause damage. The challenge is to recognize what normal network traffic looks like so that unusual network traffic can be identified as potential malicious activity. It does not matter which way you go, but it does matter that you know where you are (e.g., system and device inventory) and where you are going (e.g., network mapping and enumeration or "discovery"). The objective is to use near-real-time threat actor detection before a system is compromised to achieve effective cybersecurity. To do so requires knowledge of the system that cross-cutting systems engineering and configuration management (CM) practices provide. It also requires knowing when an indicator of attack (IOA) actually occurs. The non-proprietary more situational awareness for industrial control systems (MOSAICS) framework offers a comprehensive, vendor-agnostic, threat-informed, and integrated approach to control system defense. More importantly, utilization of the framework supports the identification and maintenance of accurate inventories at the lowest field device level, which is foundational to securing critical infrastructure in a cyber world. MOSAICS is an integrated operational capability that supports baselining tools and programmable logic controllers (PLCs) to enhance situational awareness and defense of task-critical assets in industrial control systems (ICS) (Rich Scalco, 2018). An accurate inventory with safe and active enumeration is necessary to gain visibility into the operational technology environment found in critical infrastructure. Systems engineering can help achieve effective cybersecurity through iterative system requirements analysis, which helps remove system deficiencies. The systems engineering process, encompassing requirements elicitation, analysis, specification, verification, validation, and requirements management, is valuable for addressing stakeholder needs and expectations, thereby achieving effective cybersecurity and system resiliency.

System requirements form the basis of system architecture and design (INCOSE, 2021). System requirements management is used to understand the main components of physics-based systems that define the actual physical processes (e.g., field devices such as valves and pumps), local controllers such as distributed control systems (DCS), programmable logic controllers (PLCs), remote terminal units (RTUs), and intelligent devices for sensing physical processes (e.g., process sensors and actuators). Since requirements drive everything, the time to discuss or argue over cybersecurity approaches is before anything is built. For example, a Zero Trust strategy requires authentication for an actor (e.g., a person or device) attempting to access network resources (CISA, 2023). An operational requirement to test is access control mechanisms. Organizations in sectors such as water and wastewater treatment often rely on limited internal resources and depend on physical access control systems to manage their operations. These physical access control systems can also connect to the Internet today. Examples include keycard systems, smart locks, mobile access control systems, and other visitor management systems that give access to a site. Questions to ask include: "[Does] our organization have an employee security vetting process?" and "[Does] another outside organization [have] access to our organization's industrial control system (ICS) and supervisory control and data acquisition (SCADA) capabilities?" (Scalco and Simske, 2023) Clear procedures must be established to test upgrades prior to widespread deployment. Remember to phase out any legacy credentials that allow an actor access. Remember that the main components of operational technology (OT) systems are located at the hardware and field device levels, which are discussed in Chapter 6, "Cyber Vulnerabilities Start at the Hardware Level." Controls that limit access to physical assets at the field device level must be identified, at a minimum. Cyberattacks are targeting programmable logic controllers (PLCs), which can compromise the functionality of critical infrastructure.

Another operational requirement is the situational awareness of system operations and relevant events. Personnel may lack the expertise to implement and integrate innovative solutions while maintaining the security necessary to protect the systems. Hosting technology in cloud applications may have advantages, but it may not be as cost-effective if additional costs are associated with accessing support expertise for implementation. Every system element has to be in scope. The systems engineer brings a holistic systems view to the target environment. One of the valuable roles of the systems engineer is to help identify stakeholders and key roles such as the system architects, operators, integrators, testers, and vendor partners. Addressing the control system at the control system field device level is crucial and requires engineering expertise. Facility technicians and engineers are critical. These professionals on the team must be engaged.

A question to pose might be how vulnerabilities and newly discovered threat information are shared. System requirements should be traceable to the Zero Trust authentication capability. The more situational awareness for industrial control systems (MOSAICS) is a reference framework that provides non-proprietary, vendor-agnostic, reusable requirements traceable to the desired goal and objective. MOSAICS is illustrated in Chapter 8, "Use Cases and Operational View (OV) Illustrations," and discussed in more detail in Chapters 16, "Modernization Strategies," and 17, "Architecture Approaches." Technical requirements are mapped and traceable to operational and functional requirements. Knowledge about the control system is needed for safe operations. Knowledge of the system's functions, operations, and other technical aspects is also required. The MOSAICS reference framework supports notification for account creation or modification in an automated fashion when the account creation or modification is performed in conjunction with other anomalous activity (Rich Scalco, 2018; Vermilye, 2021):

1. Control system shall review host log files for user account changes from the baseline.
2. Control system shall review host log files for changes to the user account from the baseline.
3. The control system shall generate alerts for detected attempts to escalate or elevate account privileges.
4. The control system shall detect changes in domain-level user accounts.
5. The control system shall display an alert for detected attempts to escalate or elevate account privileges.

Some of the ideas presented here are based on earlier published research (Scalco, 2022). Wolfgang Schwab and Mathieu Poujol surveyed the state of industrial cybersecurity in 2018. Most aspects have remained unchanged in the state of industrial cybersecurity teams since then. Operational technology (OT) and IT professional domains alike (e.g., technicians, engineers, business enterprise planners, and IT network operators) must be fully aligned. Cybersecurity teams across both OT and IT are often a "coalition of the willing" only after something unexpected or undesirable occurs. Schwab and Poujol's findings indicated that most companies surveyed considered cybersecurity for operational technology (OT)/industrial control systems (ICS) a significant priority. While most thought the company would likely be a target of a cybersecurity attack in the operational technology (OT) market, only 23% were compliant with the minimal mandatory industry standards or government guidelines and regulations (Schwab and Poujol, 2018). While half of the participants stated that the company had not experienced a cyber incident in the previous 12 months, Schwab and Poujol observed that participants may not have recognized if they had, as most could not detect an event or track it. The data showed that 8% revealed they do not know, and 10% do not measure cybersecurity incidents involving control systems. Most companies surveyed had already started their digital transformation, thus increasing the surface area for cyberattacks. Only 30% of the participants responded that their organization is required to report industrial security breaches and incidents to a regulatory body. Schwab and Poujol cited the low maturity of cybersecurity for operational technology (OT) industrial control systems (ICS), which is further limited by a skills gap and a lack of collaboration between OT and IT professionals. A finding is that the skills gap and a lack of collaboration between OT and IT professionals often result in varying goals, processes, tools, and even language (Schwab and Poujol, 2018).

Furthermore, 58% of the companies surveyed reported that hiring industrial control system (ICS) cybersecurity personnel with the right skills is a significant global concern. Few individuals possess the insight and experience in both the operational technology (OT) and IT realms to address cyber vulnerabilities in critical infrastructure effectively. Using systems engineering practices can help by effectively focusing the entire team.

5.4 Market Size

The operational technology (OT) market was valued at USD 210.06 billion in 2024. The market is expected to grow at a compound annual growth rate (CAGR) of 9.6% from 2025 to 2030 (Research, 2024). The operational technology (OT) market is expected to reach a value of USD 230.87 billion by 2025 and USD 364.74 billion by 2030 (Research, 2024). Recall from the study that in 2018, only 30% of participants in Wolfgang Schwab and Mathieu Poujol's research responded that their organization is required to report industrial security breaches and incidents to a regulatory body, which represents a significant percentage that manages critical infrastructure such as power, water, facilities, and manufacturing (Schwab and Poujol, 2018). The scale and scope of the operational technology (OT) market are substantial, as the tools, technologies, principles, and challenges are integral to critical infrastructure industries. Every element of the critical infrastructure system must be included in the organization's security strategy from the ground up. The plan must consist of people, processes, and tools, as well as physical and cyber access to all technology connecting devices and the target environment site. All elements of products and services rely on these underlying operational technology (OT) elements to manage and optimize complex systems effectively. The average OT components are essential in critical infrastructure. Factors driving the growth in the OT market include business demand for near real-time data analytics, predictive maintenance, and the adoption of modern cloud-based solutions to enhance visibility and control over operations (Research, 2024).

5.5 Principled Leadership

Principled leadership is vital to steering a successful team. The principled leader creates an environment that empowers everyone on the team to contribute to the success of the mission and the organization. Professionals already working in the field, those practitioners who want to up-skill, or students starting studies who wish to enter the field need foundational knowledge of circuit topologies relevant to embedded systems (i.e., computer hardware and software designed for a programmable or fixed-function to perform a task within a device), knowledge of Internet of Things (IoT) architecture, hardware enablers, sensors, and physical principles in communication components such as wireless local area network (WLAN) communication systems. Knowledge is also required to design and analyze synchronous sequential systems, multiplexers, digital circuits, and state machines, as well as to implement a programmable logic array (PLA) to construct reconfigurable digital circuits. Knowledge of computer networking and operations principles is required, including the ability to configure and operate routers, as well as specialized skills in hardware and software design.

Additionally, industry-specific domain knowledge is required for whatever critical infrastructure sector is supported, such as power and energy management and optimization for the power and energy industry. Equally, cyber fundamentals are needed. Ultimately, principled leaders who are comfortable working with cross-functional teams are essential to overcoming the "cyber paradox." All competency professionals must be engaged in the cybersecurity design and development of the system, as well as the continuous improvement and management of the system's cyber risk. Risk management essentially defines the concept of cyber-physical security and the paradox between IT and OT teams. Small critical infrastructure organizations must have a process in place to revert to manual operations in the event of a cyber-related system failure and support recovery actions to maintain a mission-capable state, including maintaining system backups and other essential resources. A key question to ask is whether the organization maintains system backups and how often these backups are tested (e.g., every 30 days, every 90 days, or at other intervals). Know this: an organization must have procedures in place to minimize the exploitation of vulnerabilities and collect and archive incident data to support incident response. The systems engineer helps establish the direction for specific discipline domains to meet business needs by establishing a collaborative environment that leverages the expertise and resources from across the organization.

In an organization with cyber business operations, which is the norm today, assume that team members are doing their best. But are they being asked to do the right things? In Lewis Carroll's book, eventually, the Cheshire Cat leads Alice to her final garden destination. There, Alice encounters three living playing-card characters named Two, Five, and Seven, painting the roses on a white rose bush red. The bush was supposed to be red, but the gardeners mistakenly planted a white bush. "*If the Queen was to find out, we should all have our heads cut off, you know. So, you see, we're doing our best.*" Painting the white roses red is analogous to asking someone to perform manual system log reviews to monitor, troubleshoot, and audit device operations. There is a better way. Alice's experience in the garden is not as idyllic as she thought when she first peered through the keyhole in the small door, nor is the functionality found in the cyber domain as idyllic. It comes with new risks to the system.

Similarly, in the cyber domain, some professionals are doing their best within an overall flawed approach, thereby contributing to an organization's technical debt, which makes it challenging to evaluate the effectiveness of security measures. Understanding the role of different disciplines ensures that professionals work together effectively to achieve the organizational vision. Cyber influences "systems" trade-off decisions in the design and development of critical infrastructure systems. Knowledge of critical cyber concepts, such as IOC and IOA, used to detect, identify, and respond to cybersecurity incidents requires a combination of knowledge in IT and OT. There is an important relationship between phases and trade-off decisions throughout the system's lifecycle. The integration of security solutions requires a high level of knowledge, expertise, and skills that no single practice, person, or vendor can provide. Principled technical leaders who are comfortable working with cross-functional teams can help an organization prevail over the "cyber paradox."

5.6 Summary

This chapter gives an overview of the "cyber paradox" and the complexity of systems that interface in the cyber medium. Systems engineering practices can help reduce complexity by effectively focusing the entire team on integrating new technologies and capabilities into control system operational procedures and management practices. Systems engineering principles and practices, such as requirements management, configuration management (CM), work breakdown structure (WBS), and model-based systems engineering (MBSE), are used to achieve effective cybersecurity and develop principled technical leaders who are comfortable working with cross-functional teams to overcome the "cyber paradox." In Wonderland, the character of the Queen, who leads a strict rank hierarchy, invites Alice to play a game of croquet without rules or an endpoint, which quickly descends into chaos. The systems engineer aligns values and perspectives that guide the team regardless of the discipline. Recall from Chapter 1, "Understanding the Game," that an essential cyber skill is collaboration. Independent of cross-functional teams, organizations may not recognize if they have indicators of compromise (IOC), as most cannot detect an event or track it in the absence of the knowledge that a diverse, experienced team brings. The Cheshire Cat also said, "*Every adventure requires a first step. Trite, but true, even here.*"

References

(CISA), C. A. I. S. A. 2023. Zero Trust Maturity Model Version 2.0. *In:* DIVISION, C. (ed.). Cybersecurity and Infrastructure Security Agency (CISA).

Carroll, L. 1865. *Alice's Adventures in Wonderland,* Oxford, Clarendon Press.

Incose 2021. *Incose Systems Engineering Handbook 5th Edition - Prototype Draft*, Incose.

Perkins, E., Rueb, D., Moran, M., Steenstrup, K. & Zimmerman, T. 2016. Predicts 2017: IT and OT Convergence Will Create New Challenges and Opportunities. Gartner.

Research, G. V. 2024. Operational Technology Market Trends.

Rich Scalco, D. B. W., Jorge Lacoste, John Andrews, Bill Beary, Ross Roley, 2018. *More Situational Awareness for Industrial Control Systems (MOSAICS) Joint Capability Technology Demonstration (JCTD)* [Online]. IACD Automate: Johns Hopkins University Applied Physics Laboratory (JHU APL). Available: https://www.iacdautomate.org/may-2018-integrated-cyber [Accessed October 2019 2019].

Scalco, A. & Simske, S. 2023. A model for measuring multi-concern assurance of critical infrastructure control systems. *Systems Engineering*, 12.

Rich Scalco, D. B. W., Jorge Lacoste, John Andrews, Bill Beary, Ross Roley, 2018. *More Situational Awareness for Industrial Control Systems (MOSAICS) Joint Capability Technology Demonstration (JCTD)* [Online]. IACD Automate: Johns Hopkins University Applied Physics Laboratory (JHU APL). Available: https://www.iacdautomate.org/may-2018-integrated-cyber [Accessed October 2019 2019].

Scalco, A. 2022. *Measuring Disagreement in Segments of the Cybersecurity Profession as a Means of Identifying Vulnerabilities,* Colorado State University.

Schwab, W. & Poujol, M. 2018. The State of Industrial Cybersecurity 2018. *In:* LAB, K. (ed.). Munich, Germany: CXP Group.

Vermilye, C. More Situational Awareness for Industrial Control Systems (MOSAICS) Requirements. MOSAICS Industry Day #3/TechConnect, October 19, 2021 2021 Washington, D.C.: MOSAICS JCTD, 13.

6

Cyber Vulnerabilities Start at the Hardware Level

"A ship in harbor is safe, but that is not what ships are built for"
– John A. Shedd (1859–1928),
American author and professor (Shedd, 1928)

"Humility is the solid foundation of all virtues"
– Confucius (551 BC–479 BC), Chinese philosopher

Abstract

Cybersecurity vulnerabilities start at the hardware level. In fact, cybersecurity vulnerability can be viewed as an issue even at the most rudimentary component level, that is, at the level of individual components. In electronics systems, these components include resistors, capacitors, diodes, transistors, and more complicated components such as integrated circuits (ICs). Even the component level poses a threat due to the rapid advancement of artificial intelligence (AI) technologies, which make the authentication, and even forensic-level authentication, of different components, chips, and devices feasible. Leaking emanation attacks have also been facilitated by the rapid development of advanced drone technologies and imaging (e.g., light detection and ranging (LIDAR)) and sensing (e.g., next-generation infrared (IR) sensors). Most existing hardware – and virtually all new hardware – is not "air-gapped" from the internet, but is instead connected and ready, willing, and able to share information with its surroundings. Hardware serves to move around bits, which can be information, code, or commands for devices or any combination thereof. "Moving around" refers to the fact that these bits are either input or output (I/O) from the hardware. Therefore, any threat agent interested in compromising the hardware may only need to monitor this I/O to determine the best approach to compromise the system. Monitoring for compromise is an area where AI is also particularly important: various AI algorithms can be used to uncover anomalous behavior in either the input or output streams, or both.

DOI: 10.1201/9788743809128-6

6.1 Introduction

Cybersecurity vulnerabilities start at the hardware level. Unsurprisingly, cybersecurity vulnerability can be viewed as an issue even at the most rudimentary component level, that is, at the level of the components. In electronics systems, these components include resistors, capacitors, diodes, transistors, and more complicated components such as integrated circuits (ICs). Depending on the circuitry, the power involved, and the nature of the application, these components can provide so-called leaking emanations such as electrical and radio signals, along with vibrations and sounds, which can be used by adversaries for a variety of purposes (up to being able to decode information in the circuit). These can be viewed as rudimentary physical security threats. Still, given the electronics involved in the attempts to spy on these signals, rather than the electronic parts themselves, they are cyber-physical security threats.

Gone are the days when we can assume that an isolated device, and especially an air-gapped one, has no vulnerability surface to attackers. Even at the component level, there is a threat due to the rapid advancement of artificial intelligence (AI) technologies, which make the authentication, and even forensic-level authentication, of different components, chips, and devices feasible. Leaking emanation attacks (e.g., being able to deduce communication information based on emanated electromagnetic spectrum energy) have also been made easier by the rapid development of advanced drone technologies and advanced imaging (e.g., light detection and ranging (LIDAR)) and sensing (e.g., next-generation infrared (IR) sensors).

In addition to such elaborate and component-specific surveillance, prospective vulnerability exploiters (i.e., threat agents ranging from basement hackers to advanced persistent threats, or APTs) may need to have little to no knowledge of the hardware's design, processes, or manufacturing to exploit vulnerabilities. This lack of need for specific knowledge is because most existing hardware – and virtually all new hardware – is not "air-gapped" from the internet, but is instead connected and ready, willing, and able to share information with its surroundings. Hardware serves to move around bits, which can be information, code, or commands for devices or any combination thereof. "Moving around" means that these bits are either input or output (I/O) from the hardware. So any threat agent interested in compromising the hardware may only need to monitor this I/O to determine how best to compromise the system. Gaining knowledge of compromise is also particularly important in the area of AI: various artificial intelligence (AI) algorithms can be used to uncover anomalous behavior in either the input or the output streams. Behavioral analytics can be combined with AI to analyze large datasets, enabling the monitoring and analysis of system activity to detect insider threats, advanced persistent threats (APTs), and zero-day attacks. Susceptibility to APTs is addressed in Chapter 7, "Threats to Information-intensive Systems," and in Chapter 22, "Behavioral Analytics." Unlike traditional enterprise IT systems used for general-purpose business operations, industrial control systems (ICS) operational technology (OT) is used for uniquely specific control system functions that involve sensors, actuators, programmable logic controllers (PLCs), and intelligent devices sensitive to specific timing and synchronization for maintaining operations. Precise timing is critical in process control systems found in high-speed, automated environments. Behavioral analytics can be used to detect anomalous behavior, insider threats, and advanced persistent threats (APTs) by establishing a baseline of known good users, devices, and applications on the system, and then identifying anomalies in near real-time using artificial intelligence (AI), machine learning, heuristics, and/or rule-based methods. This approach is particularly practical in identifying zero-day attacks that exploit previously unknown software flaws, which are cyber vulnerabilities. A question to ask is whether timestamp data is configured accurately, synchronized, and managed across the system. A follow-up question to ask is whether timestamp data is used to reference persistent time-based trends in the process control system.

With these comments in mind, the following hardware elements are examined, and some of the hardware-related concerns that necessitate a zero-trust approach are addressed. For each specific hardware element, an example of how one might corrupt the element is included. Remember that we are not threatening your system by revealing a possible attack to the world. In accordance with Kirchhoff's principle, there is no security through obscurity, and so these potential vulnerabilities need to be shared since otherwise, those who would protect your information may not know about them, but assuredly, those who would attack your network do. A corollary to Kirchhoff's principle is what we will call "The Principle of Cybersecurity Humility": If you can envision a way of attacking your system, you must assume that threat agents have already thought of the same way, and so you must protect your network from that attack.

With that, we start with hardware chips and cover different elements of hardware, leading up to and culminating with computer peripherals. In so doing, the breadth and depth of potential hardware vulnerabilities will be illustrated, at least superficially.

6.1.1 Chips

By "chips," we mean integrated circuits, or ICs, which themselves are systems of hardware. That is, most chips contain a large set of micro-, nano-, or pico-scale transistors and associated electronic components, including miniaturized capacitors and resistors, to help create logic circuitry in a single component that can be plugged into a larger computer board.

How might I, as a rogue agent, use a chip for nefarious purposes? If the chip sends data somewhere, either to another part of a system or as an output stream from the system, then I would try to use the chip for exfiltration of data by "fake" encrypting it to make it appear as if it were legitimately encrypted. An interesting possibility might be to use a trusted platform module (TPM) as the threat-exploiting hardware. A TPM is a microcontroller IC that stores, presumably securely, various artifacts used for authentication, including certificates, encryption keys, passwords, and the like. Thus, a TPM is integral to verifying and validating that a platform is both historically and currently trustworthy. The "verification and validation" (a usual part of any systems-engineered platform) element is called attestation and consists of the TPM providing a digital signature for specific measurements stored in the hardware using a key (on the TPM) that is certified by a certificate authority (CA). Since TPMs can be used in personal computers, mobile phones, and various types of network equipment (such as bridges, hubs, modems, network interface cards, routers, and switches), the threat surface of TPM compromise is wide indeed.

Based on this attack possibility, it is clear that a legitimate exploit by a rogue agent involves creating a counterfeit trusted platform module (TPM). Remember that the sketchiest supplier in your supply chain and the lowest-priced IC in your system may well cost you more than the rest of the system in the long run if you do not validate the entire procurement process. Characterization of chip behavior and routine testing for abnormal behavior are important parts of defeating "counterfeit chip" attacks.

6.1.2 Cache

For a long time in computer design, a buzz phrase might have been, "*Cache me if you can.*" While changes in how computing is generally performed, particularly for images, have shifted in recent years (e.g., to graphics processing units (GPUs) and to significantly increased amounts of RAM, the latter of which is discussed below), the cache remains important in many gaming computers and systems. Cache is essentially a storage layer (often a "card" plugged into a motherboard) that enables high-speed movement of data. This data is often transient, particularly when preloading the next few screens in a game. During high-speed episodes of a game, data is served to the processing unit(s) faster than would be possible from the main storage drive or RAM (cache is often referred to as high-speed static RAM, in fact).

Cache has known vulnerabilities. One vulnerability is *web cache deception*, wherein a threat agent manipulates a cache to store the user's private information, allowing the attacker to gain access to the cached data through "normal" cache access approaches. This type of attack is similar in design and process to attacks on browser caches; therefore, there is a high degree of understanding of "best practices" among both ethical and non-ethical hackers. A second major vulnerability is *web cache poisoning*, where the threat agent intentionally stores malicious data in the web cache server, where it can be distributed as malware to other parties using the cached server.

Stopping cache-based attacks can then benefit from inspecting the data in the cache before transmitting it, executing it, or using it to change settings on other elements in the computer. It is likely that one reason such thorough inspection does not occur is that it is at odds with the primary reason for having a cache in the first place: to increase the mean throughput of data between memory and processing. Approaches such as cache mirroring or cache pre-fetching might be useful for striking a balance here and are suitable for both throughput and security.

6.1.3 Memory and RAM

Random access memory, or RAM, was mentioned above and is the preferred type of memory for most processes since any address within the RAM can be accessed without having to reposition a "pointer" to memory, such as is needed for other direct-access storage media including CDs, DVDs, and hard disks, where the access time is dependent on a number of factors including size of the data structure to access, last data accessed, and the physical layout of memory on the storage element. Tricks such as "locking" data in memory (using "LockBits" or similar commands, depending on the coding language) are, in fact, ways of keeping the direct-access media from having unpredictable access times (either locking a larger data structure in RAM or a conveniently reduced neighborhood of the larger direct-access media).

Attacks on memory are well-known, with one obvious example being SQL injections that allow memory outside of a legitimately accessed data structure to be overwritten. This memory could be data, code, or control strings (including the registry), meaning a nearly limitless number of complications are possible. If I were attempting to simply mess up a computer's memory, I would attempt memory access through indirection operators such as pointers, in hopes these could perform an end-around to the normal intrusion protection methods, such as SQL validation. One means of preventing memory attacks is the use of digital signatures of sections of memory (hard drives, RAM, and even cache, as plausible) to ensure that sections presumably unchanged by recent operations are, in fact, unadulterated.

Another type of memory attack is the *cold boot attack,* in which a black-hat hacker with physical access to a computer performs a RAM dump by performing a hard reset on the target machine. A cold boot attack is a specific type of the more general side-channel attack, which broadly encompasses electromagnetic emanation attacks discussed in the Power section below.

6.1.4 Redundant array of independent disks (RAID)

RAID means a *redundant array of independent disks*, a term dating back to the era of spinning disk hard drives. A redundant array of independent disks (RAID) might more properly be thought of as a redundant array of storage elements, or a RASE, or even more generically as simply a redundant backup process. Whatever moniker you prefer, a RAID system stores copies of data in different locations on multiple physical storage devices and so provides a means of retrieving data should a hard disk, solid-state drive, or other storage element fail.

Improperly implemented, however, a RAID system may provide an enhanced threat surface, a feast for fraud, as it were. With data moving about between two or more physical elements, there are more chances for person-in-the-middle attacks, for partial network spoofing (e.g., IP spoofing), not to mention the old-school thuggery of physical theft/damage. If I were a black hat hacker, however, I might find the RAID system far more tempting as a means of infiltration, camouflage, and, ultimately, exfiltration of content. A key vulnerability point for RAID hardware is when the original data and the backup are being synchronized (or "synced"). At this point, malware may inject data (which, remember, can be any combination of information, code, or system control commands) in both streams and synchronize the two accounts with the extra data in both. Malware that has the same name as the RAID software and hides in the registry, for example, is a particularly insidious, ambitious, and long-term-minded attack. One way to defeat this malware is to log the digital signature of the data being backed up, both before and after the backup, using timestamps and one-way logging (where no data is returnable from the logging location) approaches. As usual, multi-factor authentication (MFA) of the logging request is a good idea.

6.1.5 Communications

Communications within hardware are often empowered by a bus (or "data bus"), which is a series of connections between components inside a computer or between computers. A good mnemonic for control of the bus, with tongue firmly in cheek, might be ROSA PARKS: Real-time Operating System Access Protocol Allows Requests to Key Subsystems. That is, access to the bus is requested by the hardware element through a request (specific command word or sequencing of bits) called an interrupt request (IRQ). It is called an IRQ because, if authenticated and authorized (validated and allowed), the IRQ causes the interrupt controller to send the interrupt request (INTR) to the CPU, interrupting its current activity in order to process the INTR (an override command, if you will). Note that a graphics processing unit (GPU) can send an INTR to the CPU (see "GPUs, graphics, and displays" below).

As expected, there is a tight coupling between communications hardware and software used to control. Device drivers, for example, are software programs specific to a particular hardware device attached to a computer (see "Peripherals" below) and "translate" between the language of the hardware and the operating system and/or its applications that call the hardware. The CPU, on the receiving end of the bus, typically performs a fetch-execute process (instruction cycle), wherein the fetch loads the bits from the current program counter into the instruction register, and the *execute* performs the requested commands associated with the bits in the instruction register. Validation of the fetch bit string prior to fetching (pre-fetch validation) is one way to prevent malicious instruction registers from being loaded into the processor. As an attacker, of course, the communications bus offers a lot of opportunities. Internal denial of service, for example, could be mediated by a rogue chip that continues to send interrupt requests. Suppose such a chip is connected to the bus. In that case, it may be programmed to remain hidden (with or without ongoing data collection or data mining for additional illicit purposes) until a specific time, at which it can cause the maximum financial damage. Such "sleeper malware" is particularly insidious since the effective infiltration may take place long before the attack is initiated.

6.1.6 Power (including batteries)

Power is increasingly fungible as more and more computing of value is performed by mobile devices. This mobility poses a threat for several obvious reasons, including the usual vulnerabilities associated with connecting to public Wi-Fi spots and shoulder surfing. However, the power use itself may provide a threat surface. Sensitive measurement of power use within a system can potentially be used to classify what the system is being used for at any given moment, and this form of "meta-espionage" is likely to become increasingly precise in the future, driven by the profound ongoing advancements in AI.

Several years back, we were able to classify attention states during learning (Liu, 2017) and recognize different human activities (Liu, 2016) based on body surface potentials. Power use by mobile computing equipment also gives off a "surface" electromagnetic potential and is thus susceptible to the leaking emanation attacks described above. The advancement of highly programmable and adaptable drone technologies, capable of carrying advanced imaging (e.g., LIDAR) and sensing (e.g., next-generation IR sensors) technologies, augments the existing repertoire of imaging and sensing equipment that can discern, with increasing resolution, power-use-relevant electromagnetic emanations. You may be thinking that such emanations will not be sufficient to determine what a processor is doing, and that is probably correct for the moment. However, if it is established that the computer user is using a particular application (like an email application), then the power emanations may be indicative of the size of different emails and their attachments, which could be used to compromise business intelligence. Since keyboard use has been known to be susceptible to this approach since 2009 (Vuagnoux and Pasini, 2009), it is clear that future advances in this field will focus on discerning information about the software running on the devices (Callan, 2016).

6.1.7 BIOS

The BIOS, or basic input/output system, consists of firmware (software supported on specific hardware device memory) that provides boot-up initialization of hardware and, thereafter, runtime services for the computer system via the corresponding operating system and its associated programs and services. Modern systems generally use a unified extensible firmware interface (UEFI) in place of BIOS, per se, but the security concerns are essentially the same. Given that BIOS is on-chip software, the attack vector on BIOS that I would choose is to mimic the BIOS install/update website and convince the user to download a rogue copy of the new or updated BIOS. This "update site spoofing" is less likely to fool a user than email impersonation attacks, mainly because someone downloading BIOS is likely to be more sophisticated than the average email user; however, it is worth a try as a rogue agent, since it essentially scales infinitely with no added cost.

The other type of attack on BIOS/UEFI is, of course, to create a "chip with salsa"; that is, a chip that includes a little extra. Instead of a rogue installation site, a rogue supplier could be used to infiltrate the supposed safety of the BIOS. BIOS is an especially tempting target for me as a potential black hat hacker since it is at the core of the system, and a compromise in BIOS is also a potential compromise in the data, application, server, and network stacks; not to mention a threat to physical security and policies, procedures and awareness.

6.1.8 Central processing unit (CPU) and motherboard

The central processing unit (CPU) is an important element of the computer – its brain, or at least its frontal lobe if you will allow the analogy – and is often the anchor of the motherboard, using up much of the communications, memory, and control signals moving around the internals of the computer. The motherboard, of course, includes memory, adapters, expansion cards, expansion slots, and other chips, including the BIOS/UEFI just described.

As a black-hat hardware hacker, I would be interested in hacking the CPU along the lines of the UEFI chip, that is, inserting a CPU with additional functionality that, for example, allows me to dump the contents for rogue logging periodically. If the CPU itself, or any of the other chips in the motherboard chipset, can be directly compromised, then I would work to insert a motherboard "hanging chad" onto the motherboard. A threat actor can program this "extra" chip to compromise the bus on the motherboard, providing a platform for infiltration, camouflage, and ultimately exfiltration, as suits my tastes for your content.

6.1.9 Graphics processing units (GPUs), graphics, and displays

From a black-hat hacking standpoint, graphics processing units (GPUs) offer similar opportunities for hardware spoofing to those discussed for other chips, including the basic input/output system (BIOS) and unified extensible firmware Interface (UEFI), as well as central processing unit (CPU) chips. They are also increasingly integrated, so they offer greater chances for the "hanging chads" mentioned as possibilities for motherboards. As mentioned above, a GPU can send an interrupt (INTR) to the CPU so it can be used as a broad tool for denial of service (DoS) attacks on a processor/computer.

BIOS screen graphics can be used to orchestrate an attack on the BIOS or UEFI of a computer (Constantin, 2023). The splash screens used to communicate with the user as the BIOS is being loaded are interpreted by the firmware on the UEFI chip, which is associated with image parsing. This image decoding software interprets the content of the image, loading the image, data, and metadata into memory. If rogue code is part of the image or its metadata, it may be interpreted and acted upon in various, arbitrary ways, ranging from buffer overflow (access to supposedly secure memory) to full-scale infiltration and camouflage of the rogue code.

6.1.10 Input/output (I/O) and universal serial bus (USB)

Of all the hardware-related vulnerabilities, input/output (I/O) and USB, or universal serial bus, are the most familiar to security professionals at all levels of training. How many cybersecurity novices – or less kindly naïfs – have found a "perfectly good" USB stick and decided to indulge in a little free memory (or simply free "eavesdropping")? Both the USB stick and the malware attachments to your computer are disseminated for free. Perhaps in such a case, USB really means "Unplanned Surreptitious Bug."

In addition to placing malware on an innocent-looking lost puppy of a USB stick, computers to which USB attachments are possible are also guilty of creating unnecessary vulnerabilities. As noted for the massive Sellafield nuclear site in the United Kingdom, "an internal attacker or supplier could still log in via a USB flash drive or CD to perform acts of sabotage" (Fell and Stegen, 2018). Obviously, protecting both ends of the potential connection is advised. If a rogue USB is attached to the computer, it should be instantly quarantined and validated before any other action is performed.

6.1.11 Peripherals (memory, mouse, keyboard, and printers)

Device drivers, as mentioned above, are software programs specifically designed for a particular hardware device attached to a computer. Among the attacks a black-hat attacker might find tempting is keyboard snooping, which can be perpetrated through acoustic and electromagnetic emanations, as well as other methods, such as surveillance. The surveillance attack of choice in a world filled with increasingly ubiquitous and powerful imaging may be to infer keystroke sequences from video surveillance (in theory, from satellites, not just drones, with the vast continuing increase in resolution of various imaging technologies).

However, a more direct method, preferred by thugs without computer science, imaging, or AI skills, is simply to install a keystroke logging software application onto the computer (as the malware "payload" from any successful attack engendered by social engineering or more sophisticated means.

Thereafter, passwords, credit card numbers, and other personally identifiable information (PII) entered via the keyboard are sitting ducks. Other peripherals, including printers and RAID servers, are similarly excellent targets for PII mining.

6.2 Summary

This chapter gives an overview of the breadth and depth of potential hardware vulnerabilities associated with computing. It is clear from this brief introduction how software can be used to compromise different types of computer hardware. Some of the attacks overviewed here are familiar to both ethical and unethical hackers worldwide. However, this chapter is also meant to motivate you to think of your own attacks. Remember, "The Principle of Cybersecurity Humility" commands you to assume that threat agents have already thought of the same attacks as you, so you must protect your network from those attacks.

References

Callan, R. L. 2016. Analyzing software using unintentional electromagnetic emanations from computing devices. *Ph.D. Dissertation.* Georgia Institute of Technology.

Constantin, L. 2023. *LogoFAIL attack can inject malware into the firmware of many computers* [Online]. https://www.csoonline.com/article/1253397/logofail-attack-can-inject-malware-in-the-firmware-of-many-computers.html: CSO. [Accessed 28 December 2023 28 December 2023].

Fell, H.-J. & Stegen, E. 2018. *The disaster of the European nuclear industry.*

Liu, X., Lei Liu, Steven J. Simske, and Jerry Liu 2016. Human daily activity recognition for healthcare using wearable and visual sensing data. *In:* IEEE (ed.) *2016 IEEE International Conference on Healthcare Informatics (ICHI).* IEEE.

Liu, X., Pang-Ning Tan, Lei Liu, and Steven J. Simske. Automated classification of EEG signals for predicting students' cognitive state during learning. Proceedings of the International Conference on Web Intelligence, 2017. 442-450.

Shedd, J. A. 1928. *Salt from My Attic,* Maine, The Mosher Press.

Vuagnoux, M. & Pasini, S. 2009. Compromising electromagnetic emanations of wired and wireless keyboards. *USENIX security symposium.*

7

Threats to Information-intensive Systems

"It is only a matter of the 'when,' not the 'if' that we are going to see something dramatic."
– Michael S. Rogers, retired US Navy Admiral, second director of US Cyber Command and 17th director of the National Security Agency (NSA)/Central Security Service (CSS) (Committee, 2014)

"Usually, when I wielded a hockey stick, it meant somebody was going to get hurt. This is just a friendly match"
– Stan Mikita (1940–2018), Canadian ice hockey player

Abstract

Systems engineers and computer scientists think about things differently. That is not to say that a systems engineer cannot be a computer scientist, nor that a computer scientist cannot practice systems engineering. The point is that each of these roles leverages different strengths at various times in a system lifecycle, particularly when it comes to systems that support critical infrastructure. Organizations can realize a significant amount of valuable functionality by leveraging computer science in critical infrastructure systems. However, connecting critical infrastructure through the cyber medium presents unique considerations, such as human health and safety, as well as environmental impacts associated with critical infrastructure (e.g., power, water/wastewater, fuel handling, and building systems). The previous chapter gave an overview of potential hardware vulnerabilities associated with computing. This chapter examines how these potential vulnerabilities can render critical infrastructure systems susceptible to cyber-induced threats, and why both engineers and computer scientists are necessary to defend these systems. With an understanding that cyber vulnerabilities can originate at the hardware level, we can examine how control system interfaces make critical infrastructure vulnerable and why both engineers and computer scientists are needed to protect these systems. Systems engineering examines things from a

DOI: 10.1201/9788743809128-7

hierarchical perspective across systems of systems (in Chapter 11, we discuss characterizing this hierarchical perspective as the metasystem and intersystem) to understand interfaces. The systems engineer applies an integrated approach to address complexity, focusing on a big-picture vantage point and utilizing engineering principles and methods. Computer scientists build and design hardware and devices, similar to computer engineers, and develop programs and software using coding and algorithms. Both systems engineering and computer science can influence the drive for innovation through technology, and both may work in fields dealing with "the cloud" or data analytics, as well as the technologies transforming them, such as artificial intelligence (AI) and machine learning (ML). The value of their respective influence matters because, as cyber threats grow in volume and complexity, depending on any one functional role is not a practical deterrent to the threats. Later chapters explore how systems engineering can be used to better understand and overcome cybersecurity vulnerabilities from both engineering and computer science perspectives about what makes critical infrastructure systems vulnerable.

7.1 Introduction

The capability to defend industrial control systems (ICS) of critical infrastructure operations is crucial because the threat of catastrophic and damaging cyberattacks on critical infrastructure is a real concern (Committee, 2014). This chapter examines how the cyber dynamic fundamentally challenges critical infrastructure operations by introducing cyber-related vulnerabilities that can disrupt control system operations and potentially lead to catastrophic failures. As in other information-intensive systems, digital transformation has arrived in critical infrastructure environments. As access to the Industrial Internet of Things (IIoT) and discussions about cybersecurity and critical infrastructure protection expand across military, national, homeland, and civilian domains, so does the opportunity for applying transformational capabilities to various secure automation applications in critical infrastructure systems. Key drivers include increased flexibility and ease of use for operators, scalability, modularity, enhanced memory access, evolving features such as wireless connectivity, and faster speed solutions. If the digital transformation has not yet touched a critical aspect of infrastructure, stand by, because it will soon likely do so – the digital transformation of critical infrastructure is a result of a technology convergence trend affecting industrial control systems (ICS) over the past decade. The role of industrial controls in controlling machinery is changing as increased software, information, and communication technology enhance the capabilities of these machines. This change is significant because it has a profound impact on the security posture of critical infrastructure. Critical infrastructure requires the same defense network as an enterprise IT network. An environmental scan of the timeline of ICS that controls infrastructure reveals that as early as 2014, senior government leaders raised concerns that criminal actors were beginning to utilize tools previously known only to nation-states (Committee, 2014). Nation-states and newly emerging groups of criminal actors were gaining the ability to shut down operations of basic critical infrastructure, such as power generation, and hinder the ability to provide these services to citizens. Gaining such destructive capability was a game-changer.

In 2014, the United States Cyber Command (USCYBERCOM) chartered an effort to address an emergent challenge in operations in the cyber domain. The problem statement was that "[n]etwork managers supporting DOD industrial control systems (ICS) lack [tactics, techniques, and procedures] TTP to detect, mitigate, and recover from nation-state level cyber-attacks" (Walsh, 2016). The effort was the joint base architecture for secure industrial control systems (J-BASICS). The lead organizational sponsor was USCYBERCOM. The operational endorsement came from the North American Aerospace Defense Command (NORAD) and the United States Northern Command (USNORTHCOM), the Office of the Undersecretary of Defense for Acquisition and Sustainment, Technology, and Logistics (OUSD(AT&L)), OUSD Energy, Installations, and Environment (OUSD(EI&E)), United States Indo-Pacific Command (USINDOPACOM), and the United States Strategic Command (USSTRATCOM).

For decades, US national laboratories have been at the forefront of research in securing industrial control systems (ICS) against cyber vulnerabilities. Their contribution to the J-BASICS research was insignificant, given their foundational role in the subsequent development of the initial cyber defensive capability for the more situational awareness industrial security (MOSAICS) framework, as mentioned earlier in Chapter 5, "Cyber Paradox." The proposed MOSAICS solution began with concept exploration to automate the output of J-BASICS and apply integrated adaptive cyber defense (IACD) principles, which are used to defend enterprise networks, to lower levels in an architecture for critical infrastructure. It also examined alternative concepts. US national laboratories, including the Idaho National Laboratory (INL), the Pacific Northwest National Laboratory (PNNL), and Sandia National Laboratory (SNL), as well as other governmental personnel and inter-agency participants, participated in the systems engineering process of reviewing alternatives, selecting, and justifying the approach. J-BASICS gives historical context and understanding of the evolution from the declaration of cyber as a "warfighting domain" to an initial cyber mitigation solution. The DOD was the proverbial canary in the coal mine for this danger. The DOD shared useful early indicators of conditions that might become dangerous and potential hazards for all. J-BASICS' charter objective was to come up with mitigations to the emerging threat:

> "[E]mploy multi-service and other Department of Defense (DOD) agency support, personnel, and equipment to develop, test, and evaluate advanced cyber industrial control system (ICS) tactics, techniques, and procedures (TTP) to improve the ability of ICS network managers to detect, mitigate, and recover from nation-state level cyber attacks" (Walsh, 2016).

The concept was that a nation-state-level cyber-attack on military installations in Guam could degrade or delay US force projection in the Asia-Pacific region. Such an attack could disrupt the delivery of basic services, such as power, fuel, or water, rendering the disruption of Command and Control (C2) and supplies, thus making units unable to project force. A decade later, in 2024, the concept remained the same, as did many of the questions. Questions that came out of the J-BASICS effort included questions from the field elements about how the TTPs would be resourced, who owned the TTPs and processes, how these would be operationalized, and how to incorporate "Industrial Control System (ICS) Thinking" into mission planning, exercises, and across department personnel as part of the greater "cyberculture"(Walsh, 2016). The workforce approach challenges differ based on the field, sector, education, and experience of the personnel involved. Professionals will judge what is considered good or bad in systems design, just as internalized evaluations are made about work based on work orientations {Dik, 2013 #188}. The systems engineer addresses system problems by leveraging the combined knowledge and technical expertise of engineers and specialists from various disciplines and cross-functional teams, including subject matter experts (SMEs) from both physical and cyber domains. Systems engineering can help tie technologies and training to business costs {Aleksandra Scalco, 2021 #229}. We elaborate further on J-BASICS in a later chapter, specifically regarding modernization strategies and the MOSAICS framework and operator enabler. The bottom line from J-BASICS was, "*Get the word out: ICS Operators may well be on the front lines at the start of the next fight. Prepare Now*!" (Walsh, 2016).

As early as 2014, Michael S. Rogers, a retired US Navy Admiral and the second director of US Cyber Command, as well as the 17th director of the National Security Agency (NSA) and Central Security Service (CSS), shared concerns. In 2014, Rogers publicly shared three top cyber concerns at the Wall Street Journal Live Conference. His top concern was that a threat actor could use cyberattacks to damage infrastructure, the second was data manipulation, and the third was the influence and capabilities of non-state threat actors. About his infrastructure concern, he stated:

> "It is only a matter of 'when' that someone uses cyber as a tool to damage the critical infrastructure of our nation," Rogers said. "I'm watching nation-states and groups within some of that infrastructure. At the moment, it seems to be really focused on reconnaissance and attempting to understand the characteristics of the structure, but it's only a matter of time, I believe, until someone actually does something destructive" (Walsh, 2016).

In the United States, the DOD defines critical infrastructure as "Systems and assets, whether physical or virtual, so vital to the United States that the incapacity or destruction of such systems and assets would have a debilitating impact on security, national economic security, national public health or safety, or any combination of those matters." The Department of Defense (DOD) distinguishes between critical infrastructure and mission-critical infrastructure, assigning the Assistant Secretary of Defense for Homeland Defense and Global Security (ASD(HD&GS)), under the Under Secretary of Defense for Policy (USD(P)), responsibility for managing the Defense Critical Infrastructure Program (DCIP) (DOD, 2008, Change 2017) (GAO, 2009). Similarly, the UK's National Cyber Security Center (NCSC) defines critical national infrastructure as "National assets that are essential for the functioning of society, such as those associated with energy supply, water supply, transportation, health, and telecommunications" (Government, 2019, Reviewed 2023).

A US Government Accountability Office (GAO) report to Congressional Committees in July 2015, entitled "Improvements in DOD Reporting and Cybersecurity Implementation Needed to Enhance Utility Resilience Planning," highlighted the need for a department-wide cybersecurity effort to mitigate industrial control system (ICS) vulnerabilities (GAO, 2015). Department-wide efforts to enhance the cybersecurity of industrial control systems (ICS) included collaborative efforts to develop and test, through joint exercises, a set of procedures for detecting, mitigating, and responding to cyber incidents in ICS. These procedures would be used by personnel to defend Department of Defense (DOD) installations against advanced persistent threat (APT) actors. The collaborative effort, initiated in 2014 by the US Cyber Command and the Joint Test and Evaluation Program, led to the development of the joint base architecture for secure industrial control systems (J-BASICS) joint capability technology demonstration (JCTD) previously mentioned. The output of J-BASICS was the advanced cyber industrial control system tactics, techniques, and procedures (ACI TTP) for Department of Defense (DOD) industrial control systems (ICS)" (USCYBERCOM, 2018). The most poignant observations made during J-BASICS data collection were that users with cybersecurity training did not behave any differently than those ICS operators who had not received cybersecurity training when confronted with phishing attacks, suggesting humans are the weakest link in system susceptibility to advanced persistent threats (Terry Merz, 2019). MOSAICS followed J-BASICS as an initiative to automate the set of procedures developed by the department, tying the defined procedures to recently emerging security orchestration, automation, and response (SOAR) technologies and methodologies. Accidents in industrial control systems (ICS) within operational technology (OT) at that time, prior to the introduction of the MOSAICS framework, were typically attributed to the complexity of systems or component failures, rather than the user's susceptibility to cyberattacks (Scalco and Simske, 2020).

7.2 Threats to Information-intensive Systems

The adage of "security by obscurity" is a failed strategy in today's always-on, always-networked environment. The NSA and the Cyber Infrastructure Security Agency (CISA) issued joint operational technology (OT) alerts as early as July 23, 2020 (CISA, 2019). A 2020 alert urged all Department of Defense (DOD), National Security Systems (NSS), Defense Industrial Base (DIB), and US critical infrastructure facilities to take immediate action to secure their operational technology (OT) assets. Alerts regarding credible information of imminent threats to the hospital and healthcare sectors and supervisory control and data acquisition (SCADA) found in water treatment systems, followed by the Cybersecurity and Infrastructure Security Agency (CISA 2020) (Cybersecurity and Infrastructure Security Agency (CISA), 2020) and the FBI (FBI, 2021). While internet-accessible OT assets have become more prevalent across the 16 US critical infrastructure sectors, the security of legacy OT systems has struggled to keep pace with the rapid exposure to cyber vulnerabilities. Combined with readily available information that helps threat actors quickly recognize OT assets connected via the Internet, you have got a "perfect storm" of:

- Easy access to unsecured assets
- Use of common, open-source information about devices
- An extensive list of exploits deployable via common exploit frameworks
- Threat agents ready, willing, and able to exploit these assets

Critical infrastructures are highly interconnected both within an infrastructure sector and with one another. In many cases, cyber systems also provide information or control to these infrastructures. Those dependencies can lead to unexpected consequences in the event of an incident. Simulation models that account for dependencies are critical to gaining insight. United States national research laboratories develop models that provide an overview of accounting for dependencies in constructing simulation models and some associated challenges.

An example is the United States' 9-1-1 emergency communication system, which exemplifies a highly connected critical infrastructure system. Research laboratories are at the forefront of developing some of these industrial control system (ICS) tools and models. Recall that industrial control systems (ICS) are control systems that include those systems often referred to as operational technology (OT), sometimes as platform information technology (PIT), and are sometimes incorrectly used interchangeably with supervisory control and data acquisition (SCADA). SCADA is used to manage and control industrial processes. Industrial control systems (ICS) encompass SCADA and distributed control systems (DCS) used in localized control process systems, such as manufacturing lines, as well as programmable logic controllers (PLCs) that automate tasks. Examples include but are not limited to, building automation systems, energy and utility monitoring and control systems, lighting, fire and life safety systems, physical security systems, fuel handling systems, logistics systems, medical systems, manufacturing systems, and weapons systems. Damage to or compromise of any of these systems may cause a mission failure or cascading failures if the control system is used as a gateway into an organization's broader information networks.

These systems are classified as context-sensitive critical infrastructure dynamic classes (CS-CIDC). Each system and component, such as SCADA and distributed control system (DCS), as well as configurations like programmable logic controllers (PLCs), has unique functional requirements, safety requirements, and topologies specific to the target system. Cybersecurity considerations in the development and design of solutions for physical systems in the cyber domain include systems engineering activities for each target system, such as:

- Identifying and assigning values to resources that require protection and conservation.
- Creating security requirements with verification methods.
- Defining sophisticated access controls and privilege management.
- Defining secure data communications/networking, and storage.
- Developing a secure architecture implementing defense-in-depth (DiD).
- Defining a secure software development lifecycle and a process for selecting, developing, and procuring trusted system components; and
- Defining physical, procedural, and administrative safeguards.

7.2.1 Dependencies

Critical infrastructures are highly interconnected. However, the industrial control system (ICS) hardware and software technologies used are specialized from typical enterprise business IT. These interconnections exist at various levels, from within an individual infrastructure sector to across sectors. For example, in the fields of electric power and communications, numerous independent entities must collaborate to provide the necessary infrastructure. These infrastructure sectors also rely on one another. Communications systems require power to operate, while electric power systems need communications services, sometimes leased from commercial providers, for their control systems.

Dependencies may be functional, such as electric power-generating plants that require fuel to operate, or geographic, such as a petroleum facility connected to a specific electric power substation. They often include human decision-makers and contractual, treaty, or regulatory constraints. Mental models are insufficient to capture the complexities, so carefully designed simulation models are necessary. Systems engineering techniques, such as model-based systems engineering (MBSE), which is discussed in further detail in Chapter 23, "Abstraction Models," are used to create a model-based organization and are helpful in the analysis and design of complex inter-systems.

Structured MBSE methods, such as agile development, are practical techniques supporting architectural design.

For example, consider the logistics of voice over internet protocol (VoIP) phones, which are increasingly common at residential and business sites. Typically, the service provider is a non-traditional telephony carrier, such as a cable company or a traditional landline carrier, that uses fiber to connect to the home. Similar to cordless landline phones, unless there is backup power at the customer's site, these phones fail instantly when the power goes out. Additionally, they depend on connections from the customer site through the carrier's network (typically internet-based facilities) until they reach an interconnection at a wire center with a traditional landline phone company. Internet facilities will have generators for backup power, with up to two to three days of fuel on-site, although this is not a regulatory requirement. Many of these facilities will also require water for cooling. Depending on the design of the water utility, electric pumps may be needed to deliver water to the facility. The water utility may or may not have backup generation for use in the event of a power outage. The internet facility will have a limited operating time in the event of a water supply loss for cooling. Wire centers and mobile switching centers also require water for cooling in many cases. All generators will require refueling if the power outage lasts longer than the on-site fuel capacity. Refueling requires transportation assets that rely on fuel and clear roadways for their operation. Sites used to load fuel onto fuel trucks also need electric power. In a significant event, facilities may compete with one another for both generators and the fuel that supplies them. Established contractual arrangements with national fuel suppliers that can provide fuel from outside the impacted area may be key to maintaining function. Electric power restoration also requires vehicles for the work crews and clear roads. In addition, the control systems for electric power may rely on wire centers or internet facilities operating on generator backup. Restoring electric power becomes more difficult and takes longer without an operational control system in place. In this manner, the model can quickly become as complex as the real system.

7.2.2 Typical security threats

Some typical security threats to information-intensive systems include malware, hacking, data theft, ransomware, denial-of-service attacks, espionage, and physical attacks. For example, major banks and other financial institutions are vulnerable to ransomware. Financial services are prime targets for these types of attacks due to the value of customer information and the profitability of cyber criminals. Criminals get into large bank or financial enterprises and move laterally, escalating privileges and then deploying encryption that enables them to encrypt files and demand a ransom payment; an online retailer relies directly on incoming traffic for creating revenue, so distributed denial-of-service (DDoS) attacks can cause damaging disruption of service to the online retailer; a transportation company such as a railroad is vulnerable to physical attack threatening critical safety systems caused by cyberattack and hacking. One example of a risk factor is the railroad infrastructure's distributed network architecture. As with other distributed systems, increased surface area means increased vulnerability. Elements of a cyberattack by a criminal syndicate against an entity may include:

- Threat agent – a malicious actor creates a threat.
- Threat – potential exploitation of a Vulnerability by a Threat Agent.
- Asset – something of value that requires protection.
- Vulnerability – weakness that an attacker can exploit.
- Security control – a measure that mitigates a vulnerability to reduce risk; collectively, it seeks to minimize the Attack Surface.
- Exposure – a compromise resulting from the exploitation of a vulnerability.

7.2.3 Safeguards

Just as in enterprise business IT network security, critical infrastructure ICS security plans should include authentication, authorization, and accounting; access control; boundary and firewall protection; and intrusion detection and response. The goal is to overcome cyber vulnerabilities in system assets and minimize risks. Three examples of categories of cybersecurity safeguards or preventative measures include administrative, technical, and physical prevention measures (Borky and Bradley, 2019):

1. Administrative prevention (i.e., planning and initiation) – procedural measures aimed at reducing vulnerabilities that may be caused by human error, such as automatically locking computers that are not in use after a specified period or implementing basic security programs, including security leadership commitment and support from other stakeholders.
2. Technical prevention (i.e., cybersecurity architecture) – firewalls and other protective devices at system boundaries; defining a cybersecurity

architecture that includes modeling the implementation of selected security controls at every level of a layered defense structure; and technical corrective (i.e., monitoring, maintenance, and upgrading) – prompt software security patch installation, and technology refreshment of obsolescent cybersecurity components.

3. Physical prevention measures (i.e., physical barriers) – placing sensitive information or high-value assets behind restrictive security barriers, such as locked and alarmed doors, and using other access controls for authorized personnel, including badges.

7.3 Communication Protocols

Communication protocols used in critical infrastructure are unique to process and building automation, as well as industrial control systems (ICS) components. Traditional IT protocols are not employed in the components that comprise these systems. Typical ICS components include SCADA, HMI, DCS, PLCs, sensors, and RTUs. Common communication protocols used that are different from IT protocols include serial interfaces such as physical interface standard RS-232 and RS-485 that define electrical signal characteristics and mechanical interface characteristics, distributed network protocol (DNP3), open-source Modbus serial communication, and the highway addressable remote transducer (HART) protocol used in automation, Building Automation, and Control (BACnet), PROFIBUS serial protocol, and PROFINET ethernet-based protocol. Some of these, such as PROFINET, lack authentication, which makes PROFINET vulnerable to common cyber threats, including spoofing. Cybersecurity best practice is to disable unnecessary protocols. Others, such as Modbus, are susceptible to software-related problems, including invalid device addressing, unrecognized function codes, or address duplication issues with other devices on the network. Some of the common communication protocols and their locations within a conceptual framework are shown in Table 7.1.

Questions to ask are: Who (if anyone) is responsible for an information system used to control industrial processes, such as manufacturing, product handling, production, and distribution? Industrial control systems encompass supervisory control and data acquisition (SCADA) systems, which are used to control geographically dispersed assets, and who (if anyone) is responsible for an information system used to control distributed control systems (DCSs) and smaller control systems using programmable logic controllers (PLC) to control localized processes. As an example, the

Table 7.1 Communication protocols found in critical infrastructure.

Communication protocol	Overview	Conceptual framework location
Building automation and control (BACnet)	Open-source standard, designed specifically for use in building automation and control networks, access control, and heating, ventilation, and air-conditioning (HVAC)	Communicates messages between the controller and various devices
Common industrial protocol (CIP)	Services for security, control, and synchronization are vulnerable to remote attacks	Application layer protocol for ethernet/IP networks that communicates between sensors and actuators and higher-level controllers
Distributed network protocol (DNP3)	Used in supervisory control and data acquisition (SCADA), reliable for long-distance data transfer	Operates at application, transport, and data link layers for communication between remote terminal units (RTUs) and control centers
Ethernet industrial protocol (Enet/IP)	Supports near real-time communication, scalable, interoperable, high-speed data transfer	Communicates between control systems, sensors, and actuators
FOUNDATION Fieldbus	Used in a factory-type environment, offers limited to no security	Located at the base-level data layer
Highway addressable remote transducer (HART)	Open source, hybrid analog and digital industrial protocol, lower cost, simple design and implementation, vulnerable, lack authentication	Bidirectional communications between intelligent field instruments and host systems, such as programmable logic controllers (PLCs)
Inter-control center protocol (ICCP), known as telecontrol application service element (TASE.2) protocol	Bi-directional WAN communication, vulnerable, lacks authentication, belongs to the IEC 60870 set of standards for remote control and monitoring	Communication between control centers, sub-stations, various types of utility and non-utility generators
Modbus	Open-source protocol used in industrial automation lacks authentication, lacks encryption, is simple, and reliable	Communicates between control devices and field instruments.
Open platform communications (OPC)	Vendor agnostic, used for integrating industrial control systems (ICS) with higher-level business operation resources such as enterprise resource planning (ERP)	Data flow and control between automation devices and management software applications
PROFIBUS (serial protocol, lacks authentication	Near real-time communication used in automation environments for time-critical industrial applications, scalable, and flexible	Communicates between controllers, sensors, and actuators
PROFINET	Ethernet-based protocol, greater speed, and more bandwidth than PROFIBUS	Communicates between controllers, sensors, and actuators
RS-232 and RS-485	Low speed, short distance, low cost, simple design	Communicates between various devices in the architecture

DOD cyber workforce framework KSAT (knowledge, skills, abilities, and tools description) for a control systems security specialist calls out knowledge of computer networking concepts and protocols and network security methodologies:

Control Systems Security Specialist (Work Role ID: 462) and Cyber Defense Infrastructure Support Specialist (Work Role ID: 521/NIST: PR-INF-001) KSAT 22 – Knowledge of computer networking concepts and protocols and network security methodologies.

7.4 Pervasive Information-intensive Systems

Nearly every business, in some form, can have a type of control system known as an industrial control system (ICS). ICS functionality pervasively handles and manages operations, monitoring, and controlling the physical world, including measuring physical variables such as temperature, pressure, speed, and flow. An example of an ICS in everyday life is a heating, ventilation, and air conditioning (HVAC) system that moves heated or cooled air throughout a facility. Different types of devices are used as industrial control systems (ICSs), including programmable logic controllers (PLCs), remote terminal units (RTUs), human–machine interfaces (HMIs), and supervisory control and data acquisition (SCADA) systems. In a residential building, it might be uncomfortable, but it is not critical to have an HVAC system to make the ambient environment more pleasant. However, for a commercial entity such as a data center that supports critical missions or business activities, maintaining controlled temperature, humidity, and airflow is crucial, as unwanted heat can lead to equipment malfunctions and downtime.

Many critical infrastructure ICS systems in operation today were designed without cyber in mind, making them outdated in today's cyber world. While it may seem that new cyber mitigation solutions lack a predecessor system, a tremendous amount of valuable research and standards development has occurred. Inputs to systems engineering efforts for critical infrastructure include predecessor research efforts as sources of information for proposed architectures and recommended concepts of operations (CONOPs). Predecessor research provides evidence-based scientific and technical advancements that contribute to systems engineering principles and practices, delivering effective solutions. These are all topics addressed by systems engineering principles and practice.

Questions to ask include whether the organization utilizes threat detection capabilities and if threat signatures are regularly updated within the organization. As an example, the DoD cyber workforce framework KSAT for a control systems security specialist calls out knowledge of cyber threats and vulnerabilities:

Control Systems Security Specialist (Work Role ID: 462) and Cyber Defense Infrastructure Support Specialist (Work Role ID: 521/NIST: PR-INF-001) KSAT *1159* – Knowledge of cyber threats and vulnerabilities.

7.5 Summary

For a hospital that supports life-giving activities, such as incubators, temperature-controlled environments are vital to operations that support prematurely born infants. The mission or purpose of the operations, as well as the high availability and uptime requirements for industrial control systems (ICS) in a data center or hospital, compel these to be considered "critical" infrastructure. For various reasons, these systems may utilize proprietary protocols or components that were designed decades ago and are now outside the support lifecycle, lacking fundamental cybersecurity features, which makes them vulnerable to cyber failures. The bottom line is that critical infrastructure can be any system that regulates devices or systems we depend upon for availability. Critical infrastructure and information-intensive systems are pervasively susceptible to cyber-induced failure. Critical infrastructure today leverages cyber for efficient operations. Therefore, the systems must be designed to overcome all the challenges that cyber presents, including dependencies, constraints, and risks to the target system, as well as any other influences that affect the system's enterprise and elements throughout the lifecycle. The hospital example leads us to Chapter 8, "Use Cases and Overview (OV) Illustrations," where we discuss what constitutes a critical infrastructure control system operating in the cyber domain and show how malware can impact systems using various architecture approaches as examples. In subsequent chapters, we will share abstraction models, explain how we transition innovation to these older, legacy systems built to last 30 years or more, and describe how to build successful teams that are comfortable competing in this domain.

References

(CISA), C. A. I. S. A. 2019. Securing Industrial Control Systems: A Unified Initiative FY 2019—2023. *In:* (DHS), D. O. H. S. (ed.). Department of Defense.

(USCYBERCOM), U. S. C. C. 2018. Advanced Cyber Industrial Control System Tactics, Techniques, and Procedures (ACI TTP) for Department of Defense (DoD) Industrial Control Systems (ICS). *In:* Defense, D. O. (ed.).

Borky, J. M. & Bradley, T. H. 2019. Protecting Information with Cybersecurity. *Effective Model-Based Systems Engineering.* Springer International Publishing.

Committee, H. I. 2014. Hearing of the House (Select) Intelligence Committee Subject: "Cybersecurity Threats: The Way Forward".

Cybersecurity and Infrastructure Security Agency (CISA), T. F. B. O. I. F., and The Department of Health and Human Services (HHS) 2020. Alert (AA20-302A) Ransomware Activity Targeting the Healthcare and Public Health Sector. *In:* Security, U. S. D. O. H. (ed.). Cybersecurity and Infrastructure Security Agency (CISA).

DOD 2008, Change 2017. Department of Defense (DOD) Instruction 3020.45, Defense Critical Infrastructure Program (DCIP) Management. *In:* (DOD), D. O. D. (ed.).

FBI, C., EPA, MS-ISAC 2021. Compromise of U.S. Water Treatment Facility. *In:* ADVISORY, J. C. (ed.).

GAO 2009. GAO-09-740R Defense Infrastructure, Defense Critical Infrastructure: Actions Needed to Improve the Consistency, Reliability, and Usefulness of DOD's Tier 1 Task Critical Asset List. *In:* (GAO), U. S. G. A. O. (ed.). Washington, DC 20548: United States Government Accountability Office (GAO).

GAO 2015. Defense Infrastructure: Improvements in DOD Reporting and Cybersecurity Implementation Needed to Enhance Utility Resilience Planning (GAO-15-749). *GAO Defense Infrastructure.*

Government, U.K. 2019, Reviewed 2023. Cyber Assessment Framework (CAF) Version 3.1. *In:* (NCSC), N. C. S. C. (ed.).

Scalco, A. & Simske, S. 2020. Engineering and Development of a Critical Infrastructure Cyber Defense Capability for Highly Context-Sensitive Dynamic Classes — Part 2, Development. *Journal of the Homeland Defense & Security Information Analysis Center (HDIAC).*

Terry Merz, C. F., Aleksandra Scalco 2019. A Context-Centred Research Approach to Phishing and Operational Technology in Industrial Control Systems. *The Journal of Information Warfare.*

Walsh, G. 2016. Joint Base Architecture for Secure Industrial Control Systems (J-BASICS). ITEA, 16 March 2016 2016. International Test and Evaluation Association (ITEA): The International Test and Evaluation Association (ITEA).

8

Use Cases and Operational View (OV) Illustrations

"Fix your little problem and light this candle."
– Alan Shepard (1923–1998), American astronaut

Abstract

A method for establishing user interactions with a system involves creating use case scenarios and operational view (OV) illustrations. Use case scenarios clarify requirements in relation to possible user and system interactions within the intended environment, as well as the desired outcome. The use case identifies dependencies, constraints, and risks to the target system, guiding the system design and development. Use cases describe how a system is used to meet its goals, identify the stakeholders involved, and outline functional requirements from a user's perspective. This systems engineering method is a particularly useful way to review and analyze the tasks of an actor and the system. Use case scenario OV illustrations provide a representation of the virtual world that can help stakeholders focus on identifying and mitigating cyber vulnerabilities and threats. OV illustrations highlight aspects of network design, monitoring network connections, network layers, and security control techniques that can be used to mitigate cyber vulnerabilities. Use cases and complementary OV illustrations can serve as examples of developing an organization's cybersecurity strategy and informing teams about the goals and objectives of the plans. Both help to communicate the system operations, interactions, and operational concepts to stakeholders. In the previous chapters, cyber threats were examined, ranging from hardware-level attacks to those perpetrated by threat actors seeking to gain unauthorized access to information-intensive systems and manipulate system functions. This chapter shows what escalation looks like using various architectural approaches, from flat network architecture to Zero Trust network design. Architecture approaches, such as network segmentation and Zero Trust elements, can enhance a system's security against ransomware attacks and minimize the risk of compromise, as illustrated in a healthcare

DOI: 10.1201/9788743809128-8

and hospital sector example. We know there are commercial technologies that can be put in place to help us respond to a cyber event. To deploy these technologies with confidence that an attack underway can be stopped requires the ability to recognize such an event if it occurs, and the ability to integrate these technologies to defend the network realistically.

8.1 Introduction

Monitoring network connections for malware in any helpful way can be an organizational challenge. Critical infrastructure is fragile, with complex cross-sector and inter-system dependencies. For example, the loss of a service in one sector can impact critical infrastructure systems that depend on the service's availability, such as electric power for a train transportation system. The level of intersystem dependencies enhances the value of the systems engineering approach, which enables stakeholders to better understand system interfaces and cybersecurity risks. Healthcare and hospital policy decision-making is unique from other sectors because policymakers must consider services designed to protect people who work in the industry and serve the underlying needs of society during times of emergency. Cyber presents the need for a paradigm shift in engineering approaches to designing secure system solutions. We will walk through system design approaches in the following example of the Healthcare and Public Health sector. Aspects of this chapter were previously published (Aleksandra Scalco, 2021).

8.2 Use Case in the Healthcare and Hospital Vertical

Healthcare and hospital organizations require reliable communication, timely and efficient data delivery, multi-node mobility, multicast technology, and energy efficiency. Additionally, the on-time delivery of accurate patient information is crucial (Egbogah and Fapojuwo, 2011). Healthcare professionals connect and monitor patients using near-real-time sensors and equipment for remote patient monitoring. However, the need for speed, efficiency, low cost, and relative simplicity of maintenance leads to a single network, also known as a "flat network." Flat networks can potentially increase network throughput while also increasing vulnerability to cyberattacks, as illustrated in Figure 8.1 (Aleksandra Scalco, 2021). This approach to network design makes systems particularly vulnerable to potential malware attacks. A layered cyber defense security approach can be better. However, a layered approach is more expensive, challenging to maintain, and susceptible to attackers using phishing techniques (Diogenes and Ozkaya, 2019).

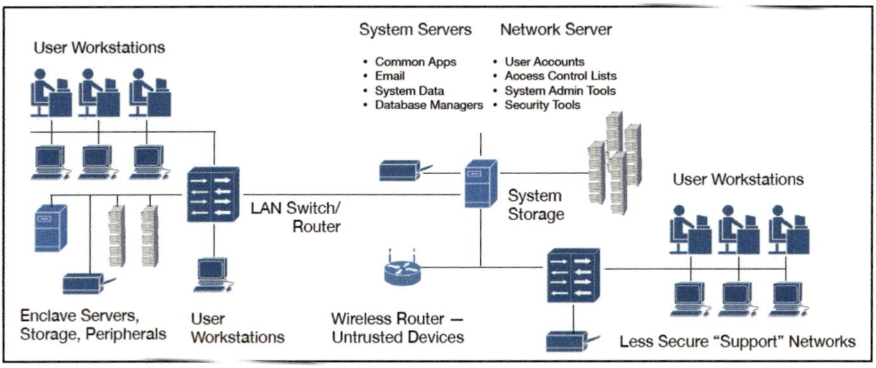

Figure 8.1 Healthcare single, "flat network" design.

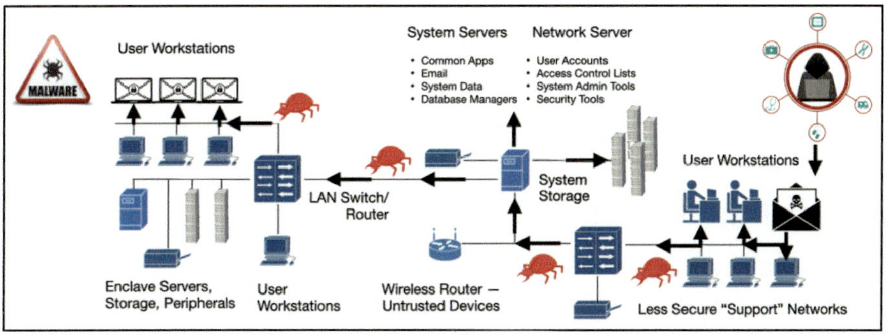

Figure 8.2 Healthcare flat network malware attack.

Monitoring these network connections for malware in any useful way is a governance challenge for any organization. Malware can spread laterally throughout the flat network design by sharing infected file attachments, as shown in Figure 8.1. As a result, any infected machine on the network can spread the malware, as illustrated in Figure 8.2. The system administrator must identify the infected workstations, disconnect the infected assets, apply patches, turn off administrative shares, remove the malware, and change the account credentials. Remediation is tedious and can force a hospital to shift network systems offline, costing precious patient treatment time (Aleksandra Scalco, 2021).

Unpartitioned, "flat" networks inherently enable insecurity. DiD, Zero Trust, and security orchestration strategies can reduce the estimated resources needed to secure a network (e.g., person-hours, training). Additionally, automation significantly improves total response time to indicators. A DiD architecture strategy suggests implementing cybersecurity at each layer of the open systems interconnection (OSI) model. Network segmentation reduces the network attack surface. Zero Trust is a network security strategy that validates everything continuously using micro-segmentation of the network. Finally, security automation integrates security processes and tools to address threats and perform mitigations rapidly. Security automation and machine learning (ML) capabilities can detect attacks and handle intrusions without requiring human intervention from subject matter experts (SMEs), operating at speeds that humans cannot match without assistance from machines (Diogenes and Ozkaya, 2019). Machine learning (ML) capabilities are integral to emerging capabilities in DiD strategies such as security orchestration. For example, ML capability can be utilized by a threat actor to analyze encrypted data traffic using algorithms that identify hidden threat patterns without requiring the decryption of the traffic. It can also identify malware based on known malware behaviors, threat reports with deeper context, and adversary tactics, techniques, and procedures (TTPs) knowledge (Aleksandra Scalco, 2021).

8.2.1 Defense-in-depth (DiD)

Defense-in-depth (DiD) security controls are implemented at each perimeter layer, and secure demilitarized zones (DMZs) using the same techniques employed in the network security layer. Network security layer enclave partitioning security controls are built on a segment of the network defined by common security, change, and service policies. Endpoint security layer controls implemented are content security (e.g., anti-virus and anti-malware) and patch management. Application security layer controls are used to protect externally facing interfaces, such as web application firewalls, and automated patch management. Data security layer controls are implemented for data classification and data/drive encryption, and mission-critical asset layer security controls are implemented as part of IT security governance, security policies, and compliance, as shown in Figure 8.3 (Aleksandra Scalco, 2021).

8.2.2 Network segmentation

Network segmentation can be accomplished physically or virtually to reduce the network attack surface. In a traditional network, all servers and workstations reside on the same local area network (LAN), which allows an attacker to move laterally from one system to another. An approach to

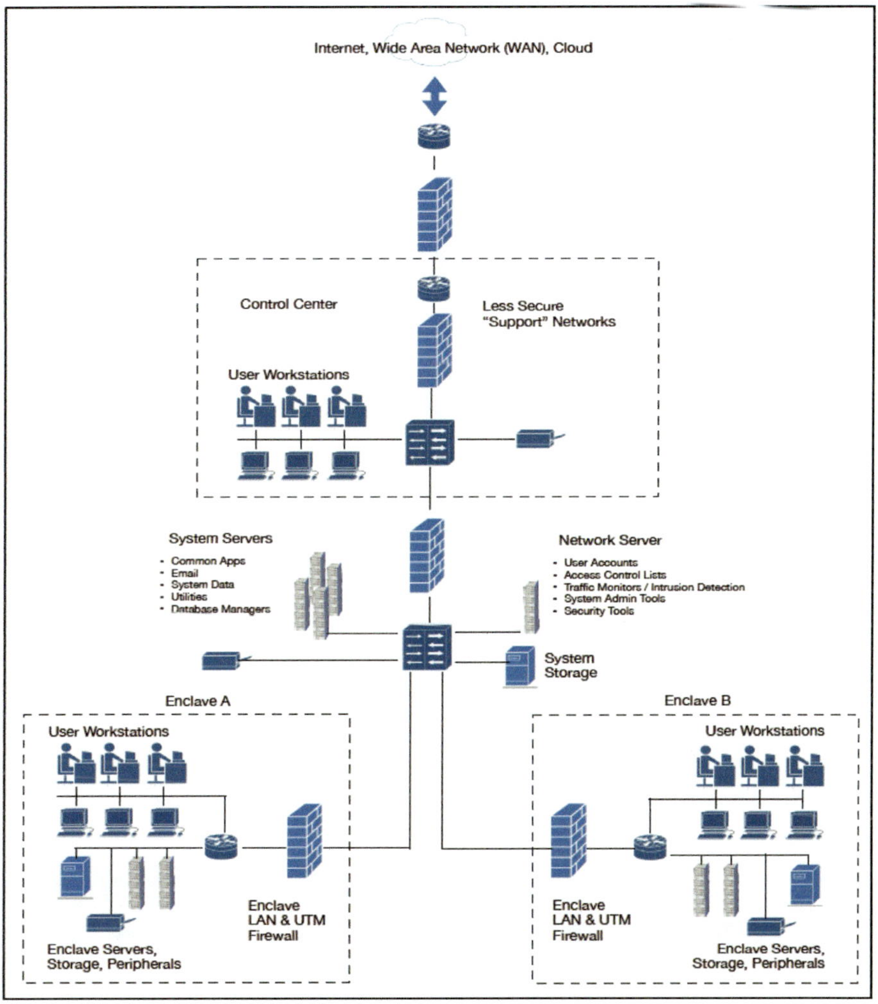

Figure 8.3 Segmented healthcare architecture design.

network segmentation involves using cryptographic segmentation and segregating networks through the application-aware defense to protect and detect cybersecurity functions as a mitigation strategy, such as application-aware firewalls, virtual local area networks (VLANs), virtual routing and forwarding (VRF) instances, and software-defined networking (SDN).

This approach can be implemented and effectively deployed on smaller, local networks, such as in a hospital's healthcare and public health sector, where network security policies can be centralized in the network control

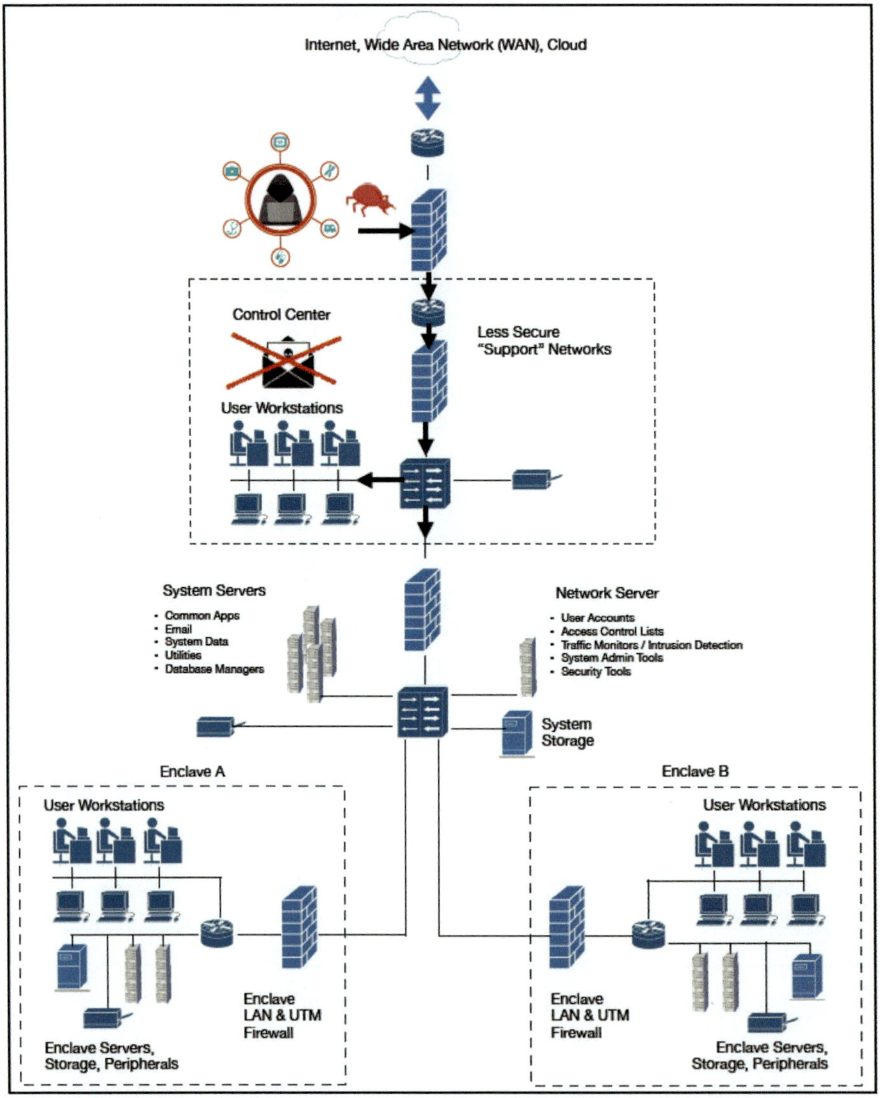

Figure 8.4 Malware attack on a segmented healthcare architecture.

plane. Application-aware network segmentation is an approach to managing cybersecurity risk and promoting a defense-in-depth security posture, which involves segmenting network traffic to reduce the attack surface. For example, one effective method to defend against a malware attack that traverses network boundaries is to segment the network into separate enclaves, as shown in Figure 8.4. An application-aware application approach allows network operations to be sorted according to the application or service the network traffic attempts to reach, and based on specific traffic contents. The first step is to identify all hospital assets, prioritize critical assets, and map the current architecture and interfaces (Aleksandra Scalco, 2021).

8.2.3 More situational awareness for industrial control systems (MOSAICS) and security automation and orchestration (SOAR)

Employing more situational awareness for industrial control systems (MOSAICS) reference framework and security zone separation architecture exemplifies how a network can be strengthened. It is worth repeating that the reference framework is non-proprietary and vendor-agnostic. Specifications for the first acquisition block have been released and are now available to the public on the Navy's website, aiming to accelerate best practices for defending critical infrastructure. Compared to traditional enterprise IT systems, essential attributes of critical infrastructure include low latency tolerance and constrained flexibility to configure system-level dynamic reconfiguration. The MOSAICS network is a necessary first step in identifying the system inventory at the lowest layers of the architecture. Key questions to ask are: Who has access to the physical network topology, and who has access to the logical network topology? What defensive countermeasures are in place to protect operations, and what defensive countermeasures are in place to detect threat activity? To answer these questions, the standard operations and activities must be characterized, and the system's inventory must be known. Integrating MOSAICS helps to achieve that. The layered approach provides security zone segmentation between operational locations and enterprise functions, as shown in Figure 8.5. MOSAICS introduces security automation to integrate security processes, applications, and infrastructure. Security orchestration can then integrate existing security tools to address incidents and perform mitigations. The MOSAICS reference framework, combined with a Zero Trust strategy, provides a superior network design framework for critical infrastructure verticals. MOSAICS provides a cyber defensive capability framework for integrating components from a control system cyber defensive solution. Zero Trust provides the security controls. The MOSAICS reference framework offers security automation and recommends commercial technologies and service integrations to minimize the mean time to repair (MTTR) following cyberattacks. MOSAICS integrates security tools and disparate systems to support security automation that prioritizes and drives response actions based on a pre-defined "Playbook" or "workflow." The security orchestration manages alerts and performs integrity checks against the detection of anomalies outside the system network boundaries, as

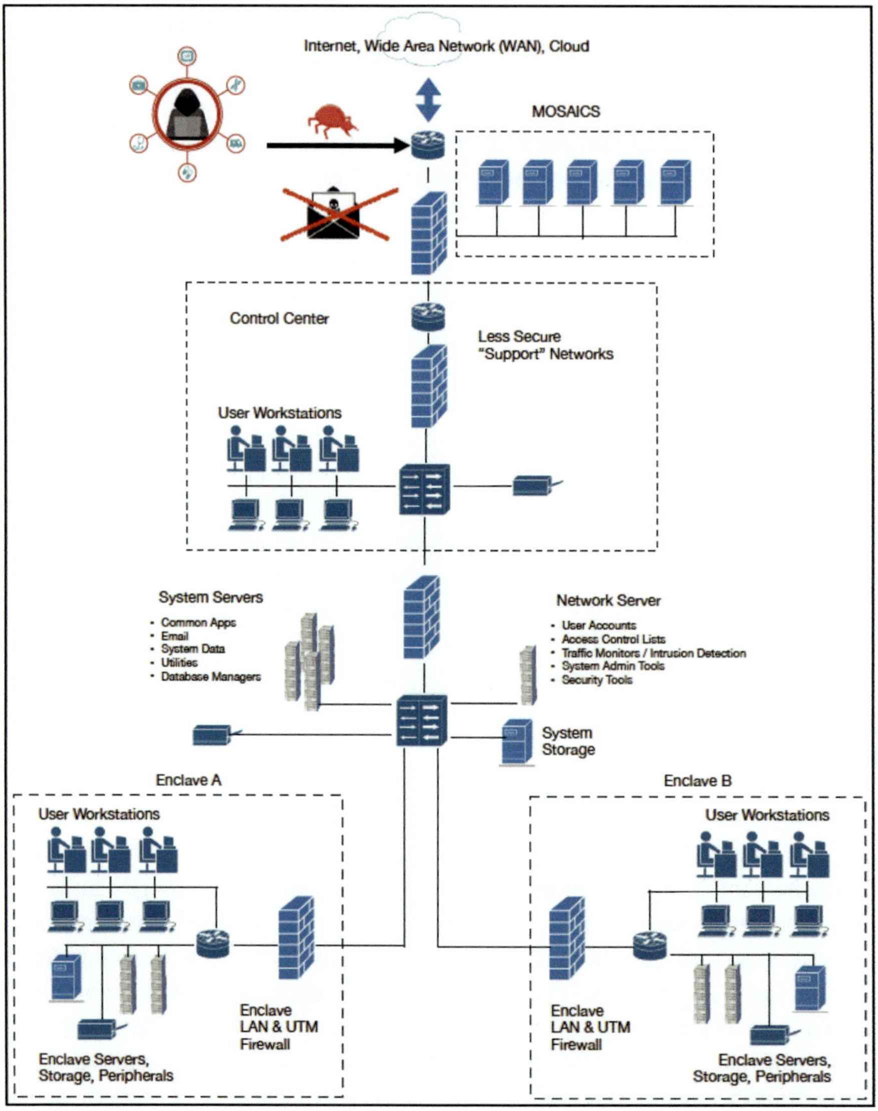

Figure 8.5 MOSAICS reference framework.

shown in Figure 8.5. Size, weight, and power constraints that impact the ability to make engineering design changes to critical infrastructure operations are observable in the reference framework. The framework's applicability encompasses delivery methods for technologies in other essential infrastructure sectors, repeatable playbooks for automated courses of action (COA), best practices, and templates for cybersecurity requirements for control systems, harmonized with concepts such as a Zero Trust strategy.

Additionally, MOSAICS introduces security orchestration, automation, and response (SOAR) capabilities to collect threat data and alerts from various security tools and disparate systems, supporting automation sequences for network defense. The automation sequences, known as "playbooks," provide immediate response efficiency for network operations and a consistent process to help administrators investigate and prioritize alert response actions (Rich Scalco, 2018). SOAR technology is a set of tools and integrated security-related processes used to support threat analysis. Zero Trust is a strategy. Both SOAR technology and a Zero Trust strategy are needed to stop adversaries from moving laterally across critical infrastructure networks. MOSAICS is a framework that describes how to integrate a system-engineered solution that leverages SOAR technology and Zero Trust strategy to deliver a capability. This approach outlines a plan to achieve optimal network security by implementing best practices for control systems. When a Zero Trust strategy is paired with the MOSAICS framework, the architecture provides a comprehensive solution for expedited and economical implementation (Aleksandra Scalco, 2021). To achieve this approach, it is necessary to understand a Zero Trust strategy, which is described next.

8.2.4 Zero Trust strategy

When implementing a Zero Trust strategy, system owners consider everything as untrusted, similar to not trusting external resource connections (SecurID, 2021). Zero Trust is a network security strategy rather than a technology. The Zero Trust strategy should permeate the organization, from the architecture through operational processes to the culture of the professionals using connected devices on the system (Simos, 2019). Everything starts from a Zero Trust level. Starting from a Zero Trust point of reference means that the trustworthiness of everything is continuously validated, including the validation process itself, from healthcare identities to applications and services used in hospitals to connect mobile devices. A Zero Trust strategy separates the control plane from external sources and utilizes trust algorithms to determine resource access to the network, such as granting, denying, or revoking access.

Additionally, internal or external threat intelligence sources provide information about new and emerging threats that can trigger automated courses of action (COAs), as outlined in the MOSAICS framework. Examples of threat intelligence include user history, behavioral analytics, IP addresses, network traffic, and locations. The Zero Trust strategy shifts the

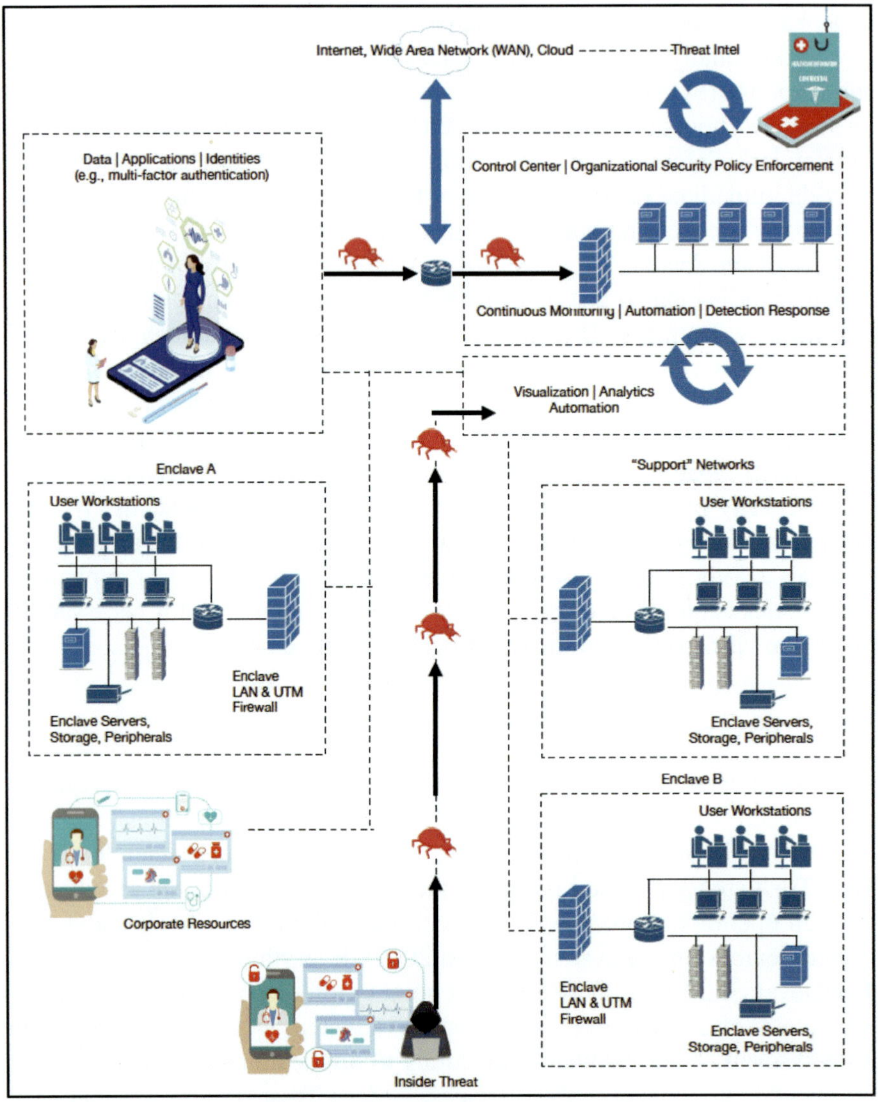

Figure 8.6 Zero Trust network strategy.

cyber defensive perimeter from a static perimeter to one that includes users, assets, services, and workflows. This more inclusive perimeter focuses on continuous monitoring, automation, and detection responses based on organizational policy and evolving threat intelligence, as shown in Figure 8.6 (Aleksandra Scalco, 2021).

Sound systems engineering practices can lead to an architecture that contributes to successful system development by providing a consistent approach to dealing with complex entities, maintaining traceability from requirements to physical components, and ensuring that all system behaviors are captured and mapped to solution elements. Systems engineering practices can provide insight into requirements refinement, prioritization, design choices, potential problems, and inform governance (Aleksandra Scalco, 2021).

8.3 Summary

Use case scenarios are a method of establishing user interactions with a system. The scenarios presented in this chapter illustrate the need to clarify requirements in relation to possible user and system interactions within the intended environment, in order to achieve the desired outcome. The identification of dependencies, constraints, and risks to the target system informs the design and development of the system. Systems use cases serve as a starting point for subsequent design and development activities specific to the critical infrastructure sector and the system of focus. Use cases provide a detailed description that enables the team to examine the system's cyber posture. A best practice for overcoming cyber vulnerabilities in critical infrastructure is to utilize systems engineering principles and practices that employ a Zero Trust strategy, coupled with a framework such as MOSAICS, to enumerate the network and automate the cyber defensive process. Doing so in an outdated environment presents an entirely different challenge, which will be discussed in Chapter 9, "Road to Commoditization."

References

Aleksandra Scalco, D. F., Steven Simske 2021. Control Systems Cyber Security Reference Architecture (RA) for Critical Infrastructure: Healthcare and Hospital Vertical Example. *Journal of Critical Infrastructure Policy,* 2.

Diogenes, Y. & Ozkaya, E. 2019. *Cybersecurity — Attack and Defense Strategies,* Birmingham, UK, Packt Publishing Ltd.

Egbogah, E. E. & Fapojuwo, A. O. 2011. A Survey of System Architecture Requirements for Health Care-Based Wireless Sensor Networks. *Sensors (Basel),* 11.

Rich Scalco, D. B. W., Jorge Lacoste, John Andrews, Bill Beary, Ross Roley, 2018. *More Situational Awareness for Industrial Control Systems (MOSAICS) Joint Capability Technology Demonstration (JCTD)*

[Online]. IACD Automate: Johns Hopkins University Applied Physics Laboratory (JHU APL). Available: https://www.iacdautomate.org/may-2018-integrated-cyber [Accessed October 2019].

Securid. 2021. *What is zero trust?* [Online]. www.SecurID.com: SecurID. Available: https://www.securid.com/en-us/blog/the-language-of-cybersecurity/what-is-zero-trust [Accessed October 30, 2021].

Simos, M. 2019. *Zero Trust strategy—what good looks like* [Online]. www.microsoft.com: Microsoft. Available: https://www.microsoft.com/security/blog/2019/11/11/zero-trust-strategy-what-good-looks-like/ [Accessed October 30, 2021].

9

Road to Commoditization

"In a commodity market, you can only be as good as your dumbest competitor."
– Peter Drucker (1909–2005), American author and educator.

"Do you believe in miracles?"
– Al Michaels (1944–), American television sportscaster, famously called the 1980 Winter Olympics men's ice hockey game between the United States and the former Soviet Union team.

Abstract

Most of the enterprise information technology (IT) industry is commoditized. IT commoditization means that much of an organization's IT is already obsolete or even irrelevant by the time a cybersecurity solution is proposed unless security operations (SECOPS) are defined and rolled out concurrently with the system rollout. What is often referred to as operational technology (OT) in critical infrastructure sectors is on the same path to commoditization. IT technology evolved into a commodity industry as efficiencies were gained from the manufacture of basic components. A component itself is not differentiated from another similar component and can be reused in an assembly across multiple manufacturers. The manufacturing footprint can be located closer to the assembly, thus saving costs. However, commoditization poses a new attack surface. Coupled with commoditization is the influence of cyber in IT and OT operations. Cyber is an IT industry disruption. Cyber can be viewed as moving further up the supply chain of component manufacturing, encompassing information technology and creating, moving, and storing finished data over its lifecycle. Cyber interfaces also move from supply chain manufacturing into process controls and critical infrastructure. Commoditization and the cyber medium create industry disruption. While commoditization and the cyber medium are both leveraged by businesses today to gain efficiencies, their use can be problematic and introduce new

DOI: 10.1201/9788743809128-9

challenges. Businesses leverage internet connectivity to perform services in the enterprise IT space and use connectivity in the OT for critical infrastructure. The business appetite for market share drives the homogenization of unique characteristics, making component units interchangeable and allowing competition on a price basis, which creates potential cybersecurity vulnerabilities.

9.1 Introduction

Recall the earlier ice hockey analogy in Chapter 1, "Understanding the Game." The skills of a young, less experienced team led to one of the greatest upset victories in hockey history. Getting inside the opponent's blue line zone is all an attacker needs to get a scoring chance. It just takes a single shot on goal to win a game against the toughest opponent. It just takes one "soft, chewy center," as the security adage made famous by John Kindervag's "No More Chewy Centers: The Zero Trust Model of Information Security," goes, to make a cyber perimeter-based model ineffective against targeted attacks (John Kindervag, 2016). All the best hockey players and all the best skaters cannot defend against a "soft, chewy center."

In keeping with the analogy of ice hockey, it was February 22, 1980, and the world watched an ice hockey game between two global giants during the Lake Placid Winter Olympics in Lake Placid, New York, United States – The Union of Soviet Socialist Republics (USSR), known as "The Soviet Union," and the United States. In the game's final moments, sportscaster Al Michaels thundered, "*Do you believe in miracles?*" about the US men's team's upset victory. There was much more behind the scenes of that disruptive, upset victory than was televised and witnessed that day. The key point is that the entire team, then and now, must work together to overcome vulnerabilities and achieve success, as anything can happen at any time. One thing to note is that with one second left in the first period, the Soviet Union's goalie gave up a point to Team US. Giving up a point made the Soviet Union's coach so mad that he pulled his goalie and put in the team's backup goalie. The coach never put the better goalie back in the game – He left a "soft, chewy center" rather than responding to the exposed vulnerability by shoring up his defensive strategy. The rest is history.

That same year, a transformation was unfolding in the global manufacturing industry. It was a disruption by design that would forever change the IT industry, leading to the commoditization of IT components and ultimately paving the way for the rise of the cloud industry. The changing landscape of the original equipment manufacturers' (OEM) supply chain relationship with the electronic manufacturing services (EMS) industry influenced the ability to convert innovative technologies and solutions into use. In this chapter, "Road to Commoditization," we share how focusing on disruption by design may best advance a cyber resiliency strategy for industrial control systems (ICSs) today and provide a roadmap for future systems. Given rapid technological advancements such as AI, most entities are fortunate to have a 5-year roadmap. Strategies to mitigate insider threats, including preventing unauthorized access to resources and achieving continuous monitoring, are discussed in further detail in Chapter 16, "Modernization Strategies."

9.2 Assume Breach

A significant takeaway from the past 20 years in enterprise IT is that we must assume a breach has already occurred in any system that is deployed. John Kindervag first codified the belief that trust is in itself a vulnerability in 2012 in what he coined as a Zero Trust Model (John Kindervag, 2012). The Zero Trust security model has since gained broad adoption. Kindervag later gave us the adage "eliminate chewy centers" as a new way of thinking about security (John Kindervag, 2012). Concepts of Zero Trust, including activities, architectural components, and organizational governance, provide valuable insights into an ever-changing ecosystem. Zero Trust gained widespread acceptance and adoption when Randy Resnick, DOD CIO Senior Advisor and Director of the Zero Trust Portfolio Management Office, took on the task of eliminating the "soft, chewy center" from DOD networks. Randy Resnick came from the National Security Agency (NSA), where he held technical leadership positions in both offensive and defensive cyber strategies. Resnick is credited with spearheading Zero Trust principles in the US government (DOD, 2024).

Additionally, many, if not most, IT products have become commoditized, which makes understanding security in the software supply chain a real challenge and a potential vulnerability. Most organizations do not adequately understand the attack surface of commoditized components that can compromise a system's cybersecurity posture. Similarly, operational technology (OT) asset owners and operators struggle to comprehend the full extent of cyber vulnerabilities in the systems that control the physical domain.

Once the IT industry became commoditized, most IT rollouts were already obsolete or irrelevant by the time a solution was proposed. It was not the integrity of the supply chain, but rather the commoditization of IT components, that was the biggest concern. Why did IT become a commodity industry, and why must we accept this fact, move past it, and anticipate the trend of cybersecurity vulnerabilities and successful attacks on systems today? IT technology became a commodity industry because, as a commodity, efficiencies were gained from the basic component manufacture itself, which is useful. The component itself is not differentiated from similar components and can be reused in assemblies across multiple manufacturers.

Additionally, the manufacturing footprint can be located closer to the assembly and the market consumer, thus saving costs. Proximity to the consumer is of significant value. This value translates into economic efficiency, as manufacturers can scale up their use for a more substantial return potential. Electronic manufacturing services (EMS) companies operate on thin margins, scaling components as commodities across product lines. Original design manufacturers (ODMs) or original equipment manufacturers (OEMs) operate on a smaller scale than the electronics manufacturing services (EMS) industry. The global EMS market size was valued at USD 626.8 billion in 2024 (Insights, 2025). The estimated compounded annual growth rate (CAGR) from 2025 to 2034 is 5.1%, attributed to the increased global demand for consumer electronics, adoption of the Internet of Things (IoT) and electronic vehicles (EVs), growth in the semiconductor industry, and the growing needs of the defense and aerospace industries. This increase puts the value projection in 2034 at USD 1.1 trillion (Insights, 2025). The trend also reflects the increasing adoption of artificial intelligence (AI) in manufacturing processes, with a growing use of robotics and automation to prevent mistakes, predictive maintenance to minimize downtime, and digital twins to reduce costs and accelerate reliable outcomes. The rise of the EMS industry is a topic addressed later in this chapter.

The rapid technological change makes it challenging for the ODM to respond quickly enough to meet the market trends that drive competition, as the EMS sector becomes more advanced in delivering end-to-end product development lifecycles and an increasingly high degree of complexity in electronic components and devices. Shifting the design and development of components and assembly to outsourced EMS companies enables the ODM to focus on the business market. At the same time, the EMS leverages technical expertise and maximizes the efficiency of manufacturing production closer to the market, thereby avoiding the costs associated with component development and warehousing. Appropriate test and development processes require increased investment in AI, automated test procedures, and digital virtual test environments, which smaller companies cannot compete with on a global scale, thereby driving consolidation in the global electronic manufacturing services market. Because commoditization poses a new attack surface, successful teams must understand the risks and better mitigate against cyber vulnerabilities. Well-established EMS companies that produce high volumes at low margins will be attracted to manufacture in countries that offer tax incentives and proximity to markets, thereby avoiding tariffs. Logistics, engineering services, testing, and development of components for critical infrastructure systems across all sectors and regions are following a similar path to commoditization as enterprise IT. This transition path presents supply chain and market disruptions that introduce cyber challenges to aging technology used in critical infrastructure systems. Key questions include: Who (or how and through what governing body) is responsible for industrial control systems (ICS), which include supervisory control systems used to control geographically dispersed assets? Who is responsible for controlling industrial processes, such as manufacturing, product handling, production, and distribution of Industrial control systems, including supervisory control and data acquisition (SCADA) systems used to control geographically dispersed assets? Who is responsible for an information system used to control distributed control systems (DCSs) and smaller control systems in the manufacturing processes that utilize programmable logic controllers (PLCs) to manage localized processes?

We can agree that parts of the DOD rely on comically atavistic technology that is difficult to maintain and replace. Overhauling the infrastructure has almost universal appeal. In 2017, US President Donald Trump invited the CEOs of America's most prominent technical companies – Apple, Amazon, IBM, Intel, Oracle, Qualcomm, Microsoft, and Google, among others – to help overhaul the US government's technical infrastructure (Solon, 2017). Some of the high-end products these companies made had become commodities, and they relied on the EMS industry to help maintain their relevance. At the same time, they were looking for strategies and partners to deliver their products and services. They had potentially commoditized themselves out of business in the pursuit of economic efficiencies. They found a willing partner in the US government. One outcome of the meeting was the drive to switch to cloud-based services, moving away from hardware, which is now, in manufacturing terms, at the end of its lifecycle and has become commoditized. According to reports about the roundtable, CEOs promoted the advantages of data analytics and cloud as a service (CaaS), which are commercial technology services sold by the companies of the CEOs giving the advice. Their hardware had already been commoditized. CaaS offered application software with users paying an on-demand use fee through paid subscriptions, bypassing the need for EMS-produced components. CaaS and other cloud solutions allowed these companies to reinvent their businesses after the EMS industry disruption.

9.3 New Business Models

As we will see shortly, the remarkably efficient EMS industry also expanded into the newly emerging markets. Apple offered a familiar business model of making product-related skills be adopted as core curriculum requirements in schools. Apple CEO Tim Cook proposed increased training on software development skills, such as coding, to advance the software-as-a-service model (Solon, 2017). Coding, as a requirement, would prime the market for advancing commercial software-based technology. Thirty-some years earlier, Apple had prepared the commercial and home consumer market for the coming home computer. Getting Apple into schools primed the market and became the business model for Steve Jobs, CEO of Apple Computer Inc. Apple had just had an initial public offering (IPO) in 1980, with a valuation of USD 1.8 billion in the United States.

The marketing strategy focused on promoting computer literacy and offering tax exemptions for computers donated to schools (Writer, 2019). In May 1983, Jobs promoted the Apple program "Kids Can't Wait" to get an Apple IIe into every primary and secondary school in California. At the same time, Apple reduced production costs by merging commercial-off-the-shelf (COTS) component functions, reducing the number of chips, and improving reliability. The move made the Apple IIe cheaper to build, thereby increasing profit margins and feeding shareholder dividends. Efforts to obtain tax credits in exchange for computers in the classroom at the federal level, as outlined in the Computer Equipment Contribution Act of 1982, failed to pass in the Senate. The "Deduction for Equipment Contributions Computer Equipment Contribution Act" of 1982 (H.R. 5573 – 97th Congress (1981–1982)) would allow for "charitable contribution income tax deduction for corporations which donate computers to qualified educational organizations (including museums and libraries)" (Watters, 2015). A valuable article about the pros and cons of special tax incentives to computer makers donating computers to schools was published in Communications of the Association for Computing Machinery (ACM) in March 1984, with Professor of Computer Science Hal Berghel arguing against the incentives as not being in the long-term public interest. California Congressman Pete Stark advocated for special tax incentives to equip schools with computers (Dallaire, 1984). Although the incentive did not pass at the federal level, similar legislation was passed at the state level. California Governor Jerry Brown signed a bill into law that provided companies donating school computers with a 25% tax credit against the retail value of the computers. Leveraging the credit gave Apple inexpensive access to market share. By the late 1980s, school computers were ubiquitous, with Apple dominating the classrooms (Vincent, 2022). The computers in the student labs used by the authors in 1987 were all Macintosh models, donated by Apple – Aleksandra Scalco at the University of Missouri in Columbia (UMC) and Steven Simske at Rensselaer Polytechnic Institute (RPI).

What happens when a company's high-end product becomes a commodity is a convincing reason that talking to the people whose business sells IT technologies may not be the best people to lead the discussion when seeking to convert innovative technologies and solutions into practical use to advance a national cyber resilience strategy. Disruptive by design is necessary to introduce disruptive technologies and creative solutions that advance a national cyber resiliency strategy. A cyber resilience strategy must be decoupled from economic commodity dependency to work. We must get out of the commodity hole we dug ourselves into. OT for ICS essentially, and predictably, follows the same technological development path of disruptive IT innovation.

9.4 Commoditization of Components

The design of components has essentially become commoditized by the EMS industry. Multi-billion-dollar revenue companies such as Sanmina, Flextronics, Jabil, and Celestica have revolutionized the modern technology of electronic component production, dramatically reduced production costs, and contributed to the standardization of manufacturing processes and component units. Standardization is a principle used in systems engineering to ensure that physical interfaces are compatible and interoperable. So, commoditization makes sense not only for manufacturing but also for fit-for-purpose use. "All physical interfaces fit because they are made to [a] standard" (Alexander Kossiakoff, 2011). The EMS industry's systems engineering contribution to the transformation of the design, development, production, and delivery of electronic components is evident in the impact on cost, reliability, and design for manufacture (D*f*M) (e.g., design feasibility), as evidenced by the reduction in the cost of standardized components. Customization of components increases the unit cost of development and production. Some design features or characteristics that contribute to laptop reliability include the operating system (OS), motherboard, power consumption unit (PCU), memory/RAM modules, graphics card, solid-state drive (SSD), cooling fan, and central processing unit (CPU).

Although the OS is not hardware, it is the single most likely factor contributing to laptop reliability (or the feature that breaks) due to malware. As discussed in Chapter 6, "Cyber Vulnerabilities Start at the Hardware Level," rudimentary component-level hardware can pose reliability and cybersecurity issues. Motherboards are one of the most complex components of the laptop and have one of the highest overall failure rates of any other laptop component. The estimated contribution to the overall computer cost of motherboards is high. The PCU contributes to laptop reliability for several reasons. Lower power consumption equates to longer battery life, and an insufficient power supply can also kill other computer components. Recall from Chapter 6, "Cyber Vulnerabilities Start at the Hardware Level," that power use can give off a "surface" electromagnetic potential and can be susceptible to emanation attacks. Symptoms of a faulty power supply include failure to power on, damage to internal components, and overheating due to fan failure, among others. The estimated contribution to the overall computer cost is low. Also, recall that cyberattacks on RAM are generally well-known. Memory or RAM modules are essential for a laptop to function properly. Relative to other components, RAM has a long lifespan. RAM is also susceptible to temperature and power surges (inadequate supply or electrical spikes can contribute to failure), and can become corrupted by both, making the power supply a potential attack vector. The estimated RAM contribution to the overall computer cost is medium. The graphics card produces the images rendered on a monitor. The estimated contribution of the graphics card to the overall computer cost is high. A solid-state drive (SSD) is the primary storage device in a laptop. SSDs use less power, are more damage-resistant, and are less likely to fail technically than spinning drives.

Spinning drives wear out faster than solid-state drives (SSDs). However, SSDs are susceptible to extreme temperatures and unreliable for data retention. The estimated contribution to the overall computer cost of the SSD is medium. The cooling fan prevents thermal throttling and overheating issues; some thinner machines omit the cooling fan and rely on alternative heat dissipation methods. Cooling fans are mechanical and are susceptible to general wear and tear. The estimated contribution of the cooling fan to the overall computer cost is low. The CPU is the controlling component of the laptop. The estimated contribution to the overall computer cost is high. In OT for ICS, the programmable logic controller (PLC) is the primary focus of control.

9.5 Industry Disruption Precedes "The Boom"

Disruptive technology or disruptive innovation is a concept that helps create a new market and value network, ultimately disrupting an existing market. The term is used in business and technology to describe innovations that make a product or service in ways that the market does not expect. One can argue that the operational technology (OT) for the critical infrastructure market did not expect the disruptive innovation of connection to the internet. "Disruptive innovation" is a more apt term, as few technologies are inherently disruptive; instead, the business model that the technology enables is what creates the disruptive impact. Clayton M. Christensen, in his bestselling book "The Innovator's Dilemma: When New Technologies Cause Great Firms to Fail," asserted that regardless of the industry, a company will be pushed aside unless its organizational leadership recognizes when and how to abandon traditional business practices (Christensen, 1997). In that sense, the disruption in the IT industry was somewhat self-inflicted, arising from the growth of the EMS industry. Christensen's provocative thoughts about innovation, first presented in 1997, remain applicable to the dynamic transformation of the critical infrastructure sector. Systems engineering rigor provides a set of principles and practices that an organization can use to capitalize on disruptive innovation without compromising its business operational edge.

Here is a timeline that is valuable to understand in the context of where OT is quickly headed. In the 1960s and 1970s, most electronics manufacturing product runs for original equipment manufacturers (OEMs) were handled by in-house assembly. Some manufacturers were contracted to assemble goods, a process known as "contract manufacturing." Essentially, when you bought a brand with few exceptions, the OEM made the product. That was true until the electronic manufacturing services (EMS) industry, discussed earlier in this chapter, was established in the late 1970s by companies such as Solectron, Flextronics, and Foxconn (also known as Hon Hai Precision Industry Co., Ltd.). The EMS industry really took off in the 1990s. The new business model for the EMS industry focused on achieving economies of scale in manufacturing for key customers, including Dell, Sony, IBM, and Apple. The EMS companies offered value-added services, including warranty, repairs, and design for manufacturing expertise. This transition to contract manufacturing helped free OEMs, as they no longer needed to manufacture and store extensive inventories of products. By the mid-1990s, the advantages of the EMS concept became more compelling, with products in nearly every market, from computer and consumer electronics to automotive, agriculture, and medical and healthcare. The emergence of surface mount technology (SMT) enabled the direct mounting of components onto the board surface, reducing the footprint size of electronic devices and increasing the resilience of parts to environmental factors. OEMs began to outsource printed circuit board assembly (PCBA) production, which is found in most electronics, on a large scale, eventually taking over the manufacturing of the actual circuit board itself. The PCB connects components and safeguards the functionality of electronic devices from potential threats that can compromise the PCB's integrity, such as moisture that can lead to corrosion or physical damage that can degrade the circuit board's functionality. This change was significant for companies such as Dell Computers.

Dell, founded in 1984, rode the same market wave as Apple Computers in a large, new commercial market. By 2000, internet sales on Dell.com had reached USD 40 million a day, making it one of the highest-volume electronic commerce (e-commerce) sites in the world. E-commerce was everywhere. Advanced PCB assembly techniques in circuit boards have made it challenging for OEMs to meet competitive market demands without the benefits of an EMS provider. For example, Dell was not manufactured by Dell, nor did Dell run the e-commerce site. The just-in-time (JIT) manufacturing wave that helped carry the internet sales was a large EMS, privately held company, SCI. You will not find that information on Dell's website or similarly on any other company's slick sheets.

9.5.1 Original equipment manufacturer (OEM) turnover to electronic manufacturing services (EMS) outsourcing

Similarly, other original equipment manufacturers (OEMs) outsourced their assembly and manufacturing to powerhouse electronic manufacturing services (EMS) companies, which leveraged high-volume, large-scale manufacturing to gain efficiencies and earn profits through sheer volume. Essentially, EMS companies assemble and integrate OEM components and parts into the finished product. The value added is increasingly shifting into systems engineering services, encompassing new product design and development, as well as test and evaluation, throughout the entire lifecycle. Control of engineering cybersecurity mitigations into products early in the lifecycle also shifts to the EMS industry. As the EMS industry grew, companies attempted to assert proprietary ownership of manufacturing lines where multiple brands were produced, ultimately leading to most divesting from the market. In 2005, IBM sold its once-heralded "Machine of the Year" personal computer (PC) division to Chinese PC maker Lenovo Group for USD 1.75 billion. IBM divested the once open-architecture personal computing machine that became an industry standard. At its peak, IBM PCs sold one PC per minute of each business day [IBM]. IBM sought more profitable software-based technology and consulting services, moving up the economic ladder. At this point, almost 40 years after the advent of the home computer, it is clear that the disruption and commoditization of IT have introduced new cybersecurity vulnerabilities. The PC itself has become a commodity.

Interestingly, less than 20 years later, after the divestiture by IBM of its commoditized PC era division to Lenovo, in an open letter to the US Navy Exchange in 2023, Chairman Mike Gallagher, a Republican from the US State of Wisconsin (R-WI) of the House Select Committee on the Chinese Communist Party (CCP) requested the removal of CCP-linked Lenovo products from the marketplace available to the service. Chairman Gallagher wrote that the affiliation with the People's Liberation Army (PLA), the Chinese Academy of Sciences (CAS), and the People's Republic of China (PRC) links to state-run cyber espionage campaigns, making the computers vulnerable to spyware and backdoors (in software, firmware, and/or hardware) that could compromise the Department of Defense (DOD) (Gallagher, 2023). The efficiencies gained from the commoditization of computers have opened up new cybersecurity vulnerabilities. As a point of reference, according to Statista, in the third quarter of 2023, Lenovo's global market share of PC shipments was 25.1% (Statista, 2021). Lenovo increased its global PC market share from less than 8% in 2009 to over 25% in approximately 15 years (Alsop, 2023). Three companies combined make up the majority of the PC market share in global sales today – Hewlett-Packard (HP), Dell, and Lenovo.

Today, the top three US PC brands—Apple, HP, and Dell—rely on Taiwan for over 90% of their components. However, this is gradually changing as some parts of their supply chain are being relocated. Apple's product brand is known for its innovative features and seamless integration across its products. HP is renowned for its reliability in a diverse range of computers, including laptops, desktops, and powerful workstations. Dell is best known for its customizable consumer configurations and competitive pricing. In 2013, Dell announced its sale to Microsoft in a leveraged buyout, with Dell retaining the largest ownership stake. By 2023, Dell announced that the company would stop sourcing component chips from China by the end of 2024 (Humphries, 2023). HP also announced its plans to diversify PC production with its EMS providers outside China in 2023. Underscoring geopolitical concerns about the cyber vulnerabilities of potential nation-state threat actors, Apple has initiated the relocation of its MacBook production to Vietnam (Shilov, 2023). Despite strategic OEM moves to production in other countries, the electronic services industry remains strong, with a projected compound annual growth rate (CAGR) of 6.8%. The estimated market size in 2021 was approximately USD 476.88 billion and is expected to reach USD 797.94 billion by 2030, providing insight into the scale and scope of the EMS market size (Skyquest, 2024). Top EMS companies today include Foxconn Technology Group and Pegatron, both headquartered in Taiwan; Jabil, with its headquarters in Florida, United States; Flextronics and Sanmina

Corporation, with headquarters in California, United States; Benchmark in Arizona, United States; Zollner in Germany; Dixon Technologies in India; Venture in Singapore; SIIX Corporation in Japan, and many others around the world.

9.5.2 IT technology timeline

Meanwhile, the developments in IT products and technologies, which we focus on in our discussions, also follow an interesting timeline that is similar to that of OT. In the 1980s, IT focused on identifying asset management; in the 1990s, the focus was on protection as demonstrated by antivirus software that could detect and automatically remove computer viruses such as McAfee Enterprise; in 2000 – Detection; in 2010 – Response; in 2020 and beyond – Recovery, Resiliency, and Elasticity. Much of the innovation and technological advancements may be directly attributed to the rise of the EMS industry. The National Institute of Standards and Technology (NIST) Cybersecurity Framework, first published in 2014, as noted in earlier chapters, is a comprehensive organizational tool used to mitigate cybersecurity risk. It follows a development timeline with five elements: identify, protect, detect, respond, and recover.

In 2024, the term "govern" was added to the NIST Cybersecurity Framework 2.0 to expand the focus to internal and external stakeholders, which the book's authors emphasize is dependent on "stakeholder alignment." NIST Special Publication (SP 800-82) expands guidance on securing OT, acknowledging the unique performance requirements needed for monitoring and controlling devices, processes, and events found in systems such as building automation systems and industrial control systems (ICS) (Stouffer et al., 2023). In a 2022 dissertation, "Measuring Disagreement in Segments of the Cybersecurity Profession as a Means of Identifying Vulnerabilities," author Aleksandra Scalco asserts that stakeholder misalignment may be greater than the overall system vulnerability (Scalco, 2022). Survey questions used in the dissertation can be found in Appendix A, "Stakeholder Alignment Survey Questions," of this book for the reader's reference and use (Scalco and Simske, 2022). While the governance function is not such a new concept, updating the NIST key tasks to include governance has a tremendous impact on an organization's cybersecurity posture. It sparked a healthy debate among experts and IT purists who did not necessarily view governance as a cybersecurity function. We had long asserted that tasks such as designing a cybersecurity strategy need to be augmented by stakeholder alignment; otherwise, the plan will never work as intended, despite all valiant efforts. This assertion is highlighted in our research about the challenges of IT and OT perspectives in a conference brief in April 2021 entitled "Championing Cultural Change and Control Systems Cyber Security" (Aleksandra Scalco, 2021) and again in 2021 in a journal article, "Model for Multi-concern Assurance in the Digital Transformation of Engineering of Critical Infrastructure Control Systems" (Scalco and Simske, 2021). In our abstract, we asserted that "the digital transformation of engineering practice, paradigms, processes, and the workforce engenders uncertainty of agreement among professionals, particularly in the critical industry control system field." We consider governance to play a key role in ensuring that system vulnerabilities can be overcome; without it, there is a greater vulnerability than the inherent system vulnerability itself.

> "Control systems are susceptible to cyber-mediated changes that can affect the control of the physical world in a way [sic] unique from data-centric information systems. Change to the system can be introduced by any proposed or forced alteration that affects the acceptability, suitability, feasibility, or resiliency to perform its intended mission, either positively or negatively. The potential impact of the cybersecurity threat on control systems is difficult to quantify. Agreement among professionals about decision authority and command and control (C2) over this threat is even more challenging to quantify. Understanding what cybersecurity entails is still not widely understood by the critical infrastructure control system workforce, and the control system assets are not widely understood by the Information Technology (IT) workforce. This research introduce[d] a model and methodology for measuring multi-concern assurance through the statistical uncertainty analysis of Likert semantic differential scales. The outcome identifies where different types of professionals do not agree about the current state of cybersecurity readiness and best practices for critical infrastructure control systems" (Aleksandra Scalco, 2021).

Cybersecurity is unmanageable without stakeholder alignment and a governance model that attaches accountability and responsibilities. The cyber medium is broader than "just IT" or "just security." That cyber is broader than just IT becomes evident when the lower levels of the open systems interconnection (OSI) model are approached, and IT touches physics-based systems. Recall from Chapter 8, "Use Cases and Operational View (OV) Illustrations," that network segmentation reduces the network attack surface. The OSI model is discussed in greater detail in Chapter 23, "Abstraction Models." The complexity of hardware and software technologies working together across boundaries in systems engineering thinking extends a system

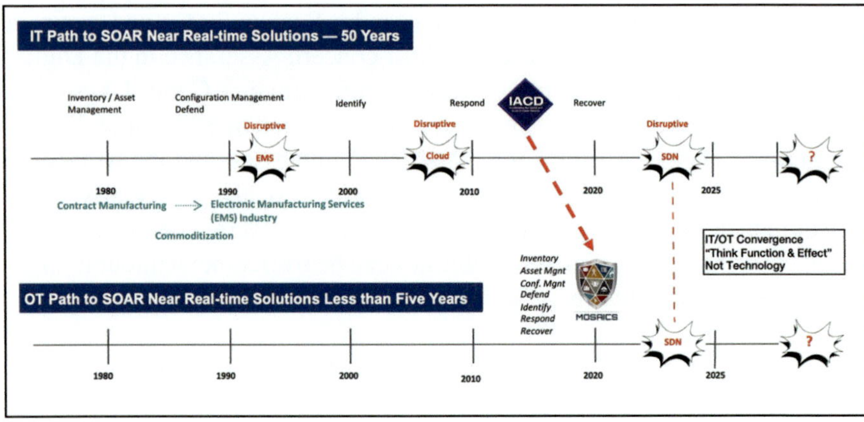

Figure 9.1 Information technology and operational technology timeline perspective (A. Scalco, 2019).

boundary beyond the higher abstraction layers to include the lowest layer at the physical level, which represents the physical and electrical aspects of the system, where computing functions control devices and sensors. The physical layer encompasses the physical channels, including copper cabling and associated processes. Everything in the OSI abstraction is a candidate for disruptive technologies and commoditization.

It has taken more than 50 years of IT investment, research, development, and experience in the commercial industry to integrate security orchestration, automation, and response (SOAR) technology into enterprise IT systems, enabling the execution and automation of tasks between people and tools as shown in Figure 9.1. The SOAR industry was heavily seeded by investments from the Department of Defense (DOD) and the Department of Homeland Security (DHS), which were realized in partnership with the Johns Hopkins University Applied Physics Laboratory (JHU APL) through the government-funded integrated adaptive cyber defense (IACD) initiative. The IACD strategy aimed to enhance the scale and speed of cyber defensive responses through automation and information sharing. Subsequently, the same transformation to OT found in critical infrastructure is being achieved in less than 3 years with significantly smaller investments by the US government through the MOSAICS efforts, which aim to realize a similar set of services that automate cyberattack prevention and response across critical infrastructure as an enterprise. All the capabilities realized by the non-proprietary, vendor-agnostic MOSAICS framework will eventually be integrated into the ICS cyber domain just as SOAR has been integrated into enterprise IT networks.

9.6 Industry Disruptions

Cyber is an industry disruption. Cyber can be viewed as moving further up the supply chain of component manufacturing, encompassing information technology and creating, moving, and storing finished data over its lifecycle. Security operations (SECOPS) brings cybersecurity teams together with operations teams by integrating monitoring tools and processes with technologies to protect organizational assets. Skilled SECOP teams can help overcome cyber vulnerabilities and enhance system resilience throughout the supply chain. System resilience encompasses monitoring and assessing risk, as well as protecting the supply chain of electronic system components, including transistors, resistors, capacitors, and integrated components that comprise a microchip or integrated circuit (IC). The microchip is a complex layering of semiconductor wafers, copper, and materials interconnected into complex assemblies. Common IC functions are used for processing and storing information in virtually all electronic devices. ICs are printed as a single unit, consuming less material and power than a single function, disparate, individual, or discrete component. The use of integrated circuits over discrete components enables large-scale integration and high production volumes of small-sized components, resulting in lower costs. It is less expensive to mass-produce, but the upfront design and development capital investment is far more significant, and the cost of building a manufacturing line is substantial. Once a factory is built, there are cost benefits and cost-effectiveness in using a shared manufacturing line for integrated circuits, which can lead to increased cost efficiency in product production. What starts at the component manufacturing level, this single production capability is one of the primary sources of competitive advancement in creating information technology capabilities: scale, proximity to market, lower-cost raw and component materials, and shared research and development costs for the design and development of innovation.

The same applies to programming languages as part of the cyber supply chain. Open-source, object-oriented programming languages, such as Python, are economical because they are free of license fees, are generally more straightforward than other languages, and generate code that can be automated, reused, and shared. However, software flaws, such as mistakes made by developers when creating code or poor coding practices, as well as the absence of source code review and security testing, can be a source of cyber vulnerabilities. Perhaps scarier at this point is what generative artificial intelligence (AI) for coding will do – without proper testing and measurement harnessing, cyber vulnerabilities may proliferate and become more difficult to detect. The benefit of converting code into standardized commodities is monetizing code at an exchange value that is interchangeable with other products and services and at a lower cost than designing and developing the code from scratch. The point here is that just as the hardware of components has become "commoditized," so has coding. Essentially, cyber is where the advantages of cost savings and hardware and software efficiencies converge into a single commoditization path.

Commoditized products simplify production and the decision-making regarding the suitability of technologies. As product components become more similar in design and production, the cost of the business model decreases. If cybersecurity is integral to a company's business proposition, then it is essential to understand the trade-offs involved in implementing business strategies and system development. Another term for commoditized could be "commercialized." The process of bringing innovation and new technologies to market is a lengthy and iterative one, encompassing investments in research and development, prototyping, and experimentation. A challenge is moving innovation into business strategy and operations. Moving innovation from science and research discovery and experimentation to doing something better, faster, or cheaper should not be rushed without considering the system lifecycle, which involves all the activities and decision points involved in system development, including cyber considerations, from entry criteria for technical reviews to contract language for commercial partners. Doing so will be regretful. As the character of Miracle Max said in The Princess Bride, "*You rush a miracle man, you get rotten miracles*" (Reiner, 1987).

There are three business timelines to follow to understand cyber vulnerabilities. One is the timeline of manufactured hardware components, another is the programming language that enables the use of computers, and the third is business efficiencies, which encompass technologies, people, and processes. There is a reason to bin "people and processes" into efficiencies that are discussed in Chapter 24, "Creating a Model-based Organization." Keep in mind that business efficiencies drive profitability. Cost reductions are calculated by subtracting the cost from the benefit. Cost reduction strategies are employed to reduce operational expenses, enhance efficiency, and minimize waste. There is a reason to discuss cyber in cost reduction strategies because, essentially, cyber is a business strategy, which we will also get to in more detail in Chapter 24. The inter-system considerations are the three timelines: hardware and software component timelines, and business efficiency. Systems engineering practice ensures that the complexity of products or services is managed from conception to operational use of physical systems, which consist of both hardware and software. The disciplined approach is holistic and requires stakeholder alignment to overcome innate system vulnerabilities. Building a successful team to overcome cyber vulnerability requires an inter-system approach, systems engineering thinking, and stakeholder alignment.

The term "manufactured components" was generally applied to hardware before Grace Murray Hopper's contribution to the design of the FLOW-MATIC computer compiler program, which had a substantial impact on the creation of COBOL. In 1959, this programming language could operate all types of computers. Programming innovation was followed in the 1970s by the introduction of the easier-to-learn Pascal language, which Apple used for software development.

Cybersecurity vulnerabilities emerged concurrently with the commoditization of industry disruptions in component manufacturing. A new idea came along, paving the way to usher in what can be called the "Cyber Era." The idea was that business and transactions need to be just-in-time (JIT), with manufacturing as closely located as possible to the end consumer market. Contract manufacturing transformed the way industries created products and goods by outsourcing aspects of production to external entities, which specialized in producing components and then purchasing these components to assemble them into the final output. The outsourcing of components gained momentum and efficiency, leading to the evolution of a new industry model within the EMS sector. It is worth noting that some of the risks of outsourcing can be obviated with careful object-oriented (OO) design, as long as the interfaces and objects are correctly designed, outsourcing software production can be made much less risky.

As a transformation unfolded in the global manufacturing industry, it was a disruption that would forever change the landscape of the IT industry, leading to the commoditization of IT components and ultimately to the rise of the cloud. The changing landscape of the OEM supply chain relationship with the EMS industry influenced the ability to convert innovative technologies and solutions into use. Focusing on disruption by design may advance a cyber resiliency strategy. A big takeaway is that we must assume breach. Additionally, many, if not most, IT products have become commoditized, which makes understanding software supply chain security a challenge and a potential vulnerability. Most organizations do not adequately understand the attack surface of commoditized components that can compromise the cybersecurity posture of a system.

Once the IT industry became commoditized, most of what we discussed regarding IT was already obsolete or irrelevant. The issue was not the integrity of the supply chain itself but rather the commoditization of the IT components. Why did IT become a commodity industry, and why must we accept this fact, move past it, and stay ahead of the trend? Because it presents a new attack surface that successful teams must understand to identify risks and better mitigate against cyber vulnerabilities.

Outsourcing also led to a shortfall in the capacity of science and technology by some of the OEMs that branded the products for final market sale. Contract manufacturers could specialize in introducing innovative technologies on their production lines, making it difficult for a single OEM to invest in new technologies and processes. In some cases, the volume of material the contract manufacturer purchased overwhelmed the available supply and consumed necessary manufacturing materials from the marketplace, preventing other manufacturers from acquiring what was required for production. New plants were built as capacity was filled at the contract manufacturing organizations. New investment and growing expertise limited the manufacturing of specialized components to those entities with multiple manufacturing lines. Investors followed the opportunities presented in a growth market sector from the 1980s through the 1990s.

Disruption has a paralyzing effect on industry leaders. With resource allocation processes designed and perfected to support sustaining innovations, they are constitutionally unable to respond. They are always motivated to go up-market and are almost never motivated to defend the new or low-end markets that the disruptors find attractive. We call this phenomenon asymmetric motivation. It is the core of the innovator's dilemma and the beginning of the innovator's solution.

The size and scale of the EMS businesses are enormous. It makes it impossible for OEMs to compete in that space and equally impossible for entities such as the DOD to influence that space. In 2017, the entire DOD addressable IT security spending (for hardware, software, and IT services only) totaled roughly USD 1.8 and 1.9 billion for civilian agencies. In 2016, it was estimated that by 2020, IT spending would grow to USD 2.5 billion for the DOD and USD 2.2 billion for civilian agencies. The US Office of Management and Budget (OMB) estimates total government IT spending in the United States at USD 74 billion, excluding the legislative and judicial branches, as well as numerous other executive branch agencies. The OMB states that of this amount, USD 54 billion is allocated to operations and maintenance, and approximately USD 20 billion to development and acquisition. Spending on legacy systems, including operations and maintenance of existing systems, accounts for approximately two-thirds of the DOD IT budget. The total available for new projects represents only a small subset of IT spending. The total Defense Department IT budget decreased from USD 35 billion in Fiscal Year 2012 to approximately USD 30 billion in Fiscal Year 2015. It is worth noting that this accounts for 0.1% of the US economy, so insurance is only USD 1 for every USD 1000 of the economy. It would be comparable to the average US citizen spending USD 70 on car insurance each year, which is unlikely. The slow wind-down of the wars in Afghanistan and Iraq contributed to the Department of Defense's (DOD) IT cutbacks. Cutbacks were most noticeable in the US Army's declining IT budget, consolidation efforts related to specific enterprise services, and migration of some IT dollars to intelligence agency solutions. Clearly, the US Department of Defense's (DOD) spending on IT solutions has taken a hit in recent years.

The EMS industry is commonly divided into tiers based on annual revenue.

- Tier 1 > USD 5B
- Tier 2 USD 500M to USD 5B
- Tier 3 USD 100M to 500M
- Tier 4 <USD 100M

Imagine, if you will, an iceberg. Nearly 90% of an iceberg is typically underwater. We do not see it. That little 10% above the waterline is what we, as consumers and government program managers, see. We know that as a market segment, the DOD is a fraction of the tip of the iceberg. Density explains why most of an iceberg is found beneath the ocean's surface. Because the densities of ice and seawater are so close in value, the ice floats "low" in the water, meaning that ice has nearly nine-tenths, or 90%, of water's density, so 90% of the iceberg is below the water's surface.

It is an iceberg, indeed. The total DOD security spending power is roughly equivalent to the revenue of one tier 2 electronic manufacturing services (EMS) company. The Worldwide IMS IT Forecast is one of the leading indicators of significant technology trends. The EMS industry is a USD 1.7 trillion market with a compound annual growth rate (CAGR) of 6.7%. Data center systems, software, devices, IT services, and communication services are all healthy growth areas for EMS companies. According to some analysts, the industry is expected to experience 12% earnings growth.

One of the EMS companies alone, Flextronics (FLEX), has a forecasted earnings growth of almost 60%. At the current rate of growth, the EMS industry is poised to surpass manufacturing in the production of all IT technology assets.

What does a trillion dollars look like? We will start with a USD 100 bill. A packet of one-hundred-dollar bills is less than one-half inch thick and contains USD 10,000. It could fit in your pocket. A little pile of 100 packets, each worth USD 10,000, is USD 1 million. You could stuff that into a grocery bag. A pallet of USD 100 million would come up to your shoulder. A billion dollars starts to look impressive. That is ten pallets. ONE TRILLION dollars. Well, it is a million million. It is a thousand billion. Twelve zeros follow it. It is a football field of pallets, up to your shoulders. Commoditization, outsourcing, new technologies, and Bitcoin are all industry disruptions.

9.7 Market Share

Cyber is an important medium used by businesses today to gain efficiency. Businesses leverage internet connectivity to perform services in the enterprise IT space and use connectivity in the OT space for critical infrastructure. The business appetite for market share drives the homogenization of unique characteristics, making component units interchangeable and allowing competition on a price basis. Market share refers to the percentage of a market that a particular brand or product controls. Companies can increase market share basically in four ways:

1. Steal market share
2. Strategic acquisition
3. Technology and innovation
4. Market to different demographics

The implication of the cyberspace medium, when viewed from a global usage perspective, reveals a stark global reliance on the internet as a catalyst for new definitions of nation-state boundaries. "An economic shift will change the way that corporations, government agencies, and other organizations, government agencies, and other organizations share storage, computing, network, and application resources." The penetration rate, expressed as a percentage of the population, is estimated at 30.2%, against an estimated global population of 6.9 billion people, making cyberspace a game-changing geopolitical system (Group, 2011). In the early history of cyberspace, in 1995, author Jean-Marie Guehenno observed how the Information Age and the resulting global community might redefine the geographical boundaries and power centers in world affairs (Guehenno, 1995). This shift is substantiated today, even as we write, by a geopolitical shift that moves national boundaries, as more business is conducted online. The globalization of economics occurs through the relocation of industrial bases of manufacturing and knowledge management to anywhere a person can access the internet. An important definition that has evolved since Guehenno published his book, which is key to understanding this declaration of a new operational domain, is the definition of cyberspace. "Cyberspace is the interdependent network of information technology (IT) infrastructures and includes the internet, telecommunications networks, computer systems, and embedded processors and controllers in critical industries" (Crowther, 2017). Some 20 years after Guehenno identified cyberspace as a game-changing influence in "End of the Nation-state," the world internet usage statistics increased from 361 million in December 2000 to more than two billion people by the end of March 2011 at a growth rate of 480% from 2000 to 2011, making the challenge how a nation responds to the many physical instantiations of a cyberspace domain a systems engineering challenge. Just as in previous land, air, and sea strategies, nations responded with their cyber strategy to promote their interests through cyberspace. For example, the United Kingdom published its "National Cyber Force: Responsible Cyber Power in Practice" in 2023 (Force, 2023 #580), and in the United States, the Department of the Navy (DON) Principal Cyber Advisor Christopher P. Cleary published the Navy's "Cyberspace Superiority Vision" a year prior in 2022 guided by three principles: "Secure, Service, and Strike" (Cleary, 2022).

9.8 Cybersecurity Commoditization Pitfalls

Commoditization, accompanied by cost-cutting in cybersecurity, can quickly become a pitfall when the intrinsic value of IT and OT is reduced to a monotechnic prioritization of profit over system resilience and security. Cybersecurity commoditization is inevitably replaced by failed cybersecurity practices that benefit economic control and access to real resources and system value. The value of critical infrastructure lies in maintaining essential operational functions, including during a cyber incident and the restoration of these functions after a breach. From a customer and asset owner's perspective, understanding the value of cybersecurity in relation to its intrinsic cost efficiency is necessary to meet business operational needs, rather than focusing solely on price. Cybersecurity commoditization presents a false sense of security. "Commoditization is the process whereby value becomes interchangeable, making it difficult to differentiate on anything other than price. From the customer's perspective, they become interested only in price and stop looking at the packaged value" (Reynolds, 2023). Central to cybersecurity is security, not price. The commercialization of cybersecurity solutions can enhance competition, leading to better products and solutions at a lower cost; however, this is not achieved through a commoditized solution itself. The differentiation is in cybersecurity. As described in Chapter 3, Cyber Policy, Regulations, and Guidelines, Section 3.4.1 Germany's Technical Sovereignty, Germany recognizes the value of cybersecurity over "monoculture" dependency on single, large vendors.

Commoditization, accompanied by constrained resources and reduced cybersecurity costs, presents a challenge but is not necessarily a doomsday scenario. Remember, the focus of this book is to share a practical systems engineering approach that can be useful to teams for successfully managing operations. Understanding what has happened to enterprise IT is valuable, as operational technologies found in critical infrastructure are taking a similar path, albeit with limited resources and a later starting point.

By the end of the 1990s and the early 2000s, many original equipment manufacturers (OEMs) sold their assembly plants to electronic manufacturing services (EMS) companies, which were vying for market share through strategic acquisitions. In 2001, when IBM sold its IBM Personal Computing Division to Lenovo, it included the right to use the ThinkPad brand name. The sale to Lenovo also gave IBM access to China's lucrative new consultancy market. A wave of consolidation followed. As aggressive EMS companies purchased other smaller EMS companies, they grabbed market share and capabilities. This strategic acquisition phase triggered a spiral of commoditization in the IT industry. By the time it was all said and done, the worldwide EMS market had become a determining force in the production of electronic products, accounting for more than 40% of all assembly. The global supply chain adds complexity that can include deliberately injected cyber vulnerabilities. "The technical, economic, political, and cultural dynamics that make it difficult, if not impossible, for individual organizations to secure themselves effectively when they are dependent on diverse, complex integrated systems" (National Academies of Sciences, 2025). Key questions to ask include: Are our upstream and downstream impacts known in the event of a data breach or cyberattack? Do our defensive countermeasures capture useful information? Have the critical assets related to control systems in our organization been identified? Can we answer these questions by tracing them back to the origin of the inventoried components and all the way to the origin of the supply chain?

Little differentiation exists among commodities taken from their natural state and, if necessary, brought up to meet minimum marketplace standards. No value is added to the commodity – all commodities of the same good sell at the same price. Ironically, one advantage to the consumer is that the margin is so thin that counterfeiters may actually avoid "competing" with these products. A product, on the other hand, can be differentiated, and the manufacturer can add value through branding and marketing. Products are made using commodities. For example, designing a system safe from cyber vulnerability is a product. Using commercial off-the-shelf (COTS) components that may be commodities does not add value. The integration of commodity components creates a branded product that can be marketed as "safe." Conventionally understood examples of commodities include crude oil and gold. Commodities are primarily traded on exchanges in the form of futures contracts, which are agreements to buy or sell a commodity at a specified future time at a predetermined price. Products are typically classified as either durable or consumable goods. Durable consumer goods, such as appliances, are generally long-lasting and purchased infrequently. Consumable goods, which include groceries, are used quickly or need frequent replacement. It is important to note that while some commodities are consumables, not all consumables are commodities.

Only three things can delay commoditization: innovation, bundling, and segmenting. Adding new technology to critical infrastructure meets the innovation delay. Bundling capabilities for operational technology (OT) into IT services can temporarily delay commoditization, and implementing a Zero Trust strategy may also support a delay in the segmentation process. For example, enterprise IT-focused products and services can be kept distinct from OT and vice versa. Eventually, commoditization at the lower levels of the architecture found in critical infrastructure will merge into a single line of effort (LOE). The way forward is systems engineering for critical infrastructure to achieve cyber vulnerability mitigation. In Chapter 25, "Stakeholder Alignment," we present a method and model for achieving valuable stakeholder alignment.

9.9 Summary

So, what happens when your high-end product becomes a commodity? And what does it have to do with critical infrastructure? How can an organization overhaul its technical infrastructure to achieve cyber resiliency? Peter Drucker famously said, "When a subject becomes totally obsolete, we make it a required course." Cyber may already be a commodity. A commodity is a basic good used in commerce as an input in the production of services or goods. One might argue, for instance, that with the container as a service (CaaS), a model whereby users manage and deploy applications in containers and collections can become a commodity service. Cloud computing itself is no longer a differentiator. It becomes a commodity-based product. Taking that assertion further, "cyber" anything is a commodity. Cyber in itself is no longer the added value. However, safe cyber at scale is. Engineered products that can be used safely in a cyber world are a value add. The uniformity of cyber is not unique as a product. The innovation is in systems engineering and the delivery and use of cyber products. Cyber is simply the use of internet-based computing. Cyber is the 21st-century modern shiny, carbon-based, black graphite pencil. Trying to make it anything other than internet-based computing is futile. Cyber is internet-based computing. That is it. Nothing more, but the use of internet-based computing is a powerful punch.

The systems engineering for critical infrastructure must be consistent with the cyber technology embedded in everything, from enterprise-level business system components to the lowest-level PLC components. This universal convergence creates potential vulnerabilities in systems, including those based on engineering and physics. The most effective way to address potential vulnerabilities is to employ a systems engineering approach, which encompasses the entire lifecycle, from system design and development to operations and maintenance. It takes a cross-functional team; although an experienced team might move more quickly, it takes time.

If cyber is taken for granted, then a key market differentiator must be cybersecurity which cannot simply be commoditized. Complacency about cybersecurity could be the end of your company, country, or culture. There are no shortcuts, only good measures. Dealing with cyber technology embedded in systems requires cybersecurity supported by a standardized, repeatable, science-based process to effectively identify vulnerabilities, meet and manage defined requirements, and adapt the people, processes, and resources to overcome vulnerabilities. Cyber, at its core, is the use of internet-based computing. It is the 21st-century, shiny equivalent of a 19th or 20th-century carbon-based, black graphite pencil. Cybersecurity is a differentiator and ensures corporate resilience.

References

(DOD), D. O. D. 2024. *Bio Senior Advisor, Zero Trust Portfolio Management Office, Department of Defense, Chief Information Officer Randy*

Resnick [Online]. https://dodcio.defense.gov/About-DoD-CIO/bios/Resnick.aspx: DOD CIO. Available: https://dodcio.defense.gov/About-DoD-CIO/bios/Resnick.aspx [Accessed 2024].

Aleksandra Scalco, S. S. 2021. Championing Cultural Change and Control Systems Cyber Security. *AIAA-INCOSE Wasatch Aerospace and Systems Engineering Mini-Conference.* AIAA Utah Section.

Alexander Kossiakoff, W. N. S., Samuel J. Seymour, And Steven M. Biemer 2011. *Systems Engineering Principles and Practice*, John Wiley & Sons, Inc. Publication.

Alsop, T. 2023. *Lenovo's personal computer (PC) unit shipment share worldwide from 2009 to 2023, by quarter* [Online]. https://www.statista.com: Statista. Available: https://www.statista.com/statistics/255306/global-pc-market-share-held-by-lenovo-since-the-1st-quarter-2009/ [Accessed April 13, 2024 2024].

Christensen, C. M. 1997. *The Innovator's Dilemma: When New Technologies Cause Great Firms to Fail,* MA, Harvard Business School Press.

Cleary, C. P. 2022. Cyberspace Superiority Vision. *In:* NAVY, D. O. T. (ed.).

Crowther, G. A. 2017. The Cyber Domain. *The Cyber Defense Review,* 2, 68–78.

Dallaire, G. 1984. Should Congress provide special tax incentives for firms donating computers to schools? *Communications of the Association for Computing Machinery (ACM),* 27**,** 185 – 194.

Gallagher, M. 2023. Letter to Navy Exchange on CCP-Linked Computers. *In:* Committee, H. S. (ed.). https://selectcommitteeontheccp.house.gov: House Select Committee on the Chinese Communist Party (CCP).

Group, M. M. 2011. *World Stats* [Online]. http://www.internetworldstats.com. [Accessed November 23, 2011].

Guehenno, J.-M. 1995. *End of the Nation-State*, University of Minnesota Press.

Humphries, M. 2023. Dell to Stop Using Chips Made in China Before the End of 2024. *PC Magazine.*

IBM. *The IBM PC* [Online]. https://www.ibm.com: IBM. Available: https://www.ibm.com/history/personal-computer#:~:text="IBM's%20share%20of%20the%20PC,in%20cash%2C%20stock%20and%20debt. [Accessed April 13, 2024].

Inisights, G. M. 2025. Electronic Manufacturing Services Market Size. *In:* Inisights, G. M. (ed.).

John Kindervag, S. B., Kelley Mak 2012. Build Security into Your Network's DNA: The Zero Trust Network Architecture. Forrester: Forrester.

John Kindervag, S. B., Kelley Mak, and Josh Blackborow 2016. No More Chewy Centers: The Zero Trust Model of Information Security. *Vision: The Security Architecture and Operations Playbook.* Forrester: Forrester.

National Academies of Sciences, E., and Medicine 2025. Cyber Hard Problems: Focused Steps Toward a Resilient Digital Future. Washington, DC: National Academy of Sciences.

The Princess Bride, 1987. Directed by R. Reiner, United States: Twentieth Century Fox.

Reynolds, M. 2023. The Pitfalls of IT Commoditization: Why Professional Advice Matters More Than Ever. Available from: https://mbrit.medium.com/the-pitfalls-of-it-commoditization-why-professional-advice-matters-more-than-ever-e4e24440fd7c.

Scalco, A. 2022. *Measuring Disagreement in Segments of the Cybersecurity Profession as a Means of Identifying Vulnerabilities,* Colorado State University.

Scalco, A. & Simske, S. 2021. Model for Multi-concern Assurance in the Digital Transformation of Engineering of Critical Infrastructure Control Systems.

Shilov, A. 2023. HP to Relocate PC Assembly to Thailand, Mexico, and Vietnam: Report. *Toms Hardware.*

Skyquest 2024. Global Electronics Manufacturing Services (EMS) Market Size, Share, Growth Analysis, By Service (Manufacturing Services, Engineering Services), By Application (Computer, Communication) - Industry Forecast 2023-2030. *In:* Skyquest (ed) *Global Electronics Manufacturing Services (EMS) Market.*

Solon, O. 2017. Trump tells tech CEOs that Washington needs to 'catch up with the revolution'. *The Guardian.*

Statistica. 2021. *Revenue of the electric power industry in the United States from 1970 to 2019* [Online]. www.statistica.com: Statistica. Available: https://www.statista.com/statistics/190548/revenue-of-the-us-electric-power-industry-since-1970/ [Accessed July 21, 2021].

Stouffer, K., Pease, M., Tang, C., Zimmerman, T., Pillitteri, V., Lightman, S., Hahn, A., Saravia, S., Sherule, A. & Thompson, M. 2023. NIST Special Publication NIST SP 800- 82r3, Guide to Operational Technology (OT) Security. *In:* Commerce, U. S. D. O. & Technology, N. I. O. S. A. (eds.). National Institute of Standards and Technology: National Institute of Standards and Technology.

Vincent, D. 2022. *The Apple IIe - Computers Of Significant History, Part 2* [Online]. Userlandia: Userlandia. Available: https://www.userlandia.com/home/2022/2/apple-iie [Accessed April 13, 2024].

Watters, A. 2015. How Steve Jobs Brought the Apple II to the Classroom. *Hack Education* [Online]. Available from: http://hackeducation.com/2015/02/25/kids-cant-wait-apple 2024].

Writer, S. 2019. *Thinking differently — growing up with the Apple computer: How Californian culture shaped the future of home computing* [Online]. Available: https://digital.nls.uk/1980s/science-technology/apple-computers/#:~:text=In%20May%201983%2C%20a%2028,get%20a%20free%20Apple%20computer. [Accessed April 13, 2024].

10

Transition of Innovation to Operations

"You rush a miracle man, you get rotten miracles."
– Miracle Max, The Princess Bride (Reiner, 1987)

"Buttercup: We'll never succeed. We may as well die here."
– The Princess Bride (Reiner, 1987)

Abstract

Critical infrastructure operational technology (OT) operations share similar business objectives with those found in enterprise IT systems. System owners and operators look for efficiency, cost optimization, and performance enhancements. Similarly, system owners and operators of OT found in critical infrastructure also face these challenges. Innovative efficiencies can lead to business success but can also involve significant change management. Innovation in operations introduces changes to organizational functions, cyber governance, and adjustments to jobs and workforce skills required to support new technologies and solutions. The transition of innovation is sometimes referred to as traversing a "valley of death." Organizational leadership must transition innovation projects from uncertainty into permanent, sustainable, and maintainable technological adaptation with minimal disruption to existing processes and operations. The successful transition of cyber solutions into critical infrastructure requires an understanding of the game, including the term "cyber," knowledge of physics-based systems, familiarity with cyber policy, regulations, and guidelines, the cyber medium, system requirements, functionality, security operations, hardware-level cyber vulnerabilities, threats, and supply chain management. Cybersecurity must be understood by the entire organization, particularly when innovation is transitioned into operations. The organization-wide understanding encompasses principles and practices, processes and activities, governance, technologies, risks, and people. The purpose of transitioning innovation into operations is to achieve improved services and efficiencies. It requires realistic and

DOI: 10.1201/9788743809128-10

accurate expectations, effective risk management, and specialized knowledge of the supported environments. Systems engineering principles and practices can help achieve the desired business objective outcomes while overcoming cyber vulnerabilities in critical infrastructure.

10.1 Introduction

Like Westley in "The Princess Bride," guiding Buttercup through the Fire Swamp, systems engineers can help teams navigate the challenges of integrating cyber solutions into critical infrastructure environments. Critical infrastructure in a cyber world is vulnerable to cyberattacks. Knowledge of the environment and risks helps the team successfully navigate the dangers encountered during the transition from innovation to operations. Westley recognizes dangers from experience, similar to leveraging shared systems engineering best practices. "*One, the flame spurt – no problem*," Westley says to Buttercup (Reiner, 1987). Discovering what new risks look like to the organization, such as cyber vulnerabilities, is a team effort. Sometimes, fear and uncertainty are rational responses to risk that warrant consideration. Transition management is not easy. As mentioned in Chapter 3, "Cyber Policy, Regulations, and Guidelines," dynamic adversary tactics necessitate dynamic tools and countermeasures to overcome these evolving adversarial tactics.

The transition of innovation is not only about integrating technology and innovation into operations. It is also about managing people who deliver functions and services and creating value, including cybersecurity requirements for legacy control system devices found in critical infrastructure throughout the device's lifecycle, from cradle to grave. As asserted in Chapter 6, "Cyber Vulnerabilities Start at the Hardware Level," and in Chapter 9, "Road to Commoditization," steering the cyber posture of the system from the start of the supply chain to the operational use is critical to system resiliency, cyber security, process safety, and predictive maintenance. Malicious compromise of process sensors can result in catastrophic failure without any indication of a cyber event. Recognizing an event as a cyber-related incident is a real problem, and perhaps the most important capability in cybersecurity (false negatives often being catastrophic). Realizing that the problem may start with the device's design is even more problematic. Process sensor data is often assumed to be uncompromised and authenticated, even when it is not. Consequently, cyberattacks are not recognized in the absence of safe, active monitoring of a control system's cybersecurity. Key questions to ask are: Does the organization use threat detection capabilities that start at the manufacture of components? Are threat signatures updated in the organization?

Who has decision authority (or how are decisions made), and for which risks are accepted when related to control processes?

Complex adaptive systems are a challenge to discuss without a model, which is discussed in greater detail in Chapter 17, "Architecture Approaches." While a particular model may represent conditions within one system, variables to user states may carry different meanings from one system to another. Several candidate approaches are used to address complexity, including seeking to understand the big picture, observing how elements within the system change, identifying the system's structure and its relationship to system behavior, and understanding test assumptions. The recommended solution architecture for a scenario involving an environment susceptible to unpredictable, high-consequence events of recursive complexity is to "design for resilience" to provide robustness and timely recovery to a minimally functional state (INCOSE, 2015). Models are helpful in facilitating the transition from integrating innovation and new product introduction (NPI) into operational use.

10.2 Uncertainty and Crossing the Valley of Death

Moving innovation into fielded operations is to traverse a "valley of death." The valley of death is uncertain and surely dangerous. A project or product can die in its first months of operation. Innovative technologies developed through science and research demonstrate the realm of the possible, without the need for readiness inspections and an operational environment. Imagining the realm of the possible is the discovery phase, and it is also the realm of scientific methodology. Understanding the landscape of modernization strategies for critical infrastructure enables organizations to prepare for emerging market growth trends and effectively respond to them. Change is not easy. Transitioning innovation into operations requires resources, skills, and a clear vision. People will be resistant to potentially encountering risks. Stakeholders may be unwilling to resource the transition if the reasons and rationale behind the innovation, as well as its benefits, are not well understood. Cybersecurity presents a differentiator. That message needs to be consistently communicated to stakeholders to manage the transition process. As described in Chapter 5, "Cyber Paradox," systems engineering practices help deal with uncertainty and complexity. There needs to be copious knowledge of a system's functional and non-technical requirements, as well as how all the different aspects of the system work together to meet these requirements. There needs to be value added to business operations, access to new markets, and opportunities for growth.

While original equipment manufacturers (OEMs) have benefited from the new market growth trends and cost efficiencies described in Chapter 9, “Road to Commoditization,” these trends and efficiencies have also allowed the electronic manufacturing services (EMS) industry to take market share from the OEMs. The market for personalized IT products and business technologies has a big appetite, as does the demand for OT components and systems. So, stay on your toes if you build your strategy around cost! You have to be clever enough to discover what “lightning sand” looks like, as Westley observed in The Princess Bride, “*so in the future, we can avoid that too*” (Reiner, 1987). Organizations that embrace cybersecurity as a key differentiator will have greater success in today’s competitive market than those that do not recognize the risks of not having cybersecurity. The risks of not having cybersecurity in critical infrastructure are substantial. Some examples include:

1. Disruption of operational functions, including power, water, fuel, and transportation
2. Physical damage to facilities and equipment
3. Threats to human life and safety
4. Environmental damage
5. Economic and financial impact, and
6. Social unrest

As an example of understanding the business value of transitioning to innovation, we consider a company from the early 2000s era that specialized in selling high-speed computer memory modules – also known as dynamic random access memory (DRAM) – used in both high-performance desktops and mobile platforms. DRAM has been around since the late 1960s. The technology significantly increased computer memory while reducing power consumption and cost. The market initially had little competition, but as numerous companies began producing DRAM modules, they became a commodity, which subsequently drove prices down. Protecting business interests is vital, as is optimizing costs while also understanding the business and service provider landscape. In the case of DRAM modules, the company reduced its prices from USD 200 to USD 20 per module over a relatively short 5-year period. This company’s founder realized they needed to completely shut down the DRAM business and start selling solid-state hard drives (SSDs), a new, faster alternative to traditional drives. There was a small problem. The market did not exist. The story ends on a positive note: the company, originally founded as Dataram in 1967, introduced an innovative product and continues in business today.

Meanwhile, by the first quarter of 2023, DRAM average selling prices dropped another 20 percent (Shilov, 2023). These types of business triggers make or break organizations. Technologies change rapidly, particularly in the cyber virtual world. Business leaders need to be able to anticipate market drivers, manage change, and operate securely in a cyber world. The operations, human life and safety, the environment, business viability, and other socio-economic factors all depend on the cybersecurity of critical infrastructure. The transition of innovation requires an understanding of the technology and innovation "valley of death" and the risks involved in traversing it or not.

10.3 Summary

A maxim that has been around for a long time, often attributed to management consultant Peter Drucker, is that "culture eats strategy for lunch." Another take on the maxim is found in Richard Draft and Thomas Lloyd's "Management," which identifies the human organizational elements leading to merger and restructuring project failures. "These expensive experiences taught change-management gurus to share slogans like: 'Culture beats strategy'" (Daft and Lloyd, 2000). Innovative efficiencies can lead to business success but can also involve significant change management. The workforce culture will beat any Zero Trust strategy if the fundamental principles and values of cybersecurity are not understood, adopted by all, and resourced. The transition of innovation to operations introduces changes to organizational functions, cyber governance, and adjustments to jobs and workforce skills required to support new technologies and solutions. Essentially, no matter how solid your business strategy is, its efficacy is bound by stakeholders who do not share the same perspectives.

Stakeholder alignment cannot be overlooked. All stakeholders must be in alignment to achieve the vision, mission, and goals (Scalco and Simske, 2022). The entire team, from technicians to system designers and resource decision-makers, must be aligned with cybersecurity innovations that are transitioning into critical infrastructure operations. Otherwise, you and your strategy will just be stomped out by the terrors of the fire swamp. Systems engineering is crucial to mitigating the risks associated with resistance to cybersecurity changes. Cybersecurity is essential for critical infrastructure systems. Having a good cybersecurity posture starts at the inception of component manufacture and continues throughout the system's lifecycle, regardless of its duration. New product and service validation ensures that changes align with the design specifications and meet the functional requirements. Risks to a successful transition include a poorly defined scope, a lack of understanding of dependencies, inadequate cybersecurity considerations in the design, and a lack of contingency plans if the sourcing partners fail to deliver their contractual obligations. Engineers knowledgeable about how to safely secure control systems used in critical infrastructure should be part of the team, along with network security personnel who may not fully understand the systems involved. Decision-makers need to hear balanced information about the challenges and benefits of transitioning innovation into operations. Systems engineering for critical infrastructure in a cyber world fosters the sharing of both good and bad news to promote informed decision-making. The purpose of validation and testing is to ensure that balanced, data-driven information is obtained.

> "*I just want you to feel you're doing well,*" Fezzes, The Princess Bride (Reiner, 1987).

References

Daft, R. L. & Lloyd, T. W. 2000. *Management*, Holt, Rinehart & Winston of Canada, Limited.

The Princess Bride, 1987. Directed by Reiner, R. United States: Twentieth Century Fox.

Incose 2015. A Complexity Primer for Systems Engineers. INCOSE.

Scalco, A. & Simske, S. 2022. Measuring Stakeholder Alignment to Overcome Control System Vulnerability. Department of Homeland Security (DHS) CISA Industrial Control Systems Working Group.

Shilov, A. 2023. DRAM Prices Dropped 20% in Q1 2023, Will Keep Falling in Q2. *Tom's Hardware.* www.tomshardware.com.

11

Lifecycle Phases

"Racing is life. Everything else is just waiting."
Steve McQueen (1930–1980), American actor as Michael Delaney, Le Mans, 1971.

"Le Mans is such a great race because you can never do anything alone. You have to work as a team member. And being a team member makes you a better person."
Tom Kristensen (1967–), Danish Racecar Driver and record holder for most wins at the 24 Hours of Le Mans.

Abstract

To understand systems engineering lifecycle phases, look to motorsports. To delve further into the sometimes brutal test of systems engineering, look to the 24 Hours of Le Mans. This 24-hour endurance race is a test of both driver and machine and a testament to the value of the interdisciplinary systems engineering field in managing complex systems throughout their lifecycle. The goals of systems engineering include organizing engineering elements into an integrated system of electronics, computer code, machine parts, and humans working together to achieve the highest levels of success in the most challenging physics-based environment at speeds of more than 200 miles per hour (mph), or 322 kilometers per hour (kph). Le Mans is the oldest and most famous endurance race in the world, held on both a race circuit, the *Circuit de la Sarthe*, and one of the world's longest public roads, the French *Route Départementale D338*, located in Le Mans, France. Since the 1920s, manufacturers of the world's finest sports cars have applied systems engineering principles to the pinnacle of competition at an annual endurance race, stirring emotions in the most dedicated enthusiasts. The most successful endurance race drivers pit themselves against machines and the clock in the time-based 24-hour endurance race. More than 60 teams representing 14 manufacturer prototypes and privateer prototypes compete in the various specification

DOI: 10.1201/9788743809128-11

classes. One common theme emerges among all teams: winning teams must collaborate to defend their position, build capacity, and improve their team's performance. Similarly, systems engineering principles and practices support the design and development from prototype through end-to-end testing, utilizing digital engineering models to test the integration of components and conducting rigorous systems engineering reviews. Commercial and government partnerships are necessary to ensure the latest commercial technologies meet the most stringent requirements and inform emerging legislation. It is a holistic race.

11.1 Introduction

The 24 Hours of Le Mans race comprises technical and sporting rules, with specific technical regulations for each class of sports car category. The Le Mans prototype class car regulations do not require a minimum number of production cars. It is what it states: a prototype, similar to a Department of Defense (DOD) Joint Capability Technology Demonstration (JCTD), is a demonstration of capability. It is not yet an operational capability. GTE-class cars raced at Le Mans have minimum production requirements. The race mix is based on technology advancements. Four categories of sports car prototypes compete in the race simultaneously against each other in their respective class and against the clock: Le Mans prototypes (LMP): LMP1, recently replaced by the new class Le Mans Hypercar, GR010, and LMP2, and the Le Mans grand touring endurance (GTE) class LMGTE-Pro, and LMGTE-Am. Le Mans GTE classes are divided into two types of production cars. The GTE-Pro class is the preserve of manufacturer teams that run professional drivers. The GTE-Am requires at least one year of production build, or build to the previous year's specifications, and has driver qualification criteria based on an FIA grading system of driver experience. Le Mans Prototype 2 (LMP2) cars are slower than Hypercars, and in 2024, a new class, LM GTE 3, will replace the GTE-Am class. The former "grand touring endurance" acronym has given way to "Gran Turismo Electric" (GTE) and the global trend of plug-in electric vehicles.

The race is an essential way for manufacturers to evaluate and demonstrate innovative technologies before integrating the technology into production cars. Similarly, the Joint Capability Technology Demonstration (JCTD) provides an opportunity to assess technology in operational environments before formally fielding it for operational use within the Department of Defense (DOD). The 24 Hours of Le Mans tests the extreme boundaries of performance in a range of conditions. Winning Le Mans builds a championship brand visibility and engages fans, which can translate into market share among consumers. Behind the scenes, before, during, and after the race, the precision orchestration of hundreds of engineers, mechanics, drivers, trainers, logistics personnel, and support staff shares information gathered at the track back to design and production. Every aspect of the system is measured at the weigh-in and scrutineering before the start, including laser measurements of the vehicle's dimensions, minimum ground clearance, and bodywork inspection. Drivers and teams are inspected and qualified, including the measurement of driver weight before the race and the monitoring of hydration during the race. Endurance car racing is no different from the engineering and team problem-solving proficiency required in systems engineering, which involves integrating technical processes and information in a high-pressure work environment to create an engineering marvel that stretches the imagination. This test of endurance achieves performance that defies the laws of physics. Similarly, critical infrastructure systems such as water utilities increasingly deploy human–machine interfaces (HMIs). The essential processes require team collaboration among various disciplines from developers, asset owners, and the entire supply chain. As described in Chapter 7, "Threats to Information-intensive Systems," system operators with cybersecurity training often do not behave any differently than those without cybersecurity training, suggesting humans might be the weakest link in susceptibility to advanced persistent threats (Terry Merz, 2019). A breach in a critical infrastructure system can result in severe damage, injuries, and even loss of life.

11.2 Concept Development through Lifecycle Phases

Like endurance racing, United States DOD projects begin with concept development, which involves exploring the realm of the possible, eliciting user requirements, conducting a needs analysis, and designing in detail. Similarly to building a GT prototype destined for the 24 Hours of Le Mans, a government product concept is followed by the implementation, integration, evaluation, testing, and fielding into operational use, followed by the post-development stage when the system is in operation and maintenance. The operations and maintenance of material solutions in the DOD are like the production of cars and the aftermarket phase. These phases in the DOD follow the Doctrine, Organization, Training, Materiel, Leadership, Education, Personnel, Facilities, and Policy (DOTMLPF-P) framework. Just as in endurance car racing, DOD projects are continually seeking funding for experimentation and research, albeit from various sources. Teams of professionals collaborate to conduct scientific and engineering experiments, applying scientific principles to design and build more efficient, faster, cheaper, and robust systems using the latest emerging technologies. Systems engineers navigate "unknown unknowns" (or more commonly called "unk-unks") to meet unknowable performance requirements and sometimes work against standards that do not yet exist. "Unk-unks" are unknown problems that could not have been foreseen. Eventually, some experimentation reaches the early prototype stage for the demonstration against the user's science and technology integrated priority list (STIPL). If the system is successfully demonstrated, it has yet another "Valley of Death" to cross. The "Valley of Death" refers to the space and time between the experimentation of science and technology and their transition into operational use.

Like the DOD, Le Mans race prototypes such as the GR010 and LMP2 are used to demonstrate innovative technologies and learn from concepts. There are also production-ready, operational systems, such as the GTE class cars, which receive recent technological upgrades. All systems require capabilities, integration, development, and engineering production documentation. In endurance car racing, drivers need licenses and team qualifications. The endurance cars are evaluated and verified against strict regulations and standards before each race. In the Department of Defense (DOD), systems are assessed against the National Institute of Standards and Technology (NIST) risk management framework (RMF) cybersecurity controls before receiving an authority to operate (ATO) on DOD networks. The DOD ATO process is currently undergoing a long-awaited transformation to modernize the traditional certification and accreditation (C&A) process, streamlining ATOs for the faster adoption of software-intensive platforms and applications. All these critical elements that define how a system is created and operated require workforce training, manuals, policies, and instructions, and most of all, the enthusiastic technical leadership and vision of the systems engineer and workforce knowledge, skills, abilities, and tasks (KSAT) of the integrated product teams (IPTs) that make up the winning team.

Systems engineers and winning teams never go it alone. The systems engineer is responsible for managing the information exchange among stakeholders and understands the roles and functions of each stakeholder within the enterprise. If a team member is reluctant to use shared working environments and tools, the systems engineer plays a key role in managing interactions between functions. Hoarding information or knowledge will not help the team to meet objectives. The team comprises a diverse range of expertise and activities. Transparency among the team members is critical, and individuals withholding information will only hinder the overall goals of mitigating cyber vulnerabilities in critical infrastructure. Systems security engineering is invaluable when discussing systems engineering for critical infrastructure, as it helps overcome cyber vulnerabilities. The systems engineer is responsible for ensuring that engineering products required by other functions are shared. Systems engineering is only one of the functions involved in successful system development. All stakeholders are responsible for the inter-system team, review processes, and peer assessments of work products in the ecosystem.

11.3 Nation State and Public–Private Partnership in Balance

These ideas were previously published as a book review (Scalco, 2024). Nations are still grappling with the right balance of law and standards when it comes to significant digital issues such as data protection and privacy rights, "technological sovereignty," IT security laws, authorities between government, military, and private stakeholders, critical infrastructure protection, economic impact, and other thought-provoking discussions for debate (Schallbruch and Skierka 2018). Governments are paying attention to the details of policies that are emerging against a broad, yet unknown, backdrop of where the boundaries lie for when the government needs to step in and mandate cyber safety, and when laws and regulations governing cybersecurity are required. Germany's approach was to focus on a public–private partnership. The government is expected to protect its citizens from negative cyber interface consequences, such as identity theft, just as private companies are expected to design and develop products or services that broadly safeguard consumers. Germany's official policy at the time of writing centered on the idea that the IT industry would reduce its dependency on untrustworthy providers and focus on the use of trustworthy solutions (Schallbruch and Skierka 2018). The German perspective emphasizes the adage, "Economic security is national security." The economic impact of cyber on national sovereignty is to ensure the country is producing economically impactful IT. National cybersecurity strategy, implementations, and tactics are essential conversations. There is a notion that cybersecurity professionals jump right in and keep everyone safe by following clear laws and statutes, while the rules and regulations are still in development. Schallbruch and Skierka effectively outline how Germany's approach to cybersecurity is evolving and what German Security policy may look like in the future.

What was most interesting was the revelation that Germany's strategic approach leaned toward being recognized globally as a country that makes and delivers safe cyber products and services. This idea of implementing technical, legal, and political measures ensures that a range of programs is available to strengthen German technologies. In the wake of the classified security information leaks, Germany went so far as to terminate contracts with some external communication providers and replace them with German providers as a measure to enhance the protection of private citizens' information (Schallbruch and Skierka 2018). Germany's approach is centered around cybersecurity as an expectation. The results of Germany's cybersecurity strategy are demonstrated by two of Germany's automotive manufacturers, Volkswagen and Porsche, discontinuing vehicle models in 2024 without indication of specific platform replacements. The cancellation of the models was announced to be directly tied to forthcoming European cybersecurity requirements. Both companies incorporated cybersecurity considerations into their engineering. The sub-theme echoes Germany's stance that commercial entities must consider cyber in engineering security risk analysis.

The balance of Germany's perspective lies between an "engineering-motivated, preventive, and technology-oriented policy" to protect personal data and the need to enable a system of consistent legal protections and the ability to investigate cybercrimes. Even years after Schallbruch's and Skierka's book was first published in 2018, the content continues to explain the methodology and thought that Germany's general population and politics have put into addressing a worldwide asymmetric threat. The result? Germany aims to become a leading exporter of technology. That is their national cybersecurity strategy. Germany can apply this strategy equally to critical infrastructure systems and their components. The discontinuation of car models by Volkswagen and Porsche demonstrates the progress made in the public–private partnership to ensure that cybersecurity aligns with this purpose in Germany (Schallbruch and Skierka 2018).

11.4 Systems Security Engineering

Common threads in sound systems engineering include taking a top-down approach with a holistic view, adopting a life-cycle orientation, making a more comprehensive effort in the initial definition of system requirements, and employing an interdisciplinary or "team approach" throughout the process (Gemmill 2011). A multidisciplinary team approach takes a tremendous amount of stakeholder identification, engagement, and team building. The discipline of systems engineering supports the development of complex systems by applying analysis and decomposing component relationships to facilitate a comprehensive understanding of the system's components and interfaces. It is essential in the introduction and employment of new operating concepts such as "cyberspace." As presented in Chapter 4, "Cyber is a Medium (Not a Domain)," when the idea of "cyber" was initially discussed in the defense industry, a standard definition remained elusive to those subject matter experts engineering the systems, building networks, and evaluating concepts to be solved for protection against adversaries. A good understanding of the idea of cyber remains elusive today. Stakeholders struggle to articulate the exact boundaries of a physical instantiation of the system's virtual world. Clarifying the system boundaries within an enterprise and identifying the constraints on a system are roles of systems engineering. People with expertise, such as software developers, often take on the role of defining system-level design problems and developing system architectures that are independent of the invisible parts of cyberspace.

The systems engineer plays a critical role by applying systems engineering techniques and methods to define the system's enterprise, functions, and interfaces between these functions. The Department of Defense (DOD) has identified gaps in its cyberspace capabilities; however, there is no comprehensive doctrine that thoroughly addresses cyberspace operations, presenting a significant challenge that needs to be addressed. The conceptual and preliminary design, as well as the detailed design and integration phases, are where the majority of the system's costs are incurred. The most significant costs incurred are typically found later in the lifecycle, during construction or production, use, refinement, and disposal of the system. It is a challenge to budget for the invisible cyber longevity in the lifecycle, given the fast-paced changes in attack methods. The systems engineering process should enable a mix of various approaches to design, development, and systems management. The expertise required encompasses management, systems engineering processes, stakeholders, and technology engineering disciplines, including software development, modeling, simulation, and analysis (Buede 2009).

The establishment of cyberspace by the DOD as a "warfighting domain" (alongside land, air, sea, underwater, and now also space) necessitates aggregating these invisible components as a required mode of system architecture. "A system is commonly defined to be a collection of hardware, software, people, facilities, and procedures organized to accomplish some common objectives" (Ibid.). For example, for aspects of the system to be integrated and functional, specialized knowledge such as software development is required. The engineer must also be conversant with interfaces, including electrical engineering and wireless communication specialties, to accomplish the system's engineering objectives. However, decomposing the challenges of cyberspace, solving complex interface problems, and reassembling solutions into an integrated solution for virtual environments requires a synthetic mode of thought, or "systems thinking." Acknowledging that certain engineering basics and technical specialty expertise needs will not be changed by the adoption of cyber, "a major characteristic of the engineering of systems is the attention devoted to the entire life cycle of the system" (Ibid.). The primary characteristic of systems engineering gives rise to the concept of systems security engineering.

Systems security engineering is an approach that focuses on a holistic security perspective throughout the system lifecycle, starting at the concept development stage. NIST Special Publication 800-160, "Systems Security Engineering: An Integrated Approach to Building Trustworthy Resilient Systems," defines the practice of systems security engineering (Ross, McEvilley et al. 2016). Subsequently, NIST SP 800-160 Volume 2 was published, focusing on cyber resiliency engineering to overcome vulnerabilities in cyber resources (Ronald S. Ross 2021). The first version of NIST Special Publication 800-160, "Systems Security Engineering: An Integrated Approach to Building Trustworthy Resilient Systems," gave insight about how the practice of systems engineering leads to more secure systems that start and built upon well-established international standards for systems engineering and well-established international standards for systems engineering as published by the International Organization for Standards (ISO), the International Electrotechnical Commission (IEC), and the Institute of Electrical and Electronic Engineers (IEEE). Systems security engineering was defined in the first impactful NIST 800-160 version, adapted from the National Aeronautics and Space Administration's System Safety Handbook, NASA/SP-210-580 (Dezfuli, Everett et al., 2011). NIST 800-160 defines systems security engineering as a specialty discipline of systems engineering. It provides considerations for security-oriented activities and tasks that produce security-oriented outcomes as part of every systems engineering process activity, with a focus on the appropriate level of fidelity and rigor in analyses to achieve assurance and trustworthiness objectives.

Principles of systems security engineering have been around for decades, predating the US Department of Defense's declaration of cyberspace as an operational "warfighting domain" in the "Department of Defense Strategy for Operating in Cyberspace" (DOD, 2011). Before that declaration, the National Aeronautics and Space Administration (NASA) had already been using systems engineering to support its philosophy that safety should be the primary consideration in a system, with the caveat that the impact on cost, schedule, and technical performance is "reasonable and acceptable" (Ibid.). Reasonable and acceptable are challenging to define in terms of safety and cyber considerations. Loss of life and loss of mission would be considered reasonable measures. NASA's handbook also addressed continuous risk management (CRM) as an integrated framework for implementing designs, plans, and processes, providing a disciplined approach to continuously assessing potential risks and implementing strategies to overcome identified issues. The focus on continuous risk management is a process that aligns with the system's baseline performance requirements. System safety is a disciplined and systematic process for identifying and evaluating risks, which must be understood and assessed in the context of the specific application.

Just as safety is paramount at the 24 Hours of Le Mans, similarly, system cybersecurity must be understood and evaluated in the context of the application. Le Mans race officials control racing for various reasons to neutralize risk on the racecourse. Depending on the type of incident encountered, race control officials can suspend racing to clear the track after an incident or enforce a speed limit to enable intervention for other safety reasons, such as a car stuck in a gravel trap. The focus must be continuous and should be a primary consideration, with the same caveats that impact cost, schedule, and technical performance, ensuring that the interest is "reasonable and acceptable." Loss of life and loss of mission would be considered reasonable measures. Cyber vulnerability mitigations must be considered and addressed throughout the system lifecycle from the onset, starting with concept development.

The systems engineering approach is holistic, requiring coordination between all stakeholders, from the engineering design team to the user, to balance the often-conflicting perspectives regarding cost, schedule, and technical performance. Systems engineering rigor ensures that quantitative data and analysis inform all design decisions and tradeoffs to support the desired outcomes. Security requirements that specify cybersecurity functions and mechanisms must be considered first-order requirements that the team cannot easily add after the fact. Accurately expressing the system with well-defined requirements that identify both functional and non-functional requirements, as well as sector-specific industry requirements, is crucial for achieving the expected performance. Examples of sector-specific critical infrastructure sectors include power, water, and fuel.

The use of the completely automated public turing test to tell computers and humans apart (CAPTCHA) is an example of a requirement for distinguishing between computers and humans. CAPTCHAs are designed to prevent bots and malware from interacting with a web page, as well as other automated attacks, such as credential stuffing. Computers are not as good as humans at reading contorted characters or otherwise distorted text. The most common form of a CAPTCHA is an image form that relies on a static database of images. Computers lack the ability humans have when it comes to processing images. Humans can pick out visual cues or associations that require interpretation. Word CAPTCHAs pose the same challenge to computers that can recognize words but do not natively understand the meaning of those words. It is challenging to teach a computer to process information in the same way humans can. However, the way data is processed is rapidly changing due to advancements in artificial intelligence (AI).

A system requirement is "[a] statement that transforms the stakeholder, user-oriented view of desired capabilities into a technical view of a solution that meets the operational needs of the user," first defined in ISO/IEC/IEEE, 2015, section 6.4.2, and later updated in ISO/IEC/IEEE 15288:2023. The system requirement definition process is also necessary. "The System Requirement Definition process creates a set of measurable system requirements that specify, from the supplier's perspective, what characteristics, attributes, and functional and performance requirements the system is to possess in order to satisfy stakeholder requirements."(ISO/IEC/IEEE 2023) It is then reasonable to consider cyber vulnerabilities in the same way one would consider safety throughout a system's lifecycle, managing risk through the involvement of all stakeholders.

11.5 Characterizing Cyber Operational Space

Systems engineering and integration support enhance the overall effectiveness of systems within cyberspace operations, enabling systems, services, and functions that are crucial in critical infrastructure sectors to mitigate cyber vulnerabilities. The application of cyber modeling and simulation is valuable for orchestrating across multiple enterprises to engineer and test the operational inter-system approach, or a "meta-system" of systems (Martin 2004). As shown in Figure 11.1, based on James Martin's work, "The Seven Samurai of Systems Engineering: Dealing with the Complexity of 7 Interrelated Systems" (Martin, 2004), systems engineering provides an approach to designing a system, recognizing that the system exists within a broader meta-system.

Before the establishment of the US Fleet Cyber Command (FLTCYBERCOM)/US Tenth (10thFLT) program offices contracted the development systems in increments, that approach often led to disparate systems designed to respond to problems in a sustainment application within the meta-system (t) with software solutions that led to potential problems in related operational and sustainment systems within the Navy meta system (t + 1). The risk analysis of software development efforts intended to address issues is often limited in perspective with respect to meta-system

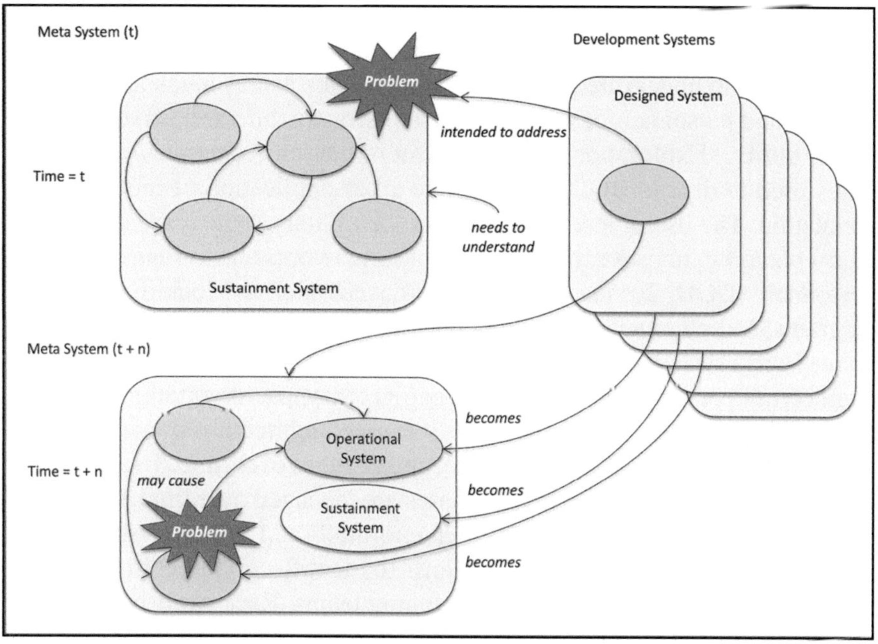

Figure 11.1 Meta-system – characterizing the cyber operational space.

dependencies. The solution, much like the DOD's path to declaring cyberspace as an operational "warfighting domain," was to change the acquisition strategy to position Navy engineering to better utilize a systems engineering approach for designed systems, both afloat and ashore.

The sheer economics of maintaining legacy systems while integrating new ones create sustainability issues and will continue to drive network architecture decisions that substantially alter the Navy's acquisition strategy. "An economic shift will change the way that corporations, government agencies, and other organizations, government agencies, and other organizations share storage, computing, network, and application resources" (Executive, 2011). "A system is commonly defined to be a collection of hardware, software, people, facilities, and procedures organized to accomplish some common objectives" (Buede 2009).

11.6 Summary

Systems engineering principles and practice support the design and development of architectures and systems specifications and criteria, including end-to-end testing and systems engineering reviews. The network links to sources of information, and the links that acquire information, the sensors and processing nodes, are part of the requirements just as much as the ships, aircraft, and personnel for the Navy, or the infrastructure and personnel for a power utility. "Exploration and refinement of tactics, techniques, and procedures used to defend DOD networks are a key application of modeling and simulation. The use of test ranges allows the military to increase awareness, adjust planning, improve the resiliency of its networks and systems, and train personnel" (DOD, 2011). Modeling approaches such as block diagrams, N^2 diagrams, and other specification language for modeling objects for requirements from a systems engineering perspective more so than from the software engineering perspective of discipline engineers whose orientation is to produce a physical instantiation of the cyberspace architecture framework help to show dependencies, real-time statue, and changes over time. To understand how systems engineering lifecycle phases are managed over time, look to the motorsports industry. To delve further into the brutal test of systems engineering, look to the 24 Hours of Le Mans. To identify common principles and practices that work best, look to the winning teams. Race car drivers rely on their team to work together to defend their position and safety, build capacity, and enhance the team's overall strength. Anything less than working together at the highest levels skirts the boundaries of potential disaster.

References

(DOD), D. O. D. 2011a. Department of Defense Cyberspace Policy Report: A Report to Congress Pursuant to the National Defense Authorization Act for Fiscal Year 2011, Section 934. *In:* (DOD), D. O. D. (ed.).

(DOD), D. O. D. 2011b. Department of Defense Strategy for Operating in Cyberspace. *In:* DEFENSE, D. O. (ed.). Washington, DC: United States Department of Defense.

Buede, D. M. 2009. *The engineering design of systems: models and methods,* Hoboken, New Jersey, John Wiley & Sons, Inc.

Dezfuli, H., Everett, C., Smith, C., Stamatelatos, M. & Youngblood, R. 2011. NASA/SP-2010-580, NASA System Safety Handbook, Volume 1, System Safety Framework and Concepts for Implementation. *In:* (NASA), N. A. A. S. A. (ed.). Washington, D.C. 20546: National Aeronautics and Space Administration (NASA).

Executive, O. O. T. N. C. 2011. Foreign Spies Stealing US Economic Secrets in Cyberspace: Report to Congress on Foreign Economic Collection and Industrial Espionage, 2009-2011. *In:* US Government, O. O. T. N. C. E. (ed.). Washington, DC: Department of Defense.

Gemmill, D. 2011. Lectures. *AerE/EE/IE 565 Systems Engineering and Analysis/Fall 2011.* Iowa State University: Douglas Gemmill.

ISO/IEC/IEEE 2023. ISO/IEC/IEEE 15288:2023, Systems and software engineering — System life cycle processes. ISO/IEC/IEEE.

Martin, J. 2004. 3.1.2 The Seven Samurai of Systems Engineering: Dealing with the Complexity of 7 Interrelated Systems. *Incose International Symposium.* Incose.

Terry Merz, C. F., Aleksandra Scalco 2019. A Context-Centered Research Approach to Phishing and Operational Technology in Industrial Control Systems. *The Journal of Information Warfare.*

Ronald S. Ross, V. Y. P., Richard Graubart, Deborah Bodeau, Rosalie Mcquaid 2021. Special Publication (N800- 160v2r100-160v2r1, "Developing Cyber-Resilient Systems: A Systems Security Engineering Approach". *In:* NIST (ed.) *Special Publications (NIST SP).*

Ross, R. S., McEvilley, M. & Oren, J. C. 2016. NIST Special Publication 800-160, Systems Security Engineering: Considerations for a Multidisciplinary Approach in the Engineering of Trustworthy Secure Systems. *In:* COMMERCE, D. O. (ed.). National Institute of Standards and Technology (NIST).

Scalco, A. 2024. *Book Review: Springer Briefs in Cybersecurity: Cybersecurity in Germany* [Online]. [Accessed].

Schallbruch, M. & Skierka, I. 2018. *Cybersecurity in Germany,* Berlin, Germany, Springer.

12

Concept Development

"Keep your head down"

– a golfer's adage

"To achieve anything in this game, you must be prepared to dabble in the boundary of disaster"

– Stirling Moss (1929–2020), British motor racing driver, and Formula One World Champion.

Abstract

The operational concept is central to the vision, mission, requirements, and scenarios in which the system is used. It helps when everyone is using the conformity of definitions and dictionaries. For example, Formula One racing adheres to a stringent set of definitions. Drivers are integral members of the design engineering team. The handoff of information between these high-performance team members is tightly coupled. However, definitions in cyberspace operations are not always uniform. The same cyberspace-related term may have one meaning to a Department of Defense (DOD) service component and an entirely different meaning to another. It is widely recognized that the concept of cyberspace operations is still evolving. Formula One racing is continually evolving to meet new safety standards and incorporate technological innovations. Most stakeholders agree that definitions of cyber warfighting and its boundaries are unclear. Yet, by 2011, more than 30 countries had already established cyber military units, and cyberspace had emerged as a potential battlefield of inevitable global proliferation (III, 2011). Cybersecurity roles and responsibilities within the Department of Defense (DOD) are distributed across various components of the DOD. The current cybersecurity organizational structure is decentralized, and many DOD components hold responsibilities (GAO, 2011). As the operational concepts are developed, the handoff between systems engineering and software engineering is pushed deeper into the systems engineering

DOI: 10.1201/9788743809128-12

"Vee." Cyberspace operations are inherently joint commands across the operational boundaries of all services and agencies. The inherent jointness of cyberspace operations makes the supporting relationships unclear, unlike the jointness of a Formula One team. This disjointedness affects the operational concept and authorities for all networks. Software and hardware engineering without systems engineering guidance in the cyberspace construct results in more significant potential for uncoordinated, conflicting, and unsynchronized architectures within the cyberspace meta-system. Enduring meta-conceptual approaches akin to design patterns, that can be used for securing data and monitoring control processes, begin with concept development. The handoff needs to be near real-time, just as cyber effects occur in near real time.

12.1 Introduction

In this chapter, we shift the sport analogy from endurance car racing to golf. One of the basics of the game of golf is sports etiquette. There are essential rules that guide player behavior on the course, following deep traditions. Playing golf requires individual skill, etiquette, and the ability to devise and execute a strategy. The etiquette rules ensure that players maintain expected traditions. The best players evaluate options and strategies for approaching the course layout, taking into account fairway conditions, weather, and any obstacles that may be present. What brings elegance to the game is knowledge of the course rules and etiquette. Rules and etiquette are where golf and cyber part ways. Cyber has few regulations, lacks etiquette, and can often appear lacking in elegance. It may not seem easy for golf beginners to know not to walk into another golfer's line of putt, but golf traditions, which can be centuries old, can be learned. Golf is a community that embraces customs, shares its traditions, and fosters a sense of community. Similarly, Formula One racers adhere to specific rules of professionalism and etiquette within their sport. For example, one does not interrupt either a golfer stepping up to the tee or a driver preparing for a race. Both the golfer and the driver require mental readiness, which is crucial in their respective competitive environments. Cyberculture is less steeped in tradition – it is more challenging to know when to support interaction between teams and when to share threat information. People generally do not want to share information about cyberattacks. Reasons vary. Vendors sell access to real threat intelligence that other organizations can use to prevent cyber incidents. There may be some adversity to sharing information that can be damaging to a reputation. Regardless of the reasons, discretionary handling of sensitive information is essential.

Camaraderie is helpful for any sports team to be successful. Professional integrity and sharing helpful information about a system's security posture, such as identifying potential cyber vulnerabilities, would be a valuable courtesy. Still, information about known vulnerabilities is often not shared early enough in the design cycle. The handoff needs to be near real-time to be successful, given the cyber interface.

12.2 External Systems Diagram

The requirement to support cyberspace operations with an architecture that is more agile, more secure, and more readily available dictates that a paradigm shift must occur in the way the acquisition strategy is executed for information technology systems. The caveat is that you have to know the boundaries of potential disaster, which leads to the challenge of characterizing cyberspace operations. A culture of systems engineering thinking is a part of the necessary holistic approach to understanding user requirements and system concept development, system performance specifications, "design-to" specifications to the handoff to software engineering for the evolution of "design-to" into "build-to," fabrication, assembly and code to "build-to" documentation, and inspection.

A systems engineering perspective is again required earlier in the integration and qualification process, including assembly and performance verification, system integration and performance verification, and demonstration and validation for the user. Examples of decisions in systems engineering at the conceptual design stage include determining if a conceptual design effort should be undertaken and what mixture of technologies should be the basis of the design. In the model of cyberspace as a warfighting operational domain, these decisions are critical to the defense of critical infrastructure networks, as there is an interplay between the three system lifecycle phases: initiation, development, and operation. The initiation phase involves planning and marketing, research, requirements gathering, and conceptual design. The development phase involves detailed design, component, and software development, research, development, test, and evaluation (RDT&E). The operational phase involves system acquisition or procurement, as well as integration and installation to facilitate its development into operational use. Both defensive computer network defense (CND) activities and offensive cyberspace computer network exploitation (CNE) and computer network attack activities are linked in a single meta-system that requires primary systems engineering steps in the definition, development, and deployment stages, supported by, but not led by, software engineering capabilities.

The simple functionalities of legacy information technologies, which operate on single inputs to meet requirements developed for independent, disparate systems, now with cyber interfaces, must be coordinated to operate on all necessary inputs within a logical architecture that defines what the system must do in a larger meta-system. Coordination of all interfaces requires decomposing the system's top-level functions into smaller, manageable components. A systems engineering approach conducted iteratively in the cyberspace operational domain, is necessary to define successively lower-level functions needed to satisfy higher-level functional requirements and to determine alternatives for sets of functional requirements that support joint command requirements within the composition of the meta-system.

12.3 The Cyberspace Meta-system

Tools a system engineer may employ to achieve the functional analysis objectives include N^2 (logic flow) diagrams, functional flow block diagrams (FFBD), integration definition for function modeling (IDEF0), modeling, timeline analysis, thread analysis, and the Hatley–Pirbhai method, which provides an integrated approach to model-based systems engineering. The handoff of the continuous integration with cyber interfaces and configuration item (CI) specifications to the software engineer occurs deeper in the engineering "Vee" after the decomposition, or "top-down" effort, has partitioned the system functions one level at a time, and identified sound definitions of all inputs and outputs. For example, the N^2 diagram represents the functional or physical interfaces between subsystem elements of the cyberspace meta-system. The N^2 is used to identify and analyze the system's interfaces with both hardware and software. From the software engineer's perspective, the composition, or "bottom-up" effort, defines functionalities that synthesize functional hierarchy from the many bottom-level functions. The best solution is to use both the decomposition and composition approaches or a "middle-out" approach. In the meta-system environment of the newly identified cyberspace operational domain, the role of the system engineer is crucial in ensuring that requirements are clearly articulated to software engineers for custodial processing, thereby maintaining the integrity of the meta-system.

In 2011, The Office of the National Counterintelligence Executive aptly predicted, "[o]ver the next 3 to 5 years, we expect that four broad factors will accelerate the rate of change in information technology and communications technology in ways that are likely to disrupt security procedures and provide new openings for collection of sensitive US economic and technology information" (Executive, 2011). These factors are a technological shift, an economic shift, a cultural shift, and a geopolitical shift. These broad factors accelerate the rate of change in information technology and communications technology and can also be expected to accelerate change at lower levels of critical infrastructure architectures. These factors change the nature of interaction and interface with process control field devices. Change to the interfaces imposes the most significant economic challenge in generations. Governments and commercial industry partners are collectively working across federal services and inter-agencies to meet the challenge, which is further complicated by the involvement of commercial partners in critical infrastructure sectors. The technology shift creates opportunities for adversaries to exploit the networks. Recall from Chapter 9, "Road to Commoditization," that commoditization and the cyber medium both create industry disruption. Original electronic manufacturers (OEMs) and IT vendors will continue to release products with shorter new product introduction (NPI) lifecycles, potentially making critical infrastructure networks even more vulnerable to adversaries. Process control systems operate in near real-time, with a latency of milliseconds, ensuring safe, reliable, and optimized processes. Shorter NPI roadmaps result in product release versions with less exhaustive testing, which further exacerbates the system-wide requirements of the cyberspace operational domain. System qualifications, interfaces, verification, validation, and acceptance cycles are all affected on a higher meta-system level than previously software-independent engineered systems.

The location where products are manufactured affects the system design solutions at higher levels in a network architecture where business-critical IT systems are located. These levels represent the segmentation of the physical processes, sensors, controls, operations, and logistics. Supply chain integrity is a concern for enterprise resource planning (ERP) systems, extending to level (or layer) three, sometimes referred to as manufacturing execution systems (MES), where a broad range of software management activities related to production take place, including maintenance, reliability, integrity, sustainability, and quality control. Architecture levels are discussed in more detail in Chapter 23, "Abstraction Models." What is essential to understand about level (or layer) three is that this is the operation and control layer that manages production. Level 3 in the Purdue enterprise reference architecture (PERA) model is where the business processes are integrated with manufacturing operations of industrial control systems found in critical infrastructure. Supply chain assurance and security controls at the field device level must be verified and validated at lower levels, as these components are also vulnerable to cybersecurity risks.

Sensors, meters, control valve functions, and motor controls are susceptible to damage. The most critical points of control occur at level 3, which serves as the backbone for computing and data sharing, encompassing data historians and manufacturing execution systems (MES); however, compromise of components can occur much earlier in the supply chain. Recall from Chapter 6, "Cyber Vulnerabilities Start at the Hardware Level," That sensors function to detect and convert signals for processing, such as temperature, light, and motion. Motor controls regulate the operation of motors, including the control of speed and direction. The integration of sensor and motor controllers enables systems to be dynamically controlled in response to environmental changes, resulting in more efficient systems suitable for applications such as industrial automation, robotics, and heating, ventilation, and air conditioning (HVAC) systems. Reliable HVAC for critical infrastructure, such as data centers, is crucial. Data centers are integral to operations that control systems in a cyber world. Components within data centers are sensitive to temperature fluctuations and changes in air pressure. Failure to regulate can result in equipment failure, and the cost of downtime for a data center that requires consistent cooling can be prohibitively high, measured in millions of dollars. A cyberattack on an HVAC system can be significant, as can unit failures resulting from the compromise of system components that may originate in the supply chain. "The globalization of the supply chain for new – and increasingly interconnected – IT products will offer more opportunities for malicious actors to compromise the integrity and security of these devices" (Executive, 2011). Supply chain integrity vulnerability was particularly true for the Department of Defense (DOD). For example, IT commodity products and solutions represent a sizable portion of the Navy's budget[1].

In the Department of Defense (DOD), prior to the standing up of the US Fleet Cyber Command (FLTCYBERCOM) and US Tenth Fleet (10thFLT), information technology and communication technology were provided through a disparate collection of acquisition methods, including component-centric contracts, task orders against general services administration (GSA) schedules, and government-wide acquisition contracts (GWACs), or interagency agreements. With the establishment of the cyberspace operational domain, the Navy eyed a more consolidated IT acquisition strategy, promoting planning consistent with the enterprise architecture (EA) initiatives between the various service components. The Navy's move to a cloud

[1] The US Department of Navy's Fiscal Year 2025 Information Technology (IT) budget is USD 6 billion.

computing paradigm posed even greater challenges. While the cloud computing approach is more cost-effective than multiple enterprise hosting environments, new security challenges have emerged that require examination from a systems engineering perspective, rather than a software code reuse case. The higher handoff in the engineering "Vee" model is less effective than the lower handoff in the system development.

12.4 Vee Model Handoff to Software Engineers

A higher handoff in the engineering "Vee" (or "V") model to software engineers does not fundamentally support the measures required to reassess network architectures and introduce new IT and OT systems. Yet, there seems to be a leaning toward handing the entirety of the cyber "problem" space over to computer scientists to solve on their own. A cultural shift is necessary in the acquisition and deployment of systems with varying engineering expectations. The systems engineer combines elements of many disciplines, such as operations research, system modeling and simulation, decision analysis, project management and control, requirements development, software engineering, industrial engineering, specification writing, risk management, interpersonal relations, liaison engineering, operations analysis, and cost estimation (Gemmill, 2011) – all necessary to minimize the impact of potential points of failure in software-intensive systems. Process sensors are designed for reliability and safety. However, software-based mechanisms with cyber interfaces can make the sensor vulnerable to cyberattacks. Recall that sensors, meters, control valve functions, and motor controls are susceptible to damage, while also forming the most critical points of control in critical infrastructure. Additionally, it is expected that technicians and operators possess the knowledge, skills, and abilities to recognize when a process sensor is compromised in near real-time due to a cyber event. The overall approach to meeting the dynamic and complex requirements of critical infrastructure should be to increase the system engineering workforce and move the handoff of activities from systems engineering to software engineering deeper into the "Vee" to achieve the strategic goal of safe systems in a cyber world.

The introduction of the concept of integrated hardware and software-intensive architectures to the US Navy's Consolidated Afloat Network Enterprise Services (CANES) afloat and next-generation enterprise network (NGEN) ashore meta-system shifts the horizontal line lower in the intersection of the "Vee" engineering model, signifying a decrease in overlap between design and integration. CANES is the Navy's single, consolidated shipboard enterprise information system. NGEN is the Navy's integration

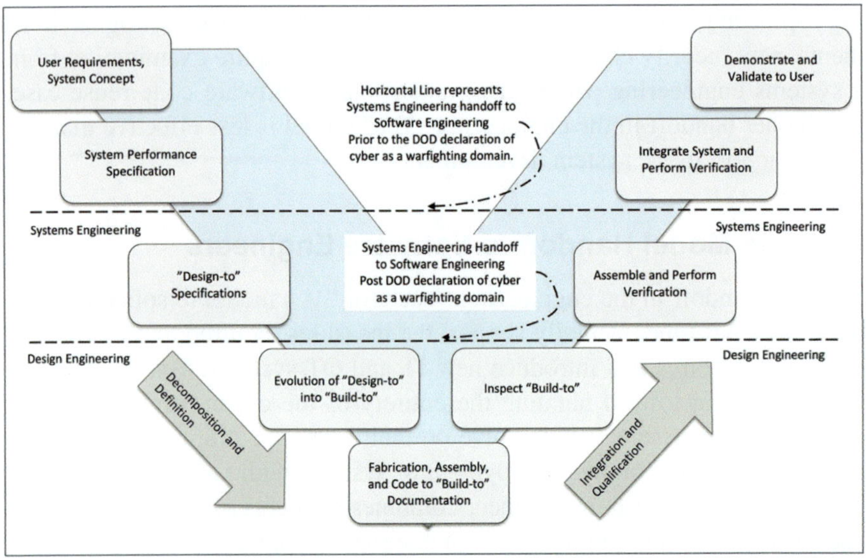

Figure 12.1 Systems engineering "Vee" (after Forsberg and Mooz [1992] (Buede, 2009)]).

of existing IT networks, services, and systems. Using modeling approaches such as block diagrams, N^2 diagrams, and other specification languages for modeling objects for requirements from a systems engineering perspective, more so than from the software engineering perspective of discipline engineers, whose orientation is to produce a physical instantiation of the cyberspace architecture framework, supports rapid comprehension of the overall mission. Both the CANES and NGEN approaches move the handoff between systems engineering activities and software engineering activities lower in the systems engineering "Vee" model. The horizontal line, as illustrated in Figure 12.1, represents the handoff between the systems engineering of the final products developed during the design process and the configuration item (CI) specifications, which are then passed to the software engineer or "discipline" design engineer (after Forsberg and Mooz [1992]) (Buede, 2009). The software engineering role in this model is to produce the physical instantiation of the design developed by the system engineer:

> "The horizontal line, drawn just under the middle intersection of the Vee, depicts the handoff of the final products of the design process, the [Configuration Item] CI specs, to discipline (or design) engineers, those engineers whose orientation is electrical, mechanical, chemical, civil, aerospace, computer science, and the like and whose job it is to produce

a physical entity. This dividing line can be drawn higher or lower to signify decreasing or increasing overlap between design and integration activities" (Buede, 2009).

12.5 Summary

Systems engineering is the most ubiquitous of the engineering skills. The intrinsic role of systems engineering is critical in designing and implementing a framework for cyberspace as an operational domain, particularly in the handoff between systems engineering and software engineering. As cyberspace plays an increasingly pervasive role in critical infrastructure operations, the role of the system engineer is a significant component of the intensifying cyber threat from adversaries. Just as a successful Formula One race team depends on near real-time feedback from the driver, so do engineers on a critical infrastructure system, who need near real-time threat information from the cybersecurity team, as well as a partnership with computer science team members. The handoff of information between these high-performance team members is tightly coupled. A top-down view of how the software interacts with the meta-system of networks is necessary to prevent malicious code from affecting networks and protect commercial networks from such threats. Mobilizing and extending advanced cyber defenses necessitates systems engineering personnel to deploy advanced architectures as countermeasures.

References

(GAO), U. S. G. A. O. 2011. Defense Department Cyber Efforts: DOD Faces Challenges in Its Cyber Activities. *In:* (GAO), U. S. G. A. O. (ed.). Washington, D.C.

Buede, D. M. 2009. *The engineering design of systems: models and methods,* Hoboken, New Jersey, John Wiley & Sons, Inc.

Executive, O. O. T. N. C. 2011. Foreign Spies Stealing US Economic Secrets in Cyberspace: Report to Congress on Foreign Economic Collection and Industrial Espionage, 2009-2011. *In:* US Government, O. O. T. N. C. E. (ed.). Washington, DC: Department of Defense.

Gemmill, D. 2011. Lectures. *AerE/EE/IE 565 Systems Engineering and Analysis/Fall 2011.* Iowa State University: Douglas Gemmill.

Iii, W. J. L. 2011. The Pentagon's Cyberstrategy, One Year Later: Defending Against the Next Cyberattack. *Foreign Affairs.*

13

Needs Analysis

"Welcome to the Grand Illusion. Come on in and see what's happening.
Pay the price, get your tickets for the show"
– Dennis De Young, American singer and songwriter,
founding member of the rock band Styx.

Abstract

Needs analysis is the most crucial phase of concept development. The new system likely must address a deficiency, the most prevalent of which is obsolescence. The systems engineering approach for conducting a needs analysis can be applied to projects of any size and scaled up for larger projects. It begins with a review of the current system's "as-is" state, followed by an analysis to determine if there is a genuine need and to identify the potential project's pros and cons. The first step is to review the total system performance of the current system and how it is used in the operational environment, given various known operating conditions. The objectives are decomposed, and system functions are identified. The constraint is evaluated within the space defined in the feasibility definition, which includes the size of the critical infrastructure site, compliance constraints, and other factors such as cost, plumbing, and heating and air-conditioning infrastructure that may affect an installed system. The primary objective is to demonstrate a valid operational need for upgrading the existing system or implementing a new one. It is essential to capture the operational requirements in sufficient detail. During this phase, we aim to verify the requirements and recognize that there are alternative solutions to a problem beyond creating a new system, such as training and material solutions. Typically, the approach is technology-driven (Kossiakoff, Seymour et al. 2020). The adage "Measure twice, cut once" is a good one to apply. The needs analysis phase is crucial.

DOI: 10.1201/9788743809128-13

13.1 Introduction

The systems engineering approach for conducting a needs analysis can be applied to projects of any size and scaled up for larger projects. For example, the analysis can be for any size system. It begins with a review of the current system's "as-is" state, followed by an analysis to determine if there is a genuine need and to identify the potential project's pros and cons. The first step is to review the total system performance of the current system and its usage in the operational environment, taking into account various known operating conditions. The objectives are decomposed, and system functions are identified. The constraint is evaluated within the space defined in the feasibility definition, which includes the size of the critical infrastructure site, compliance constraints, and other factors such as cost, plumbing, and heating and air-conditioning infrastructure that may affect an installed system. The systems engineering process can be applied to any project, regardless of its scale and scope (Scalco and Wince 2012).

13.2 Needs Analysis Objective

The primary objective of the needs analysis phase for a new system is to identify a valid operational need for the system or a major system upgrade (Kossiakoff, Seymour et al., 2020). The two principal outputs, or products, of the needs analysis phase, are:

1. An estimate of the system's operational effectiveness, which defines "what a new system should achieve to meet the identified need."
2. System capabilities "provide evidence that an affordable system capable of meeting the effectiveness target is feasible" (Kossiakoff, Seymour et al. 2020)

The primary systems engineering activities that contribute to these products are:

1. Needs validation (i.e., design validation) and technology assessment
2. Functional analysis (i.e., functional definition)
3. Feasibility definition (physical definition)
4. Operational analysis (i.e., requirements analysis)

The general products of operations analysis are a list of operational objectives and systems capabilities. The primary output of functional analysis is a list of initial functional requirements. The general product of the feasibility definition is a list of physical requirements. The general product of the need analysis validation is a list of operational validation criteria (Kossiakoff, Seymour et al. 2020). Systems studies break down the system into components and subcomponents. The most critical activity is collecting and analyzing the requirements, which allows systems engineers to perform trade-off analyses for the improvement of component elements. The goal of technology assessments is a process that researches the political, societal, economic, and legal aspects related to the system.

The operational analysis aims to identify system improvement objectives in relation to the current state, also known as the "as-is" state. For example, the needs analysis for the more situational awareness for industrial control systems (MOSAICS) framework started with a high-level objective to build an architecture and toolset to monitor and report the cybersecurity status of a DOD industrial control system (ICS) in operationally relevant, near real-time. The needs analysis phase output defined the system's operational effectiveness as enhanced automated response and remediation following the advanced cyber industrial control systems tactics, techniques, and procedures (ACI TTP) for manual responses to cyberattacks on a system (USCYBERCOM, 2016). Recall that the ACI TTP described in Chapter 7, "Threats to Information-intensive Systems," is a product of a collaborative effort by the US Cyber Command (USCYBERCOM) and the Joint Test and Evaluation (JT&E) Program under the Office of the Secretary of Defense, Director of Operational Test and Evaluation (OSD T&E). The MOSAICS technical team used the ACI TTP to inform the primary systems engineering activities that contributed to the principal outputs (products) of the needs analysis phase. Two key technologies were the map-to-model baseline tool developed by Sandia National Laboratories (Sandia) and the set of services and security orchestration, automation, and response (SOAR) tools researched by the Johns Hopkins University Applied Physics Laboratory (JHU APL). A key component of SOAR is connecting disparate tools and systems to streamline workflows and enhance system visibility. By automating ACI TTP courses of action through predefined playbooks within the ACI TTP, the system operator's manual tasks are reduced, allowing the operator or technician to focus on higher-complexity tasks. The SOAR approach has already proven valuable to the overall security posture of enterprise IT systems. The technology readiness level (TRL) assessment estimates the maturity of critical technology elements (CTE) during the needs analysis phase. This evaluation is used to determine the maturity level of an innovative technology for integration into a system. TRL is important, from the top of the architecture to the lowest-level components in critical infrastructure. Just because there is a capability for software-based monitoring of control systems does not mean

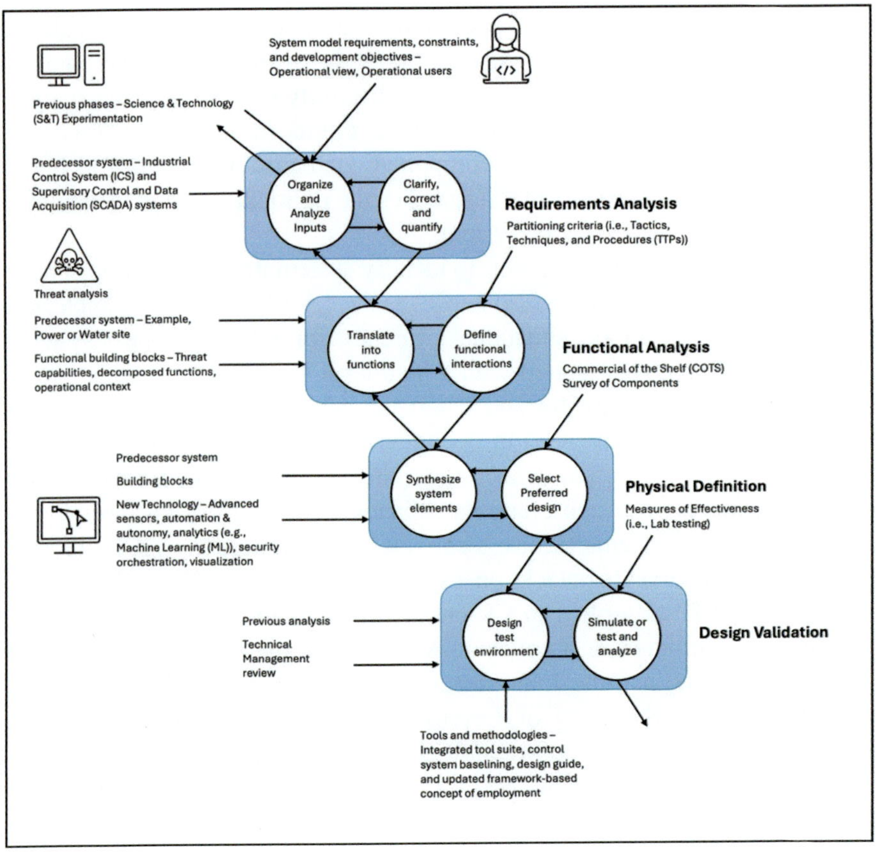

Figure 13.1(a) Generic systems engineering flow diagram.

it should be used as an interface to allow the sharing of information with business operations without a clear understanding of the data flow, latency, and any engineering, field-related issues that can be found at the sensor levels or safety functions.

Referring to Figure 13.1(a,b), the systems engineering flow diagrams illustrates the application of the systems engineering method to the "Needs Analysis Phase" flow diagram. The operational analysis phase involves a needs analysis, during which operational deficiencies are identified and addressed. For example, during this phase of MOSAICS, the technical team defined the initial capability to defend mission-critical infrastructure, system model requirements and constraints, and development objectives based on the operational views of the US Indo–Pacific Command (USINDOPACOM) and US Northern Command (USNORTHCOM). The technical team evaluated

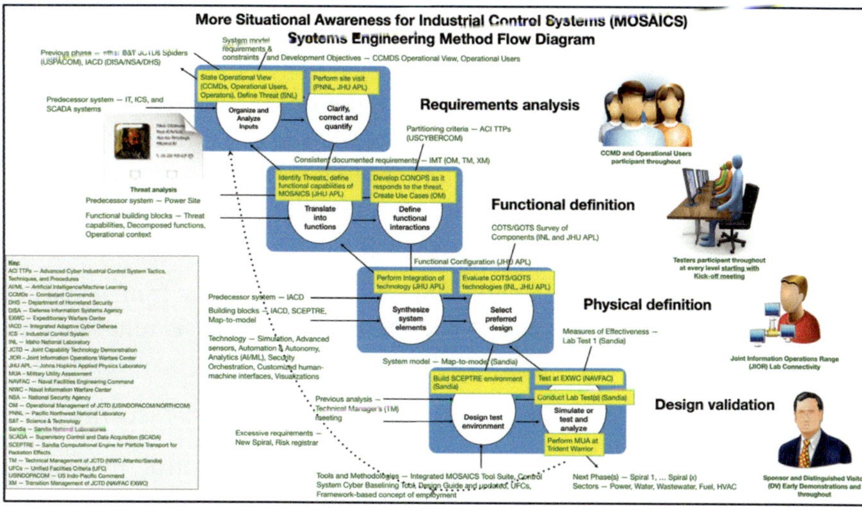

Figure 13.1(b) MOSAICS systems engineering flow diagram.

predecessor systems for deficiencies in IT, industrial control systems (ICS), and supervisory control and data acquisition (SCADA) systems. The team also assessed technology opportunities in terms of technology readiness level (TRL) and applicability. During this phase, the two processes pictured in the diagram are the analysis of projected needs and the definition of the operational approach. The significance of analyzing projected needs is that it elicits the customer's assessment of operational capability and readiness level. For example, in the case of the MOSAICS, the projected needs were characterized by USINDOPACOM and USNORTHCOM. The significance was that it was determined that there was no other such predecessor system in the entirety of the DOD. The combatant commands (COCOMs) stated their operational view, and Sandia performed a threat analysis of the requirements during the requirements analysis stage. Pacific Northwest National Laboratory (PNNL) collaborated with JHU APL to conduct site visits to clarify, correct, and quantify the requirements. The MOSAICS integrated management team (IMT) ensured consistency in the documented requirements. The information from the analysis of projected needs and the definition of the operational approach was validated through task participation, involving systems engineers, systems architects, systems analysts, needs analysts, sponsors, and operational management representing the user community. The result confirmed that the operational need existed and could not be met by another existing system solution.

13.3 Metrics

Metrics help to prioritize efforts in the design phase. For critical infrastructure in the cyber domain, metrics to consider include dwell time, mean time to detect (MTTD), mean time to respond (MTTR), time from alert to triage, and time to close. Each of these metrics assumes the operator can detect the difference between a component maintenance failure and a cyber event.

- Dwell time – The entire length of the cybersecurity event from when the attacker first enters the network to the time the attack is removed and the return to "known good."
- Mean time to detect (MTTD) – Time to discover a potential security event
- Mean time to respond (MTTR) – Time it takes to control, remediate, and eradicate a threat once it is discovered
- Time from alert to triage – Similar to dwell time, but measures the mean time from detection or alert trigger to validation and initial response action
- Time to close – Time to close is the time from opening to the close of an instance

The needs analysis phase output defined the system's operational effectiveness as enhanced automated response and remediation following the ACI TTP for DOD ICS (USCYBERCOM, 2016). The ACI TTP was a collaborative effort between the United States Cyber Command (USCYBERCOM) and the Joint Test and Evaluation (JT&E) Program, under the direction of the Director, Operational Test and Evaluation, Office of the Secretary of Defense (OSD T&E). The MOSAICS technical team used the ACI TTP to inform the primary systems engineering activities that contributed to the principal outputs (products) of the needs analysis phase.

A key output of the MOSAICS JCTD was a reference architecture (RA). Using the MOSAICS RA can help reduce development and maintenance costs, providing a better understanding of the system, ease of development, and increased productivity. "[A] RA is instantiated by the addition of relevant detail to achieve an architecture to satisfy a specific set of requirements within the domain" (Borky and Bradley, 2019).

13.4 Balancing Resilience with Risk Management

Applied systems engineering to achieve resilience and risk management for composed cyber solutions is an example of how to approach critical infrastructure (Scalco and Wince 2012). Critical infrastructure system owners require up-to-date, available, agile, and more cost-effective solutions to protect assets and property. Billions of dollars are lost to the economy each year due to cyberattacks. Traditional government cybersecurity solutions can take anywhere from three to five years to develop, certify, and deploy. In many cases, business cybersecurity solutions are just dumbed-down versions of available high-grade government solutions. By the time these security solutions are fielded, they are outdated or obsolete technology. Critical infrastructure organizations demand the latest features and capabilities found in the latest commercial products. Security-enhanced versions of these devices must incorporate cutting-edge features and applications to enhance security. A strategy is required to address what is really an applied systems engineering challenge. Utilizing the sound principles of systems engineering and analysis, we can develop a methodology to raise the bar in protecting critical infrastructure assets, thereby enhancing national security. An approach that weighs and balances security and risk while offering customers the feature-rich capabilities of commercial tools is required (Scalco and Wince, 2012).

13.4.1 Requirements analysis

The top-level function that the system must perform is to create a methodology for achieving a balanced resilience and risk management system so that integrators can use leading-edge, commercially available products without these technological advances negating the solution's security (Council 2008). The basic system assumptions are that the system will utilize commercial off-the-shelf (COTS) products in a layered approach, composed of solutions that offer feature-rich capabilities users demand and that the system shall comply with the International Society of Automation (ISA)/International Electrotechnical Commission (IEC) 62443 Standards. For the Department of Defense (DOD), critical infrastructure must comply with the National Institute of Standards and Technology (NIST) standards and the National Security Agency (NSA) directives. Although these critical infrastructure cyber domain solutions may require significant research and development resources to design initially, system owners can realize substantial savings through design reuse, such as by utilizing the MOSAICS framework. Designs and approved solutions, such as MOSAICS, are posted to a web-accessible site for reuse as part of the proposed methodology to manage the system. As an example, a critical infrastructure systems engineering approach involves combining up-to-date commercial off-the-shelf (COTS) components, achieving speed-to-field agility with NIST/NSA Suite B Cryptography, and implementing IA security that is strong enough to protect data and intellectual property from economic loss due to cyberattacks. It is essential to note that industrial control system (ICS) communications are generally not encrypted. ICS subsets of supervisory control and data acquisition (SCADA) and distributed control systems (DCS) span over large, disparate geographical regions, making these susceptible to cyberattacks. Adding encryption to troubleshooting problems, such as failed field devices, can be expensive, as can adding encryption with intrusion detection taps. Traditionally, these networks are more accessible to vendors performing maintenance, assuming that the components and connections are trusted and reliable. Recall from Chapter 5, "Cyber Paradox," that a Zero Trust strategy requires authentication for an actor (e.g., a person or device) attempting to access network resources (CISA, 2023). An operational requirement to test is the system access control mechanisms. Testing access control mechanisms become problematic for critical infrastructure systems, as system requirements must be traceable to the Zero Trust authentication capability. Encryption does not prevent a successful cyberattack on a system because there are other paths to take control of the industrial control system (ICS). For example, the zero-day Stuxnet malware attack, which was explicitly targeted at specific system software, caused physical damage to an Iranian nuclear enrichment facility in 2010. The malware was likely introduced via a compromised vendor using an infected USB flash drive, also referred to as a "thumb drive" (CISA, 2014). Stuxnet is a computer worm that checks to see if specific models of programmable logic controllers (PLCs) made by a particular manufacturer interact with machinery such as that used in uranium centrifuges. Then it alters the PLC's programming, resulting in damage to the equipment. Without behavioral analytics realized through a framework such as MOSAICS, it is challenging to diagnose any unusual activity occurring on the system, as the PLCs indicate to the control computer that the system is functioning correctly (Fruhlinger, 2017). Encryption would not resolve a malware introduction via a USB flash drive to a critical infrastructure system, for example.

However, the strategy to use commercial off-the-shelf (COTS) technology from components to product ranges combined with NIST/NSA Suite B Cryptography to compose solutions that meet security requirements is beneficial by minimizing the system design and testing expense but must be weighed against the potential risks of presenting desirable features of COTS technology that may be introduced by rapidly changing COTS technology and obsolescence, for private commercial, or government use.

The real challenge, perhaps, arises when a layered or "composed solution" is characterized from a security or risk standpoint, as the approving authority deems it "good enough," based on a set of documented statements and testing for customer usage on classified or secure networks. The challenge lies in addressing changes to this "approved" layered or composed solution once it is in the field and being used on customer platforms. These changes may be as simple as over-the-network software upgrades that may or may not affect the original characterized solution, such as when using technologies such as operational technology software-defined networking (OT-SDN). The use of SDN in critical infrastructure is briefly mentioned in Chapter 8, "Use Cases and Operational View (OV) Illustrations," as a means of segregating networks through application-aware defense to protect and detect cybersecurity functions, serving as a mitigation approach. Software-defined networking (SDN) is discussed in more detail in Chapter 16, "Modernization Strategies," and in Chapter 18, "Software Approaches." A key question in the needs analysis phase is how these changes are described in terms of their security and risk posture. At what point does this solution set need to undergo the entire process again, or can some delta testing suffice, or perhaps no further action is required? What investments are needed is the crux of the system: to devise a method for quantifying changes and determining a course of action, ensuring balanced resilience to security shortfalls caused by lifecycle changes and risk management for composed solutions, in the normal course of product development and sustainment. Any such answer must be both cost-effective and scalable (Scalco and Wince 2012).

In addition to the security shortfalls caused by lifecycle mismatches, other shortfalls offset the potential high-value return on investment (ROI) of usability in composed solutions that utilize commercial off-the-shelf (COTS) components. A 2009 study suggests that when usability evaluations are conducted, COTS equipment reveals potentially significant shortcomings in overall performance (Carl D. Thunberg, 2009). Applied systems engineering can help mitigate this performance shortfall. This potential cost savings and increased commonality from COTS procurement can offset the actual performance issues due to mismatches between equipment, users, and the changed operational environment. The objective of this study was to conduct a rapid "human-in-the-loop" evaluation of users' performance with the COTS workstation. The study demonstrated that while employing COTS components to develop composed solutions to meet requirements can be very appealing, and system owners can realize significant cost savings from resources and time saved in the research, design, and production costs of a system, this may not effectively address systems of people and equipment using the COTS composed solution (Ibid.).

Systems engineering is defined as "an interdisciplinary approach to translating users' needs into the definition of a system, its architecture, and

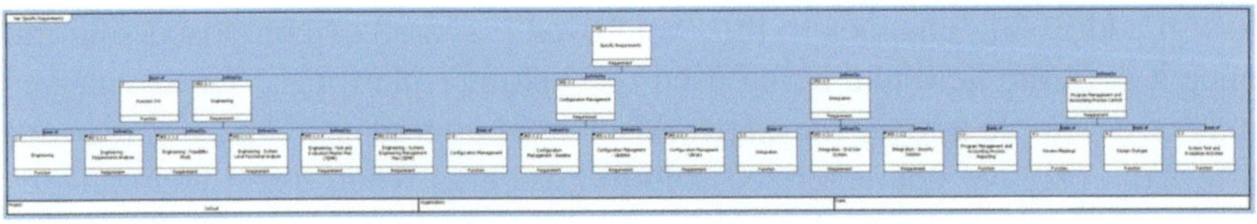

Figure 13.2 Requirements flow analysis.

design through an iterative process that results in an effective operational system. Systems engineering efforts apply over the entire life cycle, from concept development to final disposal" (DOD, 2012). Two approaches can be used as methods of applied systems engineering to address cyber challenges within: (1) a design refresh strategy and (2) a probabilistic risk analysis (PRA) strategy. The design refresh strategy is one approach to optimize the life cycle cost of a system and design refresh schedules for IT-intensive, COTS-based systems. Peter Sandborn and Pameet Singh described this strategy in a paper entitled "Forecasting Technology Insertion Concurrent with Design Refresh Planning for COTS-based Electronic Systems" (Sandborn and Singh, 2005). In this approach, "[t]he analysis methodology can be used to generate application-specific economic justifications for design refresh approaches to obsolescence management" (Ibid.). Applying a similar methodology to manage traditional cybersecurity solutions can take anywhere from 3 to 5 years to develop, certify, and deploy. System engineers lead the development of outdated or obsolete solutions to make them more resilient to technology life-cycles that are shorter than the certification process itself. The technical team can better manage the risk of deploying outdated solutions. The PRA strategy is an approach that provides a realistic assessment of risk by "providing sufficient information about the relative contribution of each component to the overall risk," allowing for optimal resource allocation (Pate-Cornell 1994).

13.4.2 Requirements flow analysis

The requirements flow analysis (Figure 13.2) in a system must show, at a minimum, all three levels of requirements and the associated functions for the top two requirements. The functions associated with the third level of requirements are illustrated in individual diagrams for each of the lower-level requirements (Scalco and Wince, 2012).

13.4.3 Mandatory requirements

Mandatory requirements are the necessary and sufficient conditions that the minimal system must have to be acceptable. These are requirements that are not susceptible to tradeoffs between requirements. Two examples of mandatory requirements and how these are traced are as follows:

Requirement (Req): The system shall have access control, where credentials must be verified and approved before access is granted.

Rationale: *Protecting the information within the system and ensuring the right personnel have access will be critical to guarding the data.*

Test method: Test

Owner: [Individual responsible]

Approved: [Date approved and by whom]

ID: ORD[1].1.1

Req: The system shall have a reliability rate of 99%.

Rationale: *For the system to be effective, it must be reliable and trusted within the community.*

Test method: Test

Owner: [Individual responsible]

Approved: [Date approved and by whom]

ID: ORD.1.2

Each solution must meet the minimum mandatory requirements specified in the requirements document. The baseline requirements are maintained as described in the configuration management (CM) plan.

ID: ORD.1.1 – Engineering

ID: ORD.1.1.1

Req: Develop a requirements analysis document

Rationale: The requirements analysis document shall, at a minimum, illuminate the system's scope, purpose, and both functional and non-functional requirements.

Owner: [Individual responsible]

---

[1] ID: ORD refers to "Identify, Define, Organize, and Document" the system's requirements, interfaces, and other critical information.

Approved: [Date approved and by whom]

ID: ORD 1.1.2

Req: Document the feasibility study results

Rationale: The feasibility study document shall identify, at a minimum, the time and resource costs, software impacts, organizational impacts, and security and privacy impacts.

Owner: [Individual responsible]

Approved: [Date approved and by whom]

ID: ORD.1.1.3

Req: Develop system-level functional analysis

Rationale: The system-level functional analysis shall enable users to view the complex system's functions as more manageable and consumable bits of information.

Owner: [Individual responsible]

Approved: [Date approved and by whom]

ID: ORD.1.1.4

Req: Develop test and evaluation master plan (TEMP)

Rationale: The test and evaluation master plan (TEMP) shall include, at a minimum, the scope, content, methodology, sequence, management, and roles and responsibilities for tasks and activities needed to perform test activities.

Owner: [Individual responsible]

Approved: [Date approved and by whom]

ID: ORD.1.1.5

Req: Develop systems engineering management plan (SEMP)

Rationale: The systems engineering management plan (SEMP) shall include, at a minimum, the systems engineering process, the systems engineering schedule, and the systems engineering process metrics.

Owner: [Individual responsible]

Approved: [Date approved and by whom]

ID: ORD.1.2 – Configuration management

ID: ORD.1.2.1

Req: Document the baseline of the composed commercial solutions capability package

Rationale: The baseline will provide a designated point in the system lifecycle where some important features of the system configuration are defined in detail (Gemmill, 2012).

Owner: [Individual responsible]

Approved: [Date approved and by whom]

ID: ORD.1.2.2

Req: Document the baseline updates of the composed commercial solutions capability package

Rationale: The baseline will provide a designated point in the system lifecycle where some essential features of the system configuration are defined in detail (Ibid.). The updates will provide a way to manage all of the changes.

Owner: [Individual responsible]

Approved: [Date approved and by whom]

ID: ORD.1.2.3

Req: Maintain a COTS composed commercial solution reference library

Rationale: The composed commercial solution library shall, at a minimum, document the configuration items, required date, classification, and a short description of the problem.

Owner: [Individual responsible]

Approved: [Date approved and by whom]

ID: ORD.1.3 – Integration

ID: ORD.1.3.1

Req: The system shall integrate a composed commercial solution capability into the end-user system.

Rationale: The system must be integrated into the end-user system so that it can be utilized by the consumers who need it.

Owner: [Individual responsible]

Approved: [Date approved and by whom]

Table 13.1 Example commercial off-the-shelf specifications.

Requirement		COTS A	COTS B	COTS C
Model	n/a			
Frequency range	20 to 3000 MHz			
Antenna size	< 22" D × 12" H			
Weight	< 20 lbs.			
Power	< 20 watts			
Price per unit	< USD 100k			
Development cost	lowest			
Availability	< 6 months			

ID: ORD.1.3.2

Req: Document system-level test results of the integrated security solution.

Rationale: It is crucial to accurately document the results of the system-level test to ensure proper integration.

Owner: [Individual responsible]

Approved: [Date approved and by whom]

ID: ORD.1.4 – Program management and accounting process reporting

Req: Conduct review meetings, coordinate design changes, and review system test and evaluation activities, at a minimum.

Rationale: The program management and accounting process reporting activities are critical to ensuring oversight and involvement from multiple angles.

Perspectives: [Individual responsible]

Approved: [Date approved and by whom]

A copy of the solution specification is maintained in a web-accessible solution library. The system owner will use the data source dashboard to monitor specification changes in commercial off-the-shelf (COTS) products used in a solution, which may impact the solution's integrity. Example data specifications are shown in Table 13.1.

13.5 Summary

Needs analysis is the most crucial phase of concept development. The systems engineering approach for performing a needs analysis can be applied to any size project and scaled to larger projects by starting with the "as is" state of the current system, followed by an analysis to determine if there is a real need and what the pros and cons are. The first step is to review the total system performance of the current system and its usage in the operational environment, taking into account various known operating conditions. Objectives are decomposed, system functions are identified, and constraints are evaluated. Metrics are established, and reliance and risk management are balanced against the need. Capturing operational requirements in sufficient detail is key to the process. "Measure twice, cut once" is a good adage to consider in this crucial phase. If you are looking to measure the value of your system, you need to know what you are measuring the system against. You do not want to be measured against things that you are not designing the system to meet, and you do not want to deliver a system that does not meet the operational needs. During this phase, requirements are verified through the systems engineering process and analysis, recognizing that alternative solutions to a problem may exist beyond creating a new system.

References

(DOD), D. O. D. 2012. *Defense Acquisition Guidebook* [Online]. https://acc.dau.mil: Defense Acquisition University (DAU). Available: https://acc.dau.mil [Accessed].

(GAO), U. S. G. A. O. 2011. Defense Department Cyber Efforts: DOD Faces Challenges in Its Cyber Activities. *In:* (GAO), U. S. G. A. O. (ed.). Washington, D.C.

(USCYBERCOM), U.S.C.C. 2016. Advanced Cyber Industrial Control System Tactics, Techniques, and Procedures (ACI TTP) for Department of Defense (DoD) Industrial Control Systems (ICS). *In:* (DOD), D. O. D. (ed.). United States Cyber Command.

Alexander Kossiakoff, W. N. S., Samuel J. Seymour, and Steven M. Biemer 2011. *Systems Engineering Principles and Practice*, John Wiley & Sons, Inc. Publication.

Borky, J. M. & Bradley, T. H. 2019. *Effective Model-Based Systems Engineering*, Springer.

Carl D. Thunberg, L. C. A. P. T. 2009. The Role of Human Systems Integration in Considering Commercial Off-the-Shelf Elements in Complex Systems. *INCOSE Insight,* 54, 4.

(CISA), C. I. S. A. 2014. ICS Advisory (ICSA-10-238-01B) Stuxnet Malware Mitigation (Update B). *In:* ICS-CERT (ed.) *ICS Advisory.* https://us-cert.cisa.gov/ics/advisories/ICSA-10-238-01B: Department of Defense.

CISA 2023. Industrial Control Systems Cybersecurity Initiative: Considerations for ICS/OT Monitoring Technologies with an Emphasis on Detection and Information Sharing. *In:* DHS (ed.). DHS.

Council, N. R. 2008. *Pre-Milestone A and Early-Phase Systems Engineering: A Retrospective Review and Benefits for Future Air Force Systems Acquisition,* Washington, DC, The National Academies Press.

Executive, O. O. T. N. C. 2011. Foreign Spies Stealing US Economic Secrets in Cyberspace: Report to Congress on Foreign Economic Collection and Industrial Espionage, 2009-2011. *In:* US Government, O. O. T. N. C. E. (ed.). Washington, DC: Department of Defense.

Fabrycky, B. S. B. A. W. J. 2011. *Systems Engineering and Analysis*, Prentice Hall.

Fruhlinger, J. 2017. What is Stuxnet, who created it, and how does it work? *CSO Online.*

Gemmill, D. 2011. Lectures. *AerE/EE/IE 565 Systems Engineering and Analysis/Fall 2011.* Iowa State University: Douglas Gemmill.

Iii, W. J. L. 2011. The Pentagon's Cyberstrategy, One Year Later: Defending Against the Next Cyberattack. *Foreign Affairs.*

Kossiakoff, A., Seymour, S., Flanigan, D. & Biemer, S. 2020. The Future of Systems Engineering. *Systems Engineering Principles and Practice.*

Ljerka Beus-Dukic, I. F. A. 2009. *Discovering Requirements: How to Specify Products and Services.*

marzolf, G. 2019. Engineering Lectures. Colorado State University (CSU): Colorado State University (CSU).

Pate-Cornell, E. M. 1994. Risk Management for the Tiles of the Space Shuttle. *Interfaces.*

Sandborn, P. & Singh, P. 2005. Forecasting Technology Insertion Concurrent with Design Refresh Planning for COTS-Based Electronic Systems. *CALCE, Department of Mechanical Engineering, University of Maryland.*

Scalco, A. W. & Wince, K. 2012. A Proposed Methodology to Achieve Balanced Resilience and Risk. IE 566 Applied Systems Engineering: Iowa State University.

14

Detailed Design

"Measure twice, cut once."

– adage, source unknown

Abstract

The system design process involves conceptual, preliminary, and detailed design based on design criteria. Design criteria are the basic design objectives for the system and its elements that must be compatible with the system operations requirements, maintenance and support concept, and data prioritization of the applicable technical performance measure (TPM) that comply with the design-to criteria to "meet all the requirements in the various applicable specifications" (Fabrycky 2011). During the detailed design phase, the system begins to take shape in response to the needs analysis requirements. An entirely new industry may emerge. Design criteria are applied in the design process through various design engineering activities, utilizing a combination of resources, including technical engineering expertise, engineering technical support, and non-technical support. The design criteria are applied at each system level throughout the system design (e.g., system configuration), preliminary design (subsystem level), detailed design and development, and system integration and testing. Resources have to be committed to deploying a design for reliability (DfR). The reliability of the solution must be validated and tested for the new use case. The maintainability of the composed solution must also be validated and tested, and the lifecycle length identified to mitigate risks. Design for maintainability (D*f*M) considers both design and maintainability concepts. Ultimately, the system design process involves conceptual, preliminary, and detailed design, all of which are based on established design criteria. Reliability analysis methods, such as failure modes, effects, and criticality analysis (FMECA) and fault tree analysis (FTA), aid in identifying and preventing design issues.

DOI: 10.1201/9788743809128-14

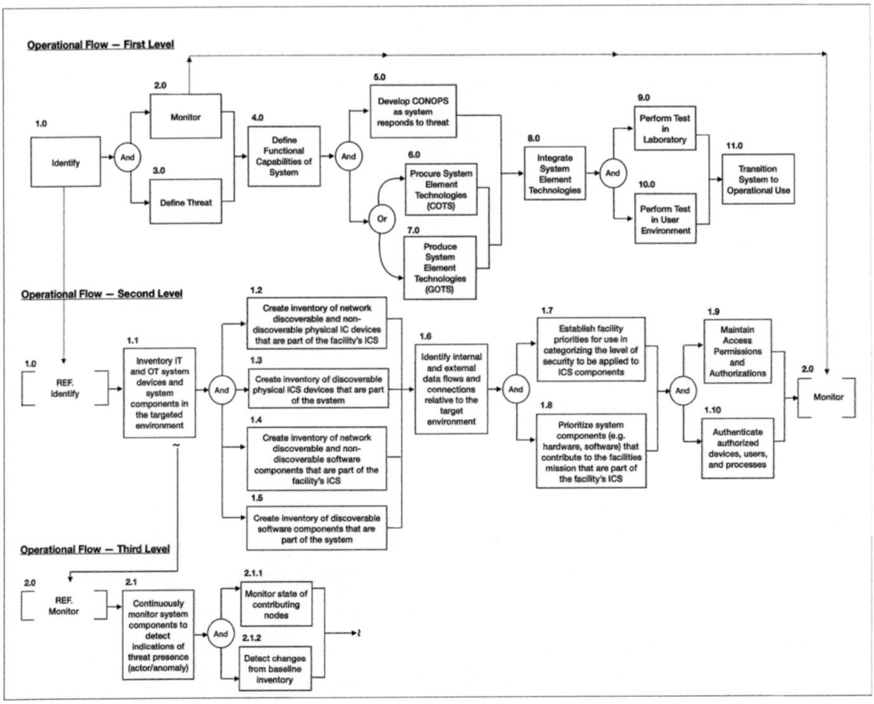

Figure 14.1 Detailed design of an electrical substation automated system network traffic flow (Scalco and Simske, 2021).

14.1 Introduction

In 2019, a new industry emerged in response to the need for continuous monitoring of data entry and exit points, as well as trust boundaries, introducing greater automation of processes and enhanced alert detection capabilities. The new industry is known as security orchestration, automation, and response (SOAR). SOAR technologies enable organizations to collect and monitor inputs in a digital workflow format. SOAR was quickly adapted for use in IT enterprise network environments and leveraged in research efforts within the power utility sector. SOAR technology serves as a stepping stone toward the adoption of more comprehensive Software-defined networking (SDN) capabilities, applicable to both IT and OT systems. SOAR offers accelerated incident response times, increased accuracy across security operations, and significant time and cost savings for response to alerts. Some commercial vendors claim that the average response time to system alerts can be reduced from 30 minutes or more to less than five minutes, which translates to approximately 83% of the time saved per alert, as shown in Figure 14.1.

The conceptual design shapes the initial cyber defensive capability, extending from the cloud to the lowest level programmable logic controller (PLC), for a critical infrastructure system. The operational need is to defend mission-critical infrastructure from cybersecurity attacks. The conceptual design process formulates how to meet the system-level requirements that flow from the operational requirements. A best practice in cybersecurity is to maintain an effective inventory baseline for both IT and operational technology (OT) networks. Without an inventory baseline, it is impossible to fully understand the system's cybersecurity posture. An operational requirement is to provide inventory for both IT and OT system devices and system components in the targeted critical infrastructure environment. A key question for system owners or operators to consider is who has access to the physical network topology. The system design process involves conceptual, preliminary, and detailed design based on design criteria. Design criteria are the basic design objectives for the system and its elements that must be compatible with the system operations requirements, maintenance and support concept, and data prioritization of the applicable technical performance measure (TPM) that comply with the design-to criteria to "meet all the requirements in the various applicable specifications" (Fabrycky 2011).

Design criteria are applied in the design process through various design engineering activities, utilizing a combination of resources, including technical engineering expertise, engineering technical support, and non-technical support. The design criteria are applied at each system level throughout the system design (e.g., system configuration), preliminary design (subsystem level), detailed design and development, and system integration and testing. For example, during the system design, design criteria are considered for performance (e.g., electrical, mechanical) and cost-effectiveness. During the preliminary design phase, design criteria considered include maintainability and survivability. Throughout the design lifecycle, qualified individuals with the right technical expertise are brought into the team to develop solutions that meet the design criteria, including flow diagrams and other tools.

The system performance requirements are the technology base that provides the framework for system development. Three principal differences between system performance requirements and system functional specifications, which are input to the concept definition phase, are examined. System functional specifications include:

1. Relate requirements to operational objectives (e.g., throughput, storage capacity, and speed of response)

2. Examine behaviors (e.g., when is availability required and at what levels of throughput)
3. Identify how the system should act under various conditions (e.g., how frequently an event occurs and how a user recovers)

The system functional specification describes what the system should do by:

1. Allocating functions to the components to meet the operational objectives
2. Specifying behavior
3. Under various conditions prescribed by the system performance requirements

For example, the outputs are defined system concepts and plans for engineering activities. "The planning outputs of this phase are usually specified to include the systems engineering management plan (SEMP), which defines in detail the systems engineering approach to be followed, the project work breakdown structure (WBS), cost estimates for development and production, test plans, and such other supporting material as may be directed" (Alexander Kossiakoff 2011). These artifacts support the realization of the system's performance requirements.

The principal criteria involved in selecting a preferred system concept are best compared using an evaluation "profile." Keeping weighted evaluations of criteria in an independent profile ensures that visibility is maintained for each concept for each critical measure of effectiveness (MOE). The rationale is that an evaluation profile is more suitable as a benchmark against which the team can measure the suitability of individual alternatives and their associated risk-to-reward ratios. For example, the profile criteria may be used to assess the impact of resource allocations or to investigate alternative technologies. The profile approach offers greater fidelity to the measures, enabling more effective resolution of conflicts and informed engineering decisions (Ibid.). An MOE is a measure of how well the objectives are achieved to determine "how well" the objectives are achieved. Example metrics include the technology readiness level (TRL) or the development time of the system. The measurement methods must be sufficiently discriminatory to ensure that the selection is the best option. The purpose of such an approach is "to retain the visibility of the complete evaluation profile of each concept (against each critical measure of effectiveness) until the final selection, rather than combining the components into a single figure of merit, a practice that is often employed but that tends to submerge significant differences" (Ibid.). A decision matrix is a valuable tool to use for testing selections and exploring the decision-making process.

The evaluation includes comparisons of operational performance, the program schedule and cost, and the risks of achieving the goals and objectives using the selected system concepts. For example, in the more situational awareness for industrial control systems (MOSAICS) joint capability technology demonstration (JCTD), the characteristics of addressing combatant command (COCOM) needs were assessed in operational demonstrations of a prototype. The technical team socialized representative concepts with the COCOMs, military services, and representative DOD agencies prior to the candidate concept's authority to proceed. Once the candidate concept is authorized, the functional allocation requirements are socialized. The development of the work breakdown structure (WBS) was a crucial step that involved subject matter expertise from all participating national laboratories and partners. During WBS development, an agreement is negotiated, and concurrence is obtained on how the team will perform the hierarchical structure and associated tasks throughout the lifecycle of the prototype development and demonstration. The WBS defines all the constituent tasks to be performed in realizing the systems engineering of the prototype.

To illustrate the systems engineering principles, consider the overall approach to designing, integrating, and fielding a cyber defensive system for critical infrastructure. The team starts with an environmental scan of the site and performs a needs analysis against available technologies. The team must consider motivation, stakeholders, requirements, use cases, and the work breakdown structure (WBS) in the needs analysis and concept exploration. The main takeaway messages are:

1. The work breakdown structure (WBS) is essential in the systems development of a program
2. Functional allocation requires brainstorming with subject matter expertise (SME), which may be "part art and part science" (Ibid.)
3. System concepts are evaluated in several ways to determine their likelihood of failure or success in achieving the system's performance objectives

14.2 Reliability Analysis

"Reliability is the probability that a system or product will perform satisfactorily for a given period when used under specified operating conditions"

(Fabrycky 2011). One of the goals is to utilize commercial off-the-shelf (COTS) products. Customers demand the latest features and capabilities found in the latest commercial products. These products and systems must be designed for reliability from the outset. However, these products and systems are designed, validated, and tested in the commercial sector for environments that may support different use objectives. The reliability of the solution must be validated and tested for the new use case. The maintainability of the composed solution must also be validated and tested, and the team must identify the anticipated lifecycle length to mitigate risks. The lifecycle length of traditional security solution development can take as long as 3 to 5 years to develop, certify, and field. By the time these security solutions are fielded, they are outdated or obsolete technology, which poses a reliability and maintainability threat to the composed solution. These factors must be considered as part of the essential system parameters during the design process (Scalco and Wince 2012).

14.2.1 Design for reliability (DfR)

The reason to commit resources for deploying a design for reliability (DfR) process is straightforward: to decrease warranty costs, increase customer satisfaction, and enhance security. The product or system should perform as intended throughout its usable life. The intended use is essential for cybersecurity solutions. Failures in the field are incredibly costly, both in terms of repair and reputation and could yield security vulnerabilities. Clearly, to be profitable, an organization's products must be reliable, and reliable products require a formal reliability process. Three critical statements summarize the best practice reliability philosophy of successful companies:

1. Reliability must be designed into products and processes using the best available systems engineering-based methods.
2. Knowing how to calculate reliability is essential, but knowing how to achieve reliability is equally, if not more, important.
3. Reliability must be designed into the system from the outset, which means that the team must initiate reliability practices early in the design process and thoroughly integrate them into the overall product development cycle.

"Reliability is the probability that a system or product will perform in a satisfactory manner for a given period when used under specified operating conditions" (Fabrycky 2011). The rapidly changing technology must be considered a major system parameter when using commercial off-the-shelf (COTS) components in a composed solution. The information assurance (IA) data and cyber process control security performance expectations may change more quickly than the certification process, which must be monitored for changes to technical performance measures that could affect the reliability of a system. The successful "cycles" between a given period of performance may be shortened and require redesign, which must also be taken into consideration as a cost factor. While it may initially appear that a COTS solution is more cost-effective, considering design modifications, validation, and test procedures, the additional costs may make this option less appealing. Measures of reliability include the reliability function, failure rate, and other factors such as availability and effectiveness. The reliability function, also sometimes called the "survival function," is one minus the failure distribution function, or unreliability function: $R(t) = 1 - F(t)$, where R is reliability, t is time, and F is failure.

The basic definition of reliability is the probability of system life $\geq t$, where t is time. During the regular operation of a mature system, it is reasonable to assume that the failure rate is constant. However, the use of COTS elements in the composed solutions increases the failure rate when changes are made to software or hardware that might affect the system's security. Recall from Chapter 6, "Cyber Vulnerabilities Start at the Hardware Level," that software can be used to compromise different types of computer hardware. Some of the attacks overviewed are familiar to both ethical and unethical hackers worldwide. You have to assume that threat agents have already thought of the same attacks as you, so you must protect your network from those attacks. The predictability of these changes is part of the challenge in designing this system. Monitoring for changes to the system using behavioral analytics is one way to protect your network from those attacks. Key questions to ask include whether the critical assets of an organization are well-defined, whether the critical infrastructure supporting these assets is known, whether single points of failure have been identified, and whether the critical infrastructure supporting these single points of failure is known (Scalco and Simske, 2023). Additionally, ask if defensive countermeasures are in place to protect operations (Scalco and Simske, 2023).

The basic component relationship of the elements that are included in the composed solution also affects the reliability of the proposed solution. This composite failure rate is one of the factors that this system will enable engineers and program management to address more effectively. The reliability of a series of components is always less than the reliability of the individual components, without exception. The lower reliability of individual components means that the team must also consider the configuration of the composed solution in the design for reliability (D*f*R). The reliability of the composed solution can be increased by a greater redundancy of the components, such as found in a parallel system. The composed solution should consider that changes in one of the components' technologies may affect the information assurance (IA) and cyber process reliability of the system. A redundancy may offer a path to keep the solution fielded and secure if one of the potential failure points is closed. To do so requires that the upstream and downstream impacts are known.

These products and systems must be designed for reliability from the outset. However, these products and systems are designed, validated, and tested in the commercial sector for environments that may support different use objectives. The team must identify the desired top-level reliability for the total system early, and reliability must be allocated down to the lower levels of the system using a composed commercial solution. The composed commercial solution is considered a subsystem of the overall system. For example, if mission reliability must be 90% ($R = 0.90\%$) with four subsystems assuming that the reliabilities are in a series, then the formula to find the total system reliability is calculated by the product of the reliability of each of the subsystems, the composed commercial solution being one of the subsystems: $(1 - w_1[1 - R_1])\ (1 - w_2[1 - R_2])\ (1 - w_3[1 - R_3])\ (1 - w_4[1 - R_4])$.

14.2.2 Failure modes, effects, and criticality analysis (FMECA)

The reliability of the composed solution must be validated and tested for the new use case. The maintainability of the composed solution must also be validated and tested, and the lifecycle length must be identified to mitigate risks. The lifecycle length of traditional government IA data security solutions can take as long as three to five years to develop, certify, and field. Cyber process control systems, similarly, can take an equally long time. By the time these security solutions are fielded, they are outdated or obsolete technology, which poses a reliability and maintainability threat to the composed solution. These factors must be considered as part of the essential system parameters during the design process. Composed commercial solutions utilize state-of-the-art technologies, offering enhanced reliability to the overall system. Getting state-of-the-art technologies is a benefit of using commercial solution methodologies to augment traditional government Off-the-shelf (GOTS) options. When selecting a composed commercial solution, the technical team must evaluate the reliability of all components prior to design acceptance, with primary emphasis on this aspect. "This includes studying the operating features, tolerances, stresses, and other characteristics of the component(s) in its intended application" (Fabrycky 2011). Reliability analysis methods include failure modes, effects, and criticality analysis (FMECA) and fault tree analysis (FTA).

FMECA is used to identify potential system weaknesses and investigate product weaknesses, examining all ways or modes in which a failure might occur and the effects of those failures. Especially important are the single-mode failures. A single-mode failure is any singular item in a particular mode that, by itself, can cause the entire system to fail. The commercial elements of a composed solution may have a high-reliability factor in the intended commercial use case. Still, the reliability factor of those components may be degraded when used in a composed DOD environment. The component reliability factor may lower the overall system reliability and may introduce a single mode of failure. FMECA is a method to identify a single mode of failure that may otherwise be difficult to detect. Using FMECA involves assessing the severity and criticality of the failure, as well as localizing the failure modes and their effects. Severity deals with the consequences of the failure mode on the overall reliability of the system. Below is a short description for each FMECA column documented in Table 14.1 (Ibid.):

1. Item: Every component or procedural step is listed.
2. Function: The function of the item is briefly described.
3. Failure modes: Describes how the item might fail. Every possible failure type is listed.

Failure effects:

1. Local effects: This is typically limited to the effects on the item exhibiting the failure mode.
2. Next higher level: This highlights the effects on the item and its larger component.
3. End effects: This column describes the effect of an assumed failure on the entire system.

Failure-detection method, testability: Describes how an item will be recognized as failing. If an item fails in single mode, you want to be sure you can detect it.

Compensating provisions: This describes how to compensate for failure modes. The use of compensating provisions is a preparation for failure, so it

Table 14.1 Failure modes, effects, and criticality analysis (Fabrycky 2011).

			Failure effects						
Item	Function	Failure mode	Local effects	Next higher level	End effects	Failure detection method, testability	Compensating provisions	Severity class	Criticality

is readily known how to detect them and how to compensate for them. The interest is specifically in single-mode failures. A single-mode failure is any failure of an item in a particular mode that, by itself, can cause the entire system to fail.

Severity class:

1. Catastrophic: Irreversible or disastrous, enduring damage. Permanent loss of property or loss of life
2. Critical: Equipment sustained disastrous but fixable damage. Although injury may occur, it is fixable. No one was seriously injured
3. Major: Serious, but no physical damage to people
4. Minor: Marginal inconvenience
5. Trivial: Inconsequential fault causing nuisance

Criticality: Expanded on in Table 14.1.

Item: Carried over from Table 14.1.

Failure modes: Carried over from Table 14.1.

Item failure rate: Describes the failure rate for the item. The item failure can come from a vendor or historical data, for example.

Mode failure rate: This column assigns the failure rate of each item to its corresponding mode. For each particular mode, what is the rate at which it will fail in that mode? It is calculated by taking the overall failure rate (lambda) multiplied by the failure mode.

At-risk time: How long does this item need to operate in order to be successful?

Loss probability: What is the probability that the failure mode will, in fact, cause an undesired system effect? It is represented by the sign Beta.

Failure effects:

Local effects: This typically refers to the effects on the item exhibiting the specific failure mode.

Next higher level: This highlights the effects on the item and its larger component.

End effects: This column describes the effect of an assumed failure on the entire system.

The basic definition of reliability is the probability of system life $\geq t$, where t is time. During the normal operation of a mature system, it is a reasonable assumption that the failure rate is constant. However, the use of COTS, or internet-connected elements in the composed solutions, increases the failure rate when changes are made to software or hardware that might affect the system's security. The predictability of these changes is part of the challenge in designing this system. The basic component relationship of the elements that are included in the solution also affects the reliability of the proposed solution. This composite failure rate is one of the factors that this system will enable engineers and program management to address more effectively. The reliability of a series of components is always less than the reliability of the individual components, without exception. The lower reliability of components means that the technical team must also consider the configuration of the composed solution in the design for reliability (D*f*R). The reliability of the solution can be increased by increasing the redundancy of components, such as those found in a parallel system. The composed solution should consider that changes in one of the components' technologies may affect the information assurance (IA) reliability of the system. Still, one also has to consider the cyber process control security of a critical infrastructure system. A redundancy may offer a path to keep the solution fielded and secure if one of the potential failure points is closed.

14.2.3 Design for maintainability (DfM)

Design for maintainability (D*f*M) considers both design and maintainability concepts. First, design is the transformation of an idea into a product, process, or service that meets both the designer's requirements and end users' needs. Second, maintainability refers to the degree to which a design can be easily, economically, and efficiently maintained or repaired. Therefore, design for maintainability can be defined as a design strategy that involves collaboration between the designer and end-user, with the following objectives: to identify and prioritize maintenance requirements, increase product availability and decrease maintenance time, enhance customer satisfaction, reduce logistics burden, and decrease life cycle costs. The effectiveness of a design for maintainability strategy can be measured using maintenance metrics and industry benchmarks. In the context of the systems engineering challenge of developing a composed solution, D*f*M would include the entirety of the composed system. If one component of the system becomes outdated, is there a commercially available replacement component that can take its place? If a new software revision is released for a component of the whole system, how will this affect the security aspects (none, some, a lot)? What needs to be done to determine this effect? Will an engineering change proposal (ECP) analysis of the changed component determine what the change affects and allow for remediation if necessary (Scalco and Wince 2012)?

14.3 Functional Analysis

14.3.1 Requirements traceability

The rationale for each composed solution package is that for the most contentious requirements, the engineered solution is partway through the lifecycle, allowing a user to gain on project time, tough design decisions are tested, and money is saved. The user starts at integration. "Decisions on requirements and design are often based on hidden assumptions" (Ljerka Beus-Dukic, 2009). Traceability enables the project team to gather relevant information about a potential composed solution, based on a set of design elements, to meet the user's needs. The set of tests to verify the requirements is part of the functional analysis process for integrating the composed commercial solution.

14.3.2 Functional flow analysis

The top-level system functions for the methodology are depicted in Figure 14.2. The top-level function was generically named "Function SYS," as noted in the top left of the picture. The top-level system functions are engineering, configuration management, integration, program management, and accounting process reporting. The engineering process begins with configuration changes made to manage updates after the baseline is created. While the solutions are engineered, integration is occurring to ensure that the solution aligns with external systems. Finally, program management and accounting process reporting are taking place to handle the managerial aspects of the solution.

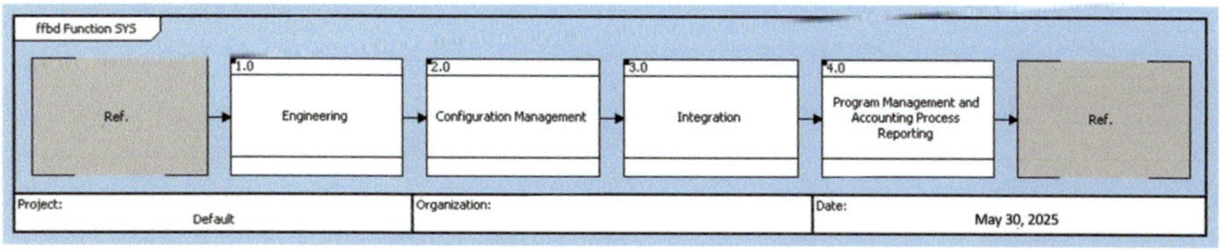

Figure 14.2 Systems engineering flow diagram.

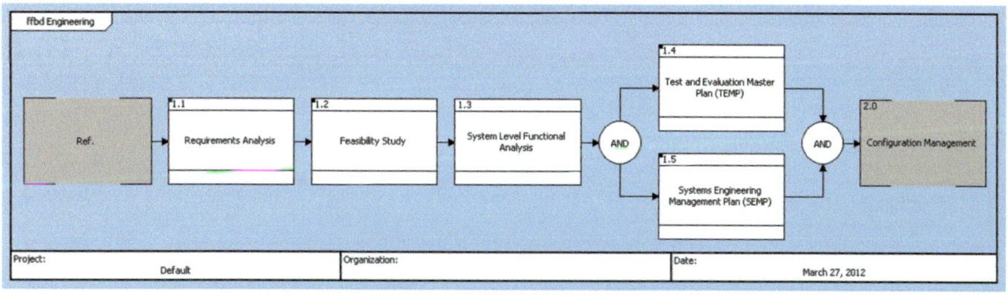

Figure 14.3 Functional flow analysis – engineering.

The top-level engineering system functions are decomposed as shown in Figure 14.3. The functional flow begins with the generation and analysis of requirements. Following this, a feasibility study is conducted to determine how realistic the requirements are. Next, a system-level functional analysis is conducted to decompose the system's functions into more manageable pieces and determine the functional, performance, and design requirements. Following the system-level functional analysis, both the test and evaluation master plan (TEMP) and systems engineering master plan (SEMP) are created. The testing process, along with the evaluation, is detailed in the test and evaluation master plan (TEMP). The systems engineering master plan (SEMP) focuses on the technical aspects of the project and the systems engineering processes to be employed in the solution.

The requirements analysis is based on the engineering functional flow analysis, as shown in Figure 14.3. Engineering functional flow analysis is further decomposed as shown in Figure 14.4. First, the scope is determined, followed by the purpose. Once the scope and purpose have been defined, the functional and non-functional requirements are concurrently created.

Figure 14.5 illuminates the functions within the feasibility study. The feasibility study shall, at a minimum, identify the time and resource costs, software impacts, organizational impacts, and security and privacy impacts. The four functions are conducted while performing the parent function, which is the feasibility study.

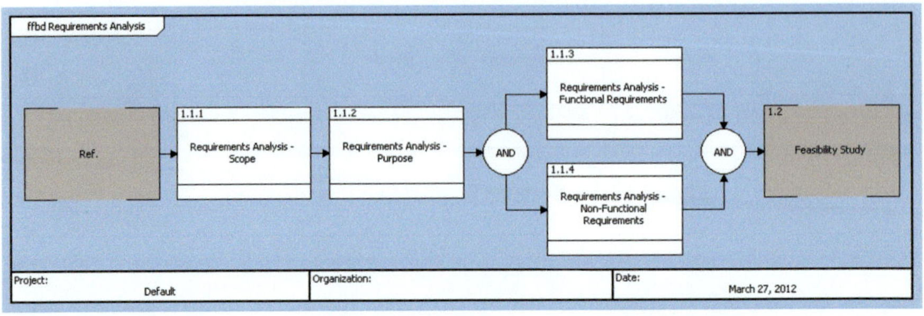

Figure 14.4 Functional flow analysis – requirements analysis.

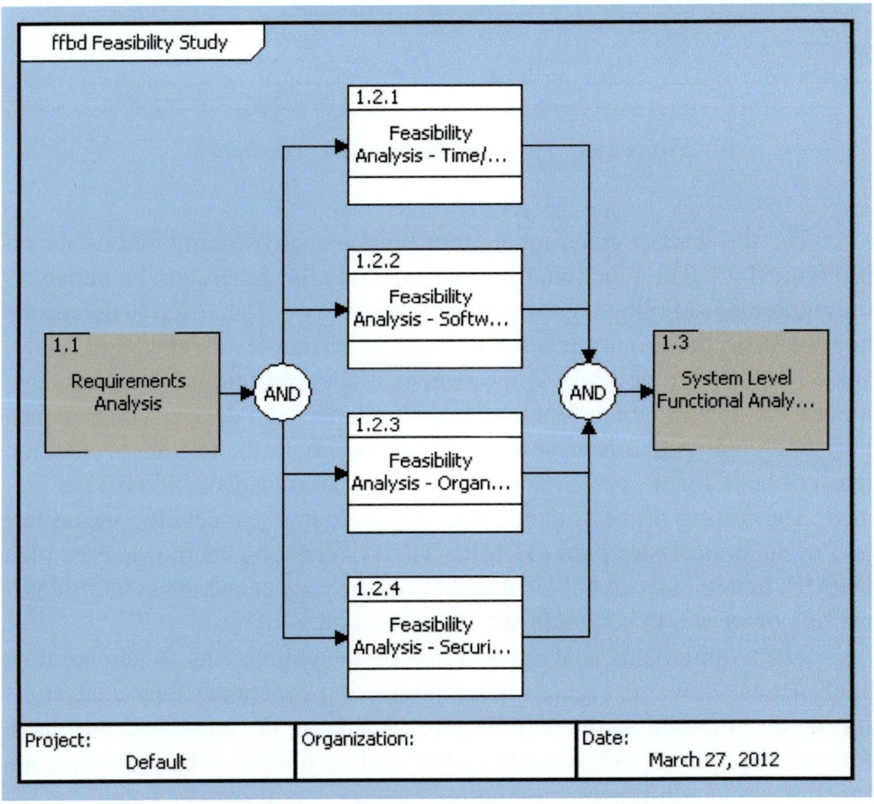

Figure 14.5 Functional flow analysis – feasibility study.

Each function associated with the test and evaluation master plan (TEMP) is shown in Figure 14.6. The scope of the TEMP is first determined, which sets the parameters for the entire test and evaluation process. Next, the content and methodology are created to determine what is being tested and

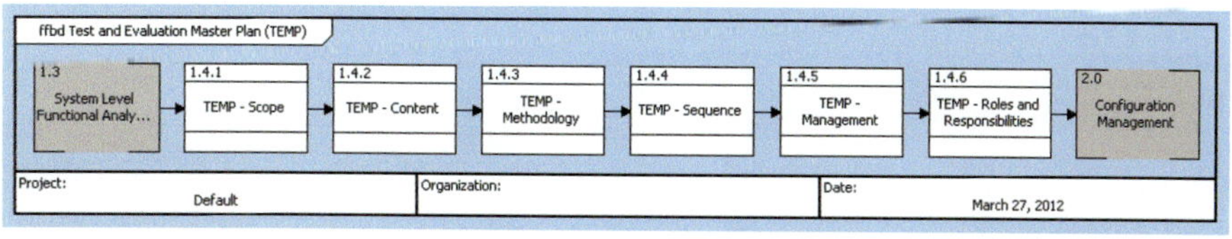

Figure 14.6 Functional flow analysis – test and evaluation master plan.

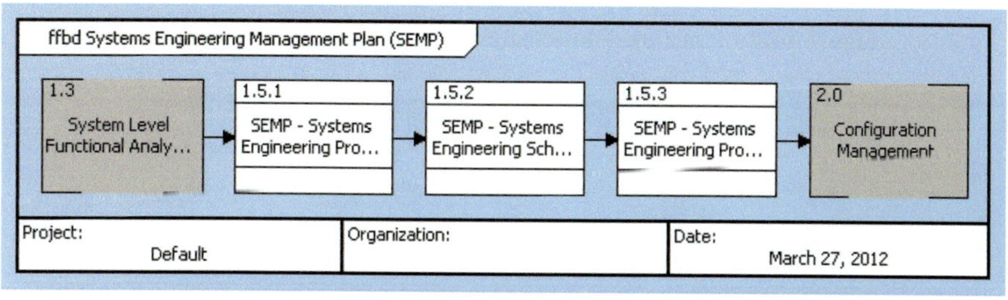

Figure 14.7 Functional flow analysis – systems engineering management plan.

how. Following the creation of the method, the sequence of testing is generated. Then, it is determined how the testing and evaluation will be managed, followed by documentation of the roles and responsibilities to ensure everyone knows what they are expected to do.

Figure 14.7 identifies the functions associated with the systems engineering management plan (SEMP). The SEMP requires creating a systems engineering process first, followed by a schedule depicting when specific systems engineering activities are expected to be completed. Finally, metrics are designed to measure the effectiveness of the systems engineering process.

Configuration management is decomposed as shown in Figure 14.8, from the top-level functional flow diagram. Configuration management deals with administering change, and the first step is to document the baseline. It is important to have a starting point that the system can revert to if needed. Next, updates to the baseline need to be captured and documented. Finally, a composed commercial solution reference library is created and maintained to contain the baseline and all the changes. A librarian is appointed to support the baseline and the composed commercial solution reference library.

The composed commercial solution reference library consists of four sequential functions, as illustrated in Figure 14.9. The first function is to document the configuration items and the required date. Next, a classification is assigned, followed by a brief description of the issue or problem that the solution resolves.

System integration is illustrated in Figure 14.10. The first function is end-user system integration. The solution must be able to integrate with the end-user systems, which is essential for its proper functioning. The second

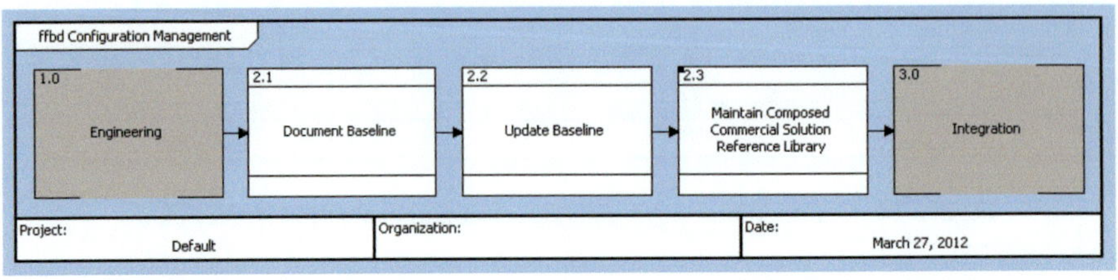

Figure 14.8 Functional flow analysis – configuration management.

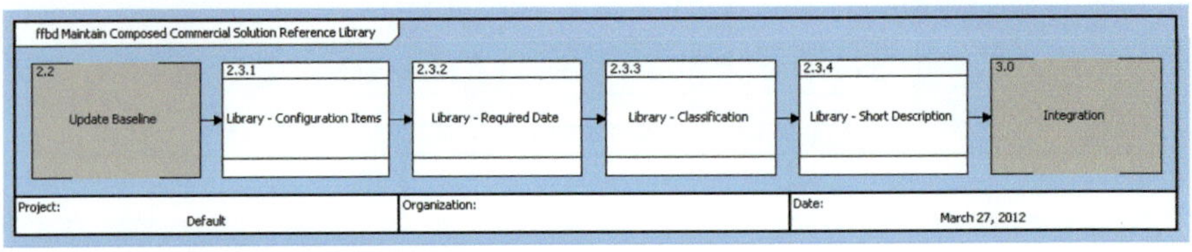

Figure 14.9 Functional flow analysis – commercial solution reference library.

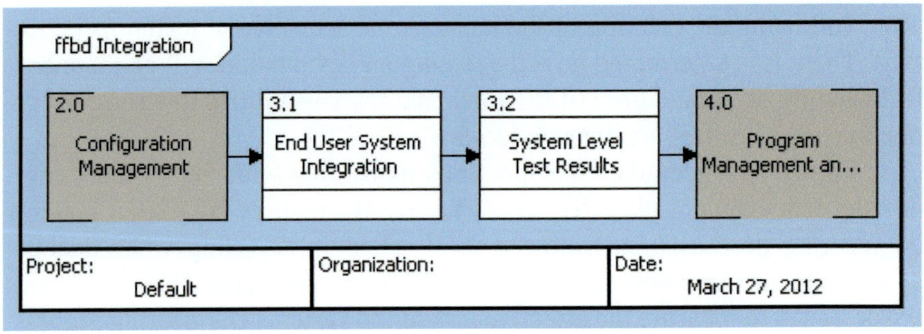

Figure 14.10 Functional flow analysis – integration.

function within integration is to document the system-level test results of the integrated security solution.

Figure 14.11 shows the decomposition of the top-level function into three lower-level functions that occur simultaneously. The systems engineering team leads the review meetings that must be conducted, coordinates design changes, and reviews system test and evaluation activities.

14.4 Summary

Systems engineering provides a standardized approach to common activities that guide system development and help understand the risks associated with design. Design criteria are the basic design objectives for the system and its

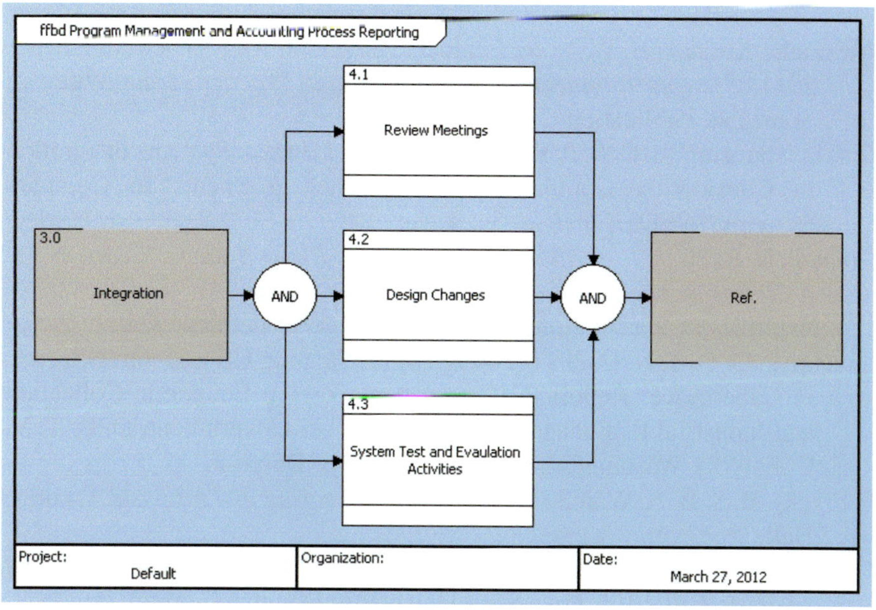

Figure 14.11 Functional flow analysis – program management and accounting process reporting.

elements that must be compatible with the system operations requirements, maintenance and support concept, and date prioritization of the applicable technical performance measure (TPM) that comply with the design-to criteria to "meet all the requirements in the various applicable specifications" (Fabrycky 2011). During the detailed design phase, the system begins to take shape in response to the needs analysis requirements.

References

(DOD), D. O. D. 2012. *Defense Acquisition Guidebook* [Online]. https://acc.dau.mil: Defense Acquisition University (DAU). Available: https://acc.dau.mil [Accessed].

(GAO), U. S. G. A. O. 2011. Defense Department Cyber Efforts: DOD Faces Challenges in Its Cyber Activities. *In:* (GAO), U. S. G. A. O. (ed.). Washington, D.C.

(USCYBERCOM), U.S.C.C. 2016. Advanced Cyber Industrial Control System Tactics, Techniques, and Procedures (ACI TTP) for Department of Defense (DoD) Industrial Control Systems (ICS). *In:* (DOD), D. O. D. (ed.). United States Cyber Command.

Alexander Kossiakoff, W. N. S., Samuel J. Seymour, And Steven M. Biemer 2011. *Systems Engineering Principles and Practice*, John Wiley & Sons, Inc. Publication.

Carl D. Thunberg, L. C. A. P. T. 2009. The Role of Human Systems Integration in Considering Commercial Off-the-Shelf Elements in Complex Systems. *INCOSE Insight,* 54, 4.

Council, N. R. 2008. *Pre-Milestone A and Early-Phase Systems Engineering: A Retrospective Review and Benefits for Future Air Force Systems Acquisition,* Washington, DC, The National Academies Press.

Executive, O. O. T. N. C. 2011. Foreign Spies Stealing US Economic Secrets in Cyberspace: Report to Congress on Foreign Economic Collection and Industrial Espionage, 2009-2011. *In:* US Government, O. O. T. N. C. E. (ed.). Washington, DC: Department of Defense.

Fabrycky, B. S. B. A. W. J. 2011. *Systems Engineering and Analysis*, Prentice Hall.

Gemmill, D. 2011. Lectures. *AerE/EE/IE 565 Systems Engineering and Analysis/Fall 2011.* Iowa State University: Douglas Gemmill.

Iii, W. J. L. 2011. The Pentagon's Cyberstrategy, One Year Later: Defending Against the Next Cyberattack. *Foreign Affairs.*

Kossiakoff, A., Seymour, S., Flanigan, D. & Biemer, S. 2020. The Future of Systems Engineering. *Systems Engineering Principles and Practice.*

Ljerka Beus-Dukic, I. F. A. 2009. *Discovering Requirements: How to Specify Products and Services.*

Marzolf, G. 2019. Engineering Lectures. Colorado State University (CSU): Colorado State University (CSU).

Pate-Cornell, E. M. 1994. Risk Management for the Tiles of the Space Shuttle. *Interfaces.*

Sandborn, P. & Singh, P. 2005. Forecasting Technology Insertion Concurrent with Design Refresh Planning for COTS-Based Electronic Systems. *CALCE, Department of Mechanical Engineering, University of Maryland.*

Scalco, A. & Simske, S. 2021. Cybersecurity Principles and Technology for a Power Utility Substation Automation System. *Journal of the Homeland Defense & Security Information Analysis Center (HDIAC) State of the Art Report (SOAR),* I and II.

Scalco, A. & Simske, S. 2023. A model for measuring multi-concern assurance of critical infrastructure control systems. *Systems Engineering*, 12.

Scalco, A. W. & Wince, K. 2012. A Proposed Methodology to Achieve Balanced Resilience and Risk. IE 566 Applied Systems Engineering: Iowa State University.

15

Design Tradeoffs

"A challenge for those who go. A dream for those who stay behind."
– Thierry Sabine (1949–1986),
Paris-Dakar Rally founder

Abstract

The Dakar Rally is one of the toughest in the world. Throughout the race, competitors and their vehicles must endure extreme conditions. A great deal of work goes into designing the race vehicles. Each vehicle must match a production model without any modification. Competitors have limited support and must be able to repair any mechanical issues, as well as replace major components, with minimal external influence. In preparing for the Dakar Rally, teams strive to keep things as simple as possible to minimize the likelihood of malfunctions. Route conditions change drastically from low-quality dirt roads to sand dunes. Drivers and vehicles must be prepared for all these conditions. While the Dakar Rally has strict rules in place regarding engines, teams can make design trade-offs in suspension, which translates into better road performance. Drivers can utilize feedback sensors and onboard electronic health checks to adjust to environmental conditions, such as remotely controlling tire pressure to respond to sand or rocky terrain. Balancing competing goals, and making choices about which goals to prioritize against competing requirements and constraints, is a design trade-off. Similarly, systems engineering for critical infrastructure in a cyber world is also dependent on design trade-offs. Improvement in one area can impact another. Low latency is essential for near-real-time systems found in industrial control systems (ICS), and high throughput is critical for data-intensive systems. ICS applications used in process controls must respond in near-real time to prevent equipment damage or ensure personnel safety. Industrial robotics requires low latency for precise movement. High latency can impact security features, making industrial control systems (ICS) vulnerable to cyber threats.

DOI: 10.1201/9788743809128-15

Choosing a network protocol for critical infrastructure involves numerous design trade-offs, such as balancing security features like encryption with robustness when considering the introduction of cybersecurity features like biometrics. Effective feedback loops and well-designed, well-executed configuration management (CM) are crucial when considering ICS, supervisory control and data acquisition (SCADA), and programmable logic controllers (PLCs), as well as industrial control communication protocols, control logic, sensors, and actuators used to control and monitor industrial operations in critical infrastructure.

15.1 Introduction

For a competitive Dakar Rally vehicle, subassemblies, suspension components, and bodywork are integrated into the frame to achieve the lowest possible center of gravity. The front and rear axles on cars are supported by independent front and rear suspension systems, which provide improved ride and handling in harsh environmental conditions. The same technologies are used in the production version of the cars. Design trade-offs are made to achieve optimal weight distribution of parts between the front and rear axles. The vehicle's technical aspects are designed with standard components, allowing them to be interchangeable with the corresponding front or rear axles. These components must be easy for the team to maintain and repair. Design trade-offs are made during the team's evaluation of approaches. Each option comes with a set of advantages and disadvantages, such as performance versus maintainability. Trade-offs are fundamental to system design. The greater the system complexity, the more challenging the iterative process of making design trade-off decisions is. There are many design trade-off decisions to be made between information technology business cost centers and operational technology, which are related to the physics-based profit centers. IT system characteristics are confidentiality, integrity, and availability of data. The National Institute of Standards and Technology (NIST) Special Publication (SP) 800-207 highlights that the objective of a Zero Trust security strategy is to "prevent unauthorized access to data and services coupled with making the access control enforcement as granular as possible" (Rose et al., 2020) (CISA, 2023). Operational technology (OT) controls physics-based systems, such as those regulating temperature, pressure, and flow. OT characteristics are safety, reliability, and functionality. Systems engineering for critical infrastructure must effectively address both when making design trade-offs in a cyber world. Modeling the system, identifying technical performance measures (TPMs), and creating robust feedback loops can help identify key considerations for infrastructure.

15.2 Modeling

Complex adaptive systems are a challenge to discuss without a model. While a particular model may represent conditions within one system, variables to user states may carry different meanings from one system to another. Several candidate approaches are used to address complexity, including seeking to understand the big picture, observing how elements within the system change, identifying the system's structure and its relationship to system behavior, and understanding test assumptions at various stages of design. The design process for the industrial control system (ICS) involves stages that follow the traditional systems engineering "Vee" model from requirement elicitation, component selection, schematic creation, build and test, integration, and system test, to maintenance. Systems engineering principles and practices are increasingly applicable in the design phase as more smart devices are connected to critical infrastructures, such as power, water, and facilities. Design trade-off decisions require industry-specific domain knowledge about industrial operation processes, instrumentation and control devices such as valves and pumps, industrial regulations and guidance, and critical infrastructure safety. Critical infrastructure safety considerations include personnel safety, critical lock-out/tag-out safety procedures that isolate machinery from all energy sources while someone is working on the equipment, safe work environments and processes, as well as safety and hazard considerations. The potential consequences of design trade-off decisions must be enduring throughout the facility's lifecycle, as changes and system upgrades in critical infrastructure do not occur as frequently as in business enterprise systems. Additional design considerations for critical infrastructure include cyber interfaces, software applications, system data, such as monitoring data flows of facility components, and governance, such as the ability to share threat information with external stakeholders.

Design trade-off decisions are made at these various design stages. The primary difference in the state of design of a system lies in the level of fidelity and detail achieved through a top-down systems engineering approach. During conceptual design, a needs analysis is performed to define the system's capabilities and effectiveness at the system level. The input to the conceptual design phase is the customer's needs. The output is the functional baseline, which is captured in the system Type A specification as a system entity. Preliminary top-level requirements are identified in the conceptual design phase (Fabrycky, 2011). In addition to understanding operational needs and control system performance metrics, potential hazards and safety requirements to prevent accidents, as well as any redundancy and safety protocols, must also be well understood.

These requirements are decomposed and expanded upon during the preliminary system design through functional analysis and the allocation of requirements to subsystems. The preliminary design process yields an allocated baseline and Type B-E specifications for the subsystems that require development. The requirements are further expanded and allocated during the preliminary design phase to define specific lower-level elements. The output of the preliminary system design is the allocated baseline. These are further defined and allocated into a detailed design. Modeling and simulation of the process and control system are used to validate the design. The output of the detailed design is the product baseline and material Type C specifications for subsystems that can be procured. The output of the production and development phase is the updated product baseline. The system specifications are used to select the appropriate sensors to monitor processes and actuators used to manipulate processes.

These system stages for the developmental lifecycle apply to the acquisition of all systems following a top-down/bottom-up, iterative systems engineering approach. The top-down approach defines requirements and decomposes and refines the system definition, while the bottom-up approach verifies and validates the system's integration and testing. The system must be thoroughly tested to ensure that the control system functions correctly, and the instruments and controllers must be calibrated to enable operators to effectively monitor and control the industrial control system.

Mock-ups, models, and prototypes support the visualization of the system, aiding the designer. "As the system design and development effort progresses, the basic process evolves from a description of the design in the form of drawings, documentation, and databases to the construction of a physical model or mock-up to the construction of an engineering or laboratory model, to the construction of a prototype, and ultimately to the production of a final product" (Fabrycky, 2011).

Some of the differences include the level of investment and time required to create them, their purposes, the level of detail, and the speed at which they can be produced or realized, as well as the ability to interact with the system, which varies between them. Mock-ups are "three-dimensional scale models constructed to provide a realistic simulation of a proposed system configuration" (Ibid.). These models can be developed to any scale and constructed from various material types. An engineering model, or a laboratory model, demonstrates the system's functions using an operational model in a research environment, such as a laboratory or an equivalent setting. A prototype can be developed as an alternative to a mock-up to provide a solid basis for design evaluation or validation of design trade-off decisions. The prototype model represents the production configuration of a system and all its elements. The exception is that the prototype has not been fully qualified for use in an operational or environmental test. As the system design progresses, tests and evaluations become more impactful to the design trade-off decision process due to the availability of actual hardware, software, and other system elements in the physical models. A model is a valuable tool in systems engineering, helping to conceptualize and understand complex ideas, processes, and protocols that are difficult to observe independently of their context. Human factors from behavioral science can add to the complexity of a system observed as a complex system, characterized by multiple stable, transient, and continuous states, or no lasting states. For example, for critical infrastructure systems, engineering models are demonstrated in laboratory environments, which allows engineers to consider the availability of test points for fault isolation, maintenance, environmental provisions, and opportunities for future growth.

Prototyping of actual hardware and software is integrated into the system for laboratory functional technical validation and verification. A second field demonstration involves operators facilitating the verification and validation of system design. During test planning and preparation, the prototype becomes real, and interface issues are resolved. Deviations from expected test results can result from deficiencies in the equipment, procedures, execution, analysis, the system under test, or excessively stringent requirements. Dealing with a test failure requires tracing it to understand the cause so that the technical team can take corrective action. Steps taken prior to, during, and after a test contribute to the diagnosis of a test failure. Before the final operational test, the systems engineering team locks down a final prototype baseline, and no further late injection of technologies will be introduced into the scenario environment.

The recommended solution architecture, given a scenario of an environment susceptible to unpredictable, high-consequence events of recursive complexity, is to "design for resilience" to "provide robustness and timely recovery to a minimally functional state" (INCOSE, 2015). An example of the "design for resilience" principle can be found in the integrated adaptive cyber defense (IACD) component of the more situational awareness for industrial control systems (MOSAICS) (APL, 2019). IACD is an extensible and adaptive framework that enhances the effectiveness of system defenses in mitigating cyber vulnerabilities. While the framework was initially created to address IT environments, it was subsequently applied to an operational technology (OT) critical infrastructure environment, utilizing systems engineering principles to enhance its effectiveness. The assumption is that if the approach is applicable to IT complexity, then it would also be relevant to OT environments. Monitoring and detection requirements include baseline comparison requirements to detect changes in the baseline inventory of components, monitoring requirements for data flows, and detection of component behavior inconsistent with normal operations. Additionally, security requirements include monitoring signatures for indications of malware and detecting violations of access rules and policies for components (Vermilye, 2021).

Systems engineering reuse has benefits in installing, maintaining, and upgrading the system throughout its lifecycle. Systems engineering principles provide a framework for integrating cyber processes and continuously monitoring the performance of the control system. Extended technology refresh cycles of 20 to 30 years or more undermine the ability to address vulnerabilities. Due to the extended refresh cycle, cybersecurity enhancements will need to meet requirements that exceed those expected from similar IT systems, as there is no existing predecessor system for OT. The extended refresh cycle highlights the use of older technologies designed for functionality requirements rather than cybersecurity requirements in OT. The convergence of OT and IT makes cyber-physical systems equally susceptible to cyberattacks, yet OT is contextually and dynamically distinct from IT. Component expertise encompasses various technological approaches to respective domains and diverse behavioral approaches and responses, particularly to human-introduced cyber vulnerabilities (HICV), which are discussed in more detail in Chapter 22, "Managing Risk."

15.3 Trade-off Considerations

Biometrics utilizes individual physical characteristics, such as fingerprints, to validate identity. Design trade-offs include lifecycle properties such as safety, reliability, producibility, and affordability. Biometrics is discussed in detail in Chapter 19, "Biometrics." Two design trade-off questions for the use of biometrics are how to group tasks into modules and how to design interfaces in the physical architecture. For example, a vital trade-off that must be considered when grouping tasks into modules is the distribution of computational tasks across different processes to enable parallel computation, as well as the increased latency resulting from data transfer between processors. Another trade-off is the cost considerations against the safety-relevant metric, or the cost-latency trade-off (Anne Collin, 2019). If all computing functions are conducted on a single processor, the cost is higher. A trade-off between cost and latency is critical.

Human factors are considered in the design of autonomous driving systems, as they perform an extensive list of tasks on behalf of humans, which naturally raises safety concerns. Each task is assigned an automotive safety integrity level (ASIL). Human factor considerations are incorporated into the details of the algorithms that underpin the functions themselves. "Increased safety can also take the form of enhanced communication between the car and the driver" (Anne Collin, 2019).

Latency is a crucial factor that is reflected in sensor architecture and sensor performance, particularly in terms of human factors such as safety and reliability. It is the maximum path length in the functional network. The ultimate goal is to minimize latency while optimizing the performance function of the environment in terms of safety and reliability. This design consideration is essential: "[I]n order to quantify the cost-latency trade-off, the optimization problem is formulated as a linear, mixed integer programming (MIP), multi-objective optimization problem, with constraints on the computational capacity of the processors" (Ibid.).

15.4 Technical Performance Measures (TPMs)

Reliability predictions in the detailed design may vary among subject matter experts (SMEs). For example, a component engineer may have a different technical personnel skill level than a cost engineer, just as a logistical engineer may have different knowledge, skills, and abilities to diagnose a system than a reliability engineer. One example of the importance of utilizing these differing SMEs effectively is the comparison between software-defined networking (SDN), which is implemented in software called a flow controller, and the traditional network control plane situated in the network appliance hardware (e.g., switches). SDN technology offers greater capabilities. However, while it is rapidly emerging in use in enterprise networks, it is still in development, expensive, and may pose security risks that are less prevalent in hardware with built-in firewalls. The comparison is further examined in the economic analysis and decision-making, which may counter initial objections based solely on price and stability attributes. Subject matter expertise (SME) for those involved in lifecycle operations and maintenance needs to be considered, just as the driver in the Dakar Rally needs to be taken into account. Who will fix the system when it fails? Design trade-offs have to be made for failure rate; mean-time-between-failure (MTBF) reliability measures for repairable items; mean-time-between-maintenance (MTBM) for repairable fielded systems; maintenance labor hours per hour, per cycle, or per action; mean-downtime (Mdt) when the system is unavailable for use due to a failure; operational availability (Ao); and operational readiness.

It is essential to assess the lifecycle cost of producing any unique product to ensure that the product's ultimate requirements will be met at or before the critical design review, including maintenance and future upgrades to the control system. Unique products may incur high acquisition, sustainment, maintenance, and component replacement costs to support systems in varying configurations. Assess the critical useful life of these items to inform maintenance and component replacement plans, maintain configuration control throughout the process, and effectively implement the design process. Design review, evaluation, and feedback are practical steps and systems engineering best practices for a continuous, ongoing, and interactive process of reviewing and evaluating day-to-day activities. Formal design reviews at specific gateways for design and system acquisition decisions are necessary to realize the applicable technical performance measure (TPM) of life-cycle cost. For example, considerations before the critical design review for systems developed for the Department of Defense (DOD) typically include either government-off-the-shelf (GOTS) products, which are created exclusively for the purpose, or commercial-off-the-shelf (COTS) products, which are commercially developed and produced. GOTS are custom-developed, which can add to the total lifecycle cost of the system.

A better option is to look for available, high technology readiness level (TRL) commercial off-the-shelf (COTS) products and down-select an appropriate commercially available product to meet the requirement specifications. Another step is to "maintain some form of configuration control (i.e., "change" control) in the process" (Fabrycky, 2011). Adherence to the baseline management can aid in maintaining the system design and acquisition process. Configuration management (CM) contributes to the integrity of the system design by supporting the vital system development baselines, which include the functional baseline, the allocated baseline, and the product baseline – all of which are essential elements throughout the system's lifecycle. "Formal change control of system-level changes is usually exercised by a designated group composed of senior engineers with recognized technical and management expertise capable of making judgments among performance, cost, and schedule" (Kossiakoff, 2011). The goal of integration is to engineer the new system into a cohesive and compelling operating whole that meets the desired objectives.

15.5 Feedback Loop

The test plan serves as a valuable reference throughout the system design, engineering, development, and production. Feedback capability is essential for recording and communicating what has been learned about the system to the design and engineering team, allowing for corrections as early in the lifecycle as possible. The correct information must be provided to the responsible engineering and management personnel. The objective is to conduct a thorough assessment of the system's performance in a user environment. Good data collection, analysis, reporting, and feedback capabilities enable the team to manage builds more effectively and deliver innovative solutions. Maintaining accurate data allows the team to compare new data with the established baseline, facilitating informed decision-making. Although some variance may be expected, conditions must be such that effective results may be achieved. The fulfillment of test objectives involves a formal, approved procedure, typically documented in manuals. Throughout the testing process, data is collected and analyzed, resulting in an assessment of the system's performance and effectiveness against the specified requirements (Fabrycky, 2011). Questions answered by good data collection and analysis are: "What is the 'true' performance and effectiveness of the system operating with all of its interfaces? What is the 'true' performance and effectiveness of the logistics and maintenance support infrastructure? Are all the initially specified TPM and related requirements being met? Is the system compatible with the other systems operating within the same SOS structure?" (Ibid.). It is essential to capture this information, as the results may indicate deficiencies that require correction in the software or another module, which may necessitate a design change.

Configuration management (CM) provides traceability throughout the system lifecycle. As such, CM is important in the implementation of the "system test, evaluation, and validation" process, as CM artifacts make changes visible to stakeholders. A baseline enables parallel development for version control, making it easier to test and create new versions. Proper CM enables designers and engineers to verify which components pass or fail test events, allowing for effective changes and corrections. For example, during unit testing, system configuration control, status accounting, and configuration audits allow for easier tracking of the tested items. Results from tests are also considered configuration items, which enable the determination of whether system requirements have been met through traceability in configuration management (CM).

The technology readiness level (TRL) is a method for estimating the maturity of essential technology elements in DoD acquisition, enabling the acquisition of sufficient technical knowledge to develop the system. SDN still needs to be demonstrated in a relevant environment. One approach to achieving this is to create a technology development strategy (TDS) that incorporates software-defined networking (SDN) into the baseline system. Categories of tests and their applications are integrated into the evaluation process of the system throughout its life cycle, iteratively and recursively. Basic categories and the associated lifecycle phase are: analytical test in the conceptual and preliminary design phases for system evaluation using design workstations and analytical models (e.g., computer-aided design (CAD), computer-aided engineering (CAE), computer-aided manufacturing (CAM), computer-aided styling (CAS)); Type 1 testing in the preliminary and detail design and development phases for evaluation of engineering and service test models, system components, breadboards, mock-ups, and rapid prototyping; Type 2 testing in the detail design and development lifecycle phase for evaluation of prototype and production models (e.g., production sampling); Type 3 testing in the production phase, which are evaluated at the test site(s); and Type 4 testing during the system utilization and life-cycle support phase for continuous evaluation of the system in operational use (Fabrycky, 2011). CAD is an essential tool for engineers that allows them to visualize the shape, dimensions, and physical characteristics of objects. CAM is used in the translation of software that controls and automates manufacturing processes into machine-readable instructions, ensuring the high precision and repeatability required for processing and producing exact parts that meet specifications. CAE is used to analyze and simulate the performance of systems and products. Integrating these tools into an interconnected workflow is valuable for system and product type testing, and crucial in the design phase, as they provide useful feedback to generate enhancements.

15.6 Summary

Modeling systems, feedback loops, and Innovative technologies, such as software-defined networking (SDN), offer solutions to update critical infrastructure in the context of a cyber world problem statement. Including cybersecurity in the model is akin to incorporating SECOPS (secure concept of operations) into the CONOPS (concept of operations) of a system; that is, it is part of the detailed design from the start. Just as the Dakar Rally can be terrifying for competitors who find themselves in extreme, unanticipated conditions such as sandstorms and other desert conditions that defy human understanding, cyber innovation can also put any organization off course. The Dakar Rally is a living endurance laboratory for manufacturers. Design trade-offs are decisions made when one feature must be sacrificed to achieve a desired outcome. A less expensive solution might introduce increased latency, which is not desirable in critical infrastructure. Business goals and priorities may introduce cyber vulnerabilities to the system, where biometric authentication might be a more effective design. Low latency is crucial for critical infrastructure to ensure timely response and maintain operations. Just as the Paris-Dakar rally is a competition of the extremes, famous for its off-road endurance and rugged terrain, critical infrastructure is a field where personnel often work in harsh environmental conditions. While anyone aged 18 years or older who holds a Fédération Internationale de l'Automobile (FIA) Rally license can enter the Paris-Dakar Rally, the vehicles raced must be approved by an accredited mass-market vehicle class in accordance with FIA standards. The scalability of rigidized vehicles and experienced operators is a challenge. Scalability in growing network traffic to monitor and control critical infrastructure versus security features is part of the design trade-offs that need consideration in modernization strategies, as discussed in Chapter 16, "Modernization Strategies."

References

Anne Collin, A. S., Yuto Imanishi, Eric Rebentisch, Taisetsu Tanimichi, Olivier L. De Weck 2019. Autonomous driving systems hardware and software architecture exploration: optimizing latency and cost under safety constraints. *Systems Engineering*.

(APL), J. H. U. A. P. L. J. 2019. *Integrated Adaptive Cyber Defense (IACD)* [Online]. https://www.iacdautomate.org: Johns Hopkins University Applied Physics Laboratory (JHU APL). [Accessed December 92019].

(CISA), C. A. I. S. A. 2023. Zero Trust Maturity Model Version 2.0. *In:* Division, C. (ed.). Cybersecurity and Infrastructure Security Agency (CISA).

Fabrycky, B. S. B. A. W. J. 2011. *Systems Engineering and Analysis*, Prentice Hall.

Rose, S., Borchert, O., Mitchell, S. & Connelly, S. 2020. Zero Trust Architecture. *NIST Special Publication 800–207*, 59.

Vermilye, C. More Situational Awareness for Industrial Control Systems (MOSAICS) Requirements. MOSAICS Industry Day #3/TechConnect, October 19, 2021 2021 Washington, D.C.: MOSAICS JCTD, 13.

16

Modernization Strategies

"I try to make it look easy, but the behind-the-scenes stuff is the challenge."

– Steph Curry (1988–),
American professional basketball player

Abstract

Modernization strategies that utilize software-based technology and architectural approaches, such as operational technology software-defined networking (OT-SDN), software-defined perimeter (SDP), machine learning (ML), and the promise of artificial intelligence (AI) technologies, are transforming the nature and art of perimeter security. These modernization strategies can help an organization improve efficiencies, just like the three-point shot changed the game of basketball. The addition of the three-point perimeter made the game more exciting. A good "three-point shooter" can force the defense to extend further out to the court perimeters, opening up the middle court for offensive players to attack the net. You do not want an adversary inside the middle court. You do not want an adversary inside your cyberspace either. New and evolving technologies enable Internet of Things (IoT) designers to use the Internet for many impactful applications and services.

The latest technologies, including autonomous systems, robotics, advanced communication, machine learning (ML), and artificial intelligence (AI), can bring about enhanced efficiency and connectivity. These technologies provide the benefits of easier, centralized remote management of networks from a single centralized dashboard. However, depending on the deployment, these technologies can also expose the system to various network layer attacks, such as the possibility of a compromised software-defined networking (SDN) controller at the control plane layer. Adhering to

DOI: 10.1201/9788743809128-16

the principles of Zero Trust[1], you do not want an adversary inside the network. SDN is a disruptive technology that separates network control from physical infrastructure, from a centralized view and control. The technology enables efficient policy enforcement. An organization's defensive teams also need to learn the mechanisms for specific configurations of the OT-SDN to strengthen the security posture and response capabilities against evolving cyber threats. Software-defined perimeter (SDP) is another software-based technology that creates secure network connections for users from any location. Software-defined perimeter (SDP) is a security architecture that is useful for remote teams that do not require access to a physical network perimeter. SDP utilizes software for access control based on the identity and context of the device, as well as access to specific resources, which reduces the potential attack surface by limiting access. While virtual private networks (VPNs) enable all users to access the entire network, software-defined perimeters (SDPs) can be used to manage specific access levels for individual users. VPNs establish inbound connections. SDPs respond to outbound-only connections, or "tunneling," that keep the SDP blocked from the Internet. SDP implements Zero Trust security principles, which, when combined with defense-in-depth (DiD) and a cyber defensive framework, such as more situational awareness for industrial control systems (MOSAICS), provides a comprehensive security approach for critical infrastructure. These strategies are particularly effective when introducing new technologies that can expose vulnerabilities to adversaries.

16.1 Introduction

The game of basketball has not always followed the same rules. Neither has the internet. The game has changed, and the rules of the road for assets connected to the Internet have changed. For basketball, the game evolved with the introduction of a new three-point basket in 1979. A three-point shot is any basket made beyond the three-point perimeter line. The three-point perimeter is the area outside an arch line at a set radius measured below the basket's center and two parallel lines on either side of the court. The distance varies somewhat at different game levels, but the principle is the same. It adds excitement by augmenting the attack surface on the periphery of the

[1] Zero Trust is sometimes referred to as "ZT." In engineering "ZT" has a specific meaning that refers to a dimensionless value to quantify efficiency of thermoelectric materials. As this book addresses critical infrastructure, the authors do not use the abbreviation to avoid confusion of terminology used between the IT and operational technology (OT) communities

defending team's defense. The Boston Celtics' Chris Ford made that first historical shot, but Larry Bird of the Boston Celtics was one of the first to truly weaponize it. Larry Bird was not initially a fan of the change, but became one of the greatest three-point shooters, helping to change how the game was played. He said of the three-point shot in his biography: "The first year I came into the NBA, the league instituted the three-point shot, but at first, it meant nothing to me. I never thought I'd be shooting one" (Larry Bird, 1990). For Larry Bird, the distance for the shot just looked too far from the basket, and he thought any two-point lead at the end of a game was enough to make any gain. The new rule changed the game, however, because it augmented the danger of peripheral system penetration by 50 percent (from two points to three points).

Similarly, ignoring cybersecurity makes it easier for hackers and puts the system at risk. Hackers look for easy targets to make significant gains, seeking the high-scoring shots that can help them score a win. Dynamic segmentation in response to changing threats enables the system operator to have greater control and achieve greater flexibility when utilizing a software-defined security architecture. Identity-based access and security at the layer of connectivity limit the potential for unauthorized access and prevent lateral movement across the network. Many cybersecurity issues, in fact, stem from human error. Simple examples are an employee clicking on a phishing link or using shared passwords. If you make a mistake in basketball, you give the opponent a free throw shot. Players can score one point for each free throw made. A free throw can be considered the equivalent of forgetting to install critical security updates. You have just given an adversary a free shot at your network, your data, or even your business. The three-point player is more efficient at scoring points and can do more damage with a single-made basket. Again, basketball today looks very different than when the three-point perimeter was introduced. Offensive and defensive strategies and plays of the game all changed. The perimeter of the offensive zone, where the three-pointers are made, is suddenly a greater risk than before the three-point shot. Moreover, in some cases, forcing defenses to employ a perimeter-based protective strategy, the three-point line can open defenses to additional threats inside the periphery. By limiting access to authorized users and resources at specific identified times and enforcing access policies, the potential for a threat actor or human-induced error to breach the system is limited.

Returning to cybersecurity, since a software-defined perimeter (SDP) authenticates both a user and their access device, the technology can be used to limit access to protected resources. Machine learning (ML) and artificial intelligence (AI) technologies also have enormous potential to change the perimeter game in cyber. In the past, systems were protected using perimeter security tools, such as firewalls, to control incoming data and block outside traffic from accessing the system. The evolving cyber-attack tactics, techniques, and procedures (TTPs), or "playbooks," changed how nefarious actors gain unauthorized access. The older perimeter approach, which uses firewalls to shut the doors to gateways that threat actors use, has changed to models that presume threat actors are already inside the network. ML and AI algorithms can detect behavioral activity and identify suspicious behavior, such as lateral movement from a user's desktop computer entry point to escalated access privileges. Threat actors use lateral movement to expand access from a compromised asset within an organization's network into deeper within the system.

Collecting and correlating data from various tools provides a comprehensive view of the organization's security posture, enabling faster detection and response. Systems engineers need to bring together subject matter expertise from all competencies who know the systems and processes involved, because active monitoring of industrial control systems (ICS) can potentially lead to compromise of the critical infrastructure's normal operations. This sensitivity to active monitoring can make the implementation of modernization strategies problematic. For example, active scanning of ICS can result in a system crash of critical operations that can be a costly error.

16.2 Zero Trust Strategy

The concept of the Zero Trust security strategy extends cybersecurity beyond the perimeter-based security model to make security ubiquitous throughout the native environment, not just at its gates. A Zero Trust security approach aims to minimize the time to respond to system attacks by limiting lateral movement across the network. Zero Trust also restricts unauthorized, unauthenticated access to networks, applications, and data. The Zero Trust model assumes that potential attackers are present both internally and externally in an organizational network. Organizations employing the Zero Trust model approach to security utilize multi-factor authentication, least-privilege access, near real-time monitoring, and a range of endpoint monitoring, detection, and response capabilities as part of their defense strategies.

As highlighted in Chapter 1, "Understanding the Game," and in Chapter 9, "Road to Commoditization," Zero Trust was first formally introduced by Forrester Research Analyst John Kindervag in a paper published in November 2012, "Build Security into Your Network's DNA: The Zero Trust Network Architecture" (John Kindervag, 2012). This concept was further elaborated in the Forrester paper, "No More Chewy Centers: Introducing the Zero Trust Model of Information Security," followed by articles about simplifying cybersecurity and creating a data security and control framework for dissecting and analyzing data (John Kindervag, 2016).

Zero Trust is used to enhance security in complex networks by limiting user or device access, regardless of the network's location or status. The principles of Zero Trust are:

- Assume a hostile environment – All network traffic is untrusted regardless of location, and all traffic is logged and inspected
- Assume breach – Presume a least privilege posture
- Least privilege access – Practice strict access control within the organization
- Persistent engagement – Collaboration is essential and is practiced continuously for improvement across all capability areas
- Multi-factor authentication with contextual assessment – Authorization beyond assurance of a user's identity to include insights into contextual information such as login location and known device to validate access attempts (DISA and Team, 2022).

The Zero Trust pillars and capabilities include user, device, workloads, data, network/environment, automation/orchestration, and visibility/analytics (DISA, 2022). The user pillar feeds and supports visibility and analytics. It includes authentication, authorization, and privilege access management (PAM). The device pillar provides and supports visibility and analytics, as well as automation and orchestration. It includes authentication, authorization, and compliance. The network/environment pillar also feeds and supports visibility and analytics, automation, and orchestration. It includes software-defined data center (SDDC), macro segmentation, application delivery, and micro-segmentation. The application and workload pillar provides feeds and support for visibility, analytics, automation, and orchestration. It encompasses computing, application development, supply chain management, application delivery, and micro-segmentation. The data pillar also feeds and supports visibility, analytics, automation, and orchestration. It includes storage, data tagging, data loss prevention (DLP), data rights management (DRM), and machine learning (ML) (DISA and Team, 2022).

The visibility/analytics pillar feeds and supports automation and orchestration. It includes accounting and analysis, discovery and baselining, security information and event management (SIEM), machine learning (ML), and artificial intelligence (AI). Security information and event management (SIEM) allows organizations to detect, analyze, and respond to security threats by correlating near-real-time and historical security-related data from various collection sources for analysis. The automation and orchestration pillar also feeds and supports visibility and analytics. It includes application programming interface (API) standards, incident response, security orchestration, automation, and response (SOAR), as well as artificial intelligence (AI). The API is the set of rules that enable software applications to exchange data. The open application programming interface (Open API) is publicly accessible, allowing developers to interact without direct access to the application's source code, thereby enabling interoperability. Security orchestration, automation, and response (SOAR) automates and orchestrates security operations and threat detection by integrating various tools and tasks to help manage security incidents in an automated way. Encryption is used across all the Zero Trust pillars and capabilities. Recall from Chapter 13, "Needs Analysis," that encryption does not prevent a successful cyberattack on a system because there are other paths to take control of an industrial control system (ICS) and to get access to the programmable logic controllers (PLC) that can be used to damage equipment. However, encryption does protect sensitive information and is a fundamental and essential layer of security.

Zero Trust maturity stages can be categorized into five stages, ranging from pre-Zero Trust to basic, intermediate, advanced, and ongoing Zero Trust Improvement across the Zero Trust pillars and capabilities: user, device, workloads, data, network/environment, automation/orchestration, and visibility/analytics (Dransfield, 2022).

Asset owners and operators face added challenges in realizing Zero Trust in industrial control systems (ICS) found in critical infrastructure: each system is unique, often relatively old, and almost always lacks upgrade resources. Dealing with aging systems is where the modernization strategy and the more situational awareness for industrial control systems (MOSAICS) framework come into play. A more holistic approach is needed: one that protects critical infrastructure from various cyber threats, both internal and external, from tampering and unauthorized access to data.

16.3 Layered Defense Solution using a DID/Zero Trust Architecture

Protecting critical infrastructure assets requires a holistic defense-in-depth (DiD) approach that uses multiple layers of defense to address various threats. An example of an overall layered defense solution strategy is to leverage the segmentation of a Purdue reference model complemented by principles of defense-in-depth (DiD) and a Zero Trust security network architecture to isolate functions throughout the factory into distinct network zones using segmentation. The goal of a layered defense structure is to achieve resilience. Resilience is the ability to operate securely in the face of network compromise. DiD layers and security controls are located at the perimeter security layer. Controls implemented include application gateways, firewalls, and secure network demilitarized zones (DMZs), utilizing the same techniques employed in the network security layer. The DMZ serves as a buffer between an internal network and external interfaces. The network security layer enclave partitioning security controls are built on a segment of the network defined by collective security, change, and service policies. The endpoint security layer controls and implements content security, including antivirus and anti-malware, as well as patch management. The application security layer controls are designed to protect externally facing interfaces, such as web application firewalls and automated patch management systems. The data security layer controls implemented data classification and data/drive encryption, and the mission-critical asset layer security controls implemented are IT security governance, security policies, and compliance. Specific security controls and procedures to be implemented at various points in the security solution using both passive observation scanning of network behavior and safe active network monitoring to protect against cyber threats and unauthorized access include:

- Policy/governance – Principle of defense-in-depth (DiD) and Zero Trust security network architecture, attribute-based access control (ABAC) with multi-factor authentication, strong passwords, insider threat detection tools and procedures, secure personnel acquisition and management practices, including background checks for job applicants, employee training to maintain security practices.
- Perimeter/boundary security – Layered physical security barriers for facility access control and protection of sensitive resources, boundary router with secure routing rules; demilitarized zone (DMZ) with proxy servers for applications, email, web applications, and others; intrusion detection/protection system (IPDS); security Information and event management (SIEM); data loss prevention (DLP); VPN for external communications; and a Honeynet.
- Network layer – Network firewall, data loss prevention (DLP), network IPDS, enclave, unified threat management (UTM) firewalls.

- Endpoint protection – Desktop firewalls, endpoint data loss prevention (DLP), endpoint intrusion detection and location-aware prevention (IDLP), Intrusion detection and prevention system (IDPS), automated patch management.
- Application layer – System administration tools and procedures to detect and mitigate vulnerabilities, database monitoring, database secure gateway, application DLP, DPI/SML firewall, extensive vulnerability scanning and penetration testing, configuration auditing, and trust-level validation of information technology components.
- Data security – Data classification/encryption, database DLP, data integrity monitoring, secure data backup, enterprise rights management, mandatory access control (MAC) for all sensitive data, principle of least privilege, and fine-grained need-to-know.

The concept of network segmentation is integral to a defense-in-depth strategy, which reduces the network attack surface to defend against network penetration from adversarial threats and protects against task overload or other potential adverse network events. An organization can utilize network segmentation either physically or virtually. In a traditional network, all servers and workstations reside on the same local area network (LAN), which allows an attacker to move laterally from one system to another. Components that do not need to communicate with each other are not permitted – by design – to do so. This approach simplifies network zones based on function, allowing firewall boundaries to effectively prohibit traffic. An example of this approach is illustrated in Figure 16.1. The approach requires strong governance practices as each functional business owner must provide current interface information for maintenance and operations. Network segmentation is achieved by implementing packet filtering or stateful firewall inspection at the zone boundary. Network segmentation ensures that only authorized traffic can enter and exit zones and that unauthorized traffic is denied access. The Zero Trust network strategy is a concept used as an example of an additive manufacturing (3D printing) facility to secure networking and to interact with other company facilities (e.g., the headquarters), materials suppliers, customers, and transportation providers. Zero Trust security assumes that there are no trusted users or system applications. This approach reduces the attack surface, thereby improving the organization's defensive posture. Innovative cybersecurity involves the continuous prevention and protection of systems through security policy enforcement, as part of a comprehensive approach to defending sensitive information and processes associated with additive

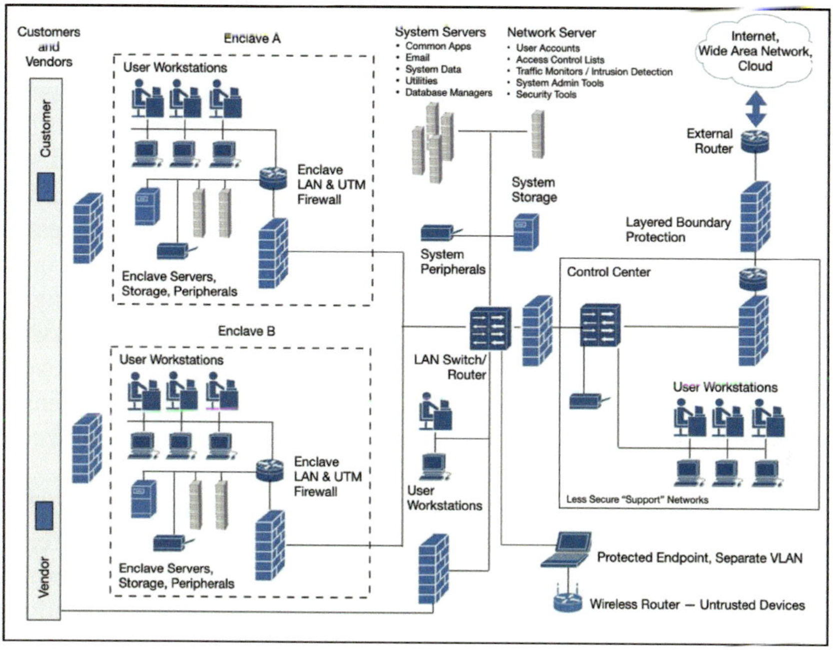

Figure 16.1 Anatomy of an additive manufacturing (3D) facility.

manufacturing, also known as 3D printing. A Zero Trust security approach utilizes a software-controlled, micro-segmentation approach to create security zones based on risk. This micro-segmentation approach can be applied across any network infrastructure that requires more robust access controls than what is achieved with traditional access control lists. The principles of Zero Trust security are based on network access that is never trusted, and threats can originate from both within and outside the system's ownership. Zero Trust security employs the principle of least privilege for all system users, and micro-segmentation facilitates access controls between security zones.

16.3.1 Access control strategy

The access control strategy is an integrated, multi-layered defense approach using intrusion detection systems, perimeter security, access control, and fire alarm systems. All of these sensors are integrated into a larger security platform monitored in a security control room. This approach enables a quick correlation of alerts, leading to faster response times. Administrative controls are implemented in a coordinated, mutually supportive solution consisting of designated restricted areas, visitor identification (ID)/logging/ badging/escort policy and enforcement/and associated paperwork, no expectation of privacy, audit trails and access logs, asset classification and control, emergency procedures, general housekeeping, and pre-and post-employment procedures (i.e., employee check-in and check-out). Appropriate security controls and safeguards include up-to-date access control lists based on the security technical implementation guide (STIG). The STIG is a set of technical guidelines that provide specific configurations for software and hardware. Access control protocols are used at all points of connection. The default system configuration should not be used, as it allows access to the system without a password. Administrative roles should be strictly managed and limited without exception. Access credentials and roles need to be regularly maintained to ensure that sensitive information and processes are protected from unauthorized or malicious access.

Next-generation technologies, such as operational technology software-defined networking (OT-SDN), used in critical infrastructure, are discussed in greater detail in Chapter 18, "Software Approaches." The operational technology software-defined networking (OT-SDN) enables administrators to control the network, provision resources, and modify network configuration settings from a centralized user interface, all without requiring additional hardware. The network management functionality is excellent, and the technology effectively implements the organization's work security strategy. The OT-SDN separates network device control from the data being transported and separates the switching software from the network hardware. However, without a rigorous access control strategy, the organization can quickly run into serious problems by implementing OT-SDN as a solution. Giving administrative access to the flow control also provides full access to the network control plane and the access control list (ACL). Recall that a good "three-point" player can force the defense further out to the court perimeter, opening up the midrange for offensive players to attack the net. You don't want an adversary open in the midrange, either.

Technologies such as operational technology software-defined networking (OT-SDN) can provide the benefits of easier centralized remote management of networks from a single centralized dashboard. However, depending on the deployment, it can also expose the system to various network layer attacks, such as the possibility of a compromised software-defined network (SDN) controller at the control plane layer. Attacking an access control list (ACL) poses a significant threat to system security. Packet filtering, or "stateless" inspection, allows or denies traffic based on the transmission control protocol (TCP), user datagram protocol (UDP), internet control message protocol (ICMP), and internet protocol (IP) headers of packets using an ACL. It is fast, inexpensive, and unsophisticated, vulnerable to IP or DNS address spoofing. It cannot counter application-specific vulnerabilities or spoofed addresses due to its limited logging capabilities. Technology is not the only vulnerability to networks. Insider threat is another.

16.3.2 Personnel security

Personnel security involves managing actions, resources, people, and processes, while employing and supporting systems in a secure and protected manner, to ensure that an acceptably low level of security risk is maintained over time. Personnel security in hiring practices is also a critical consideration in systems engineering for infrastructure. The systems engineer informs the human resources staff of risks associated with personnel hiring policies and actions. The human resources (HR) staff are also stakeholders in the system. The role of HR staff is to protect the organization from lawsuits, of which cybersecurity is a potential source. Steps a factory HR staff should take to minimize security risks associated with organizational personnel include:

- Adequate employee background checks and clearances (i.e., checking references, employment data, legal and other records, credit, and drug testing).
- Periodically re-screening personnel with sensitive access.
- Maintaining employment agreements and documenting disciplinary consequences of violations
- Managing effective hiring and termination practices (i.e., recovering keys and badges, changing locks and passwords, deactivating accounts, sending notifications)
- User monitoring of personnel, especially those with special privileges (i.e., direct observation, random testing, attendance audits, workstation monitors, explicitly declaring a "no expectation of privacy" in policy)
- Establishing an unambiguous bias for security responsibilities and practices in job descriptions

A badge is used to associate an identity with an individual, and the system is used to authenticate an individual. A badge is an item used as a basis for authentication. A numerical access code is information that can be shared and may also be easily guessed. An access badge is a credential that identifies the holder as having access to a specific facility area or system network. The access badge contains an associated number, which the card reader reads based on credential information. This number can be used to generate a report that displays the date and time the badge was used to access a controlled area, along with the access PIN used to secure a facility door at the entrance. Access requires that a person hold both the badge and know the numerical PIN to gain entry to the site. Hiring and termination practices should include recovering keys and badges, changing locks and passwords, deactivating accounts, and sending appropriate notifications to ensure a seamless transition.

Changing key codes to doors ensures that all spaces are secure long after an employee leaves (i.e., from angry or disgruntled terminated employees, and accurate logging and tracing of door entry). Changing key codes is part of a comprehensive approach to managing and implementing access control to facility areas and resources, such as insider threats. Systems using key codes for doors provide convenient access programming and can be easily changed each time an employee leaves or is reassigned to different access points.

One drawback is that someone with key code access to a door also has access to enter the area. An organization can implement the most leading-edge technical solutions, but without adequate personnel security and access control, the technical solution is rendered ineffective. All stakeholders need to be vigilant of the "sin of omission." A "sin of omission" is a practice that we knew we should have put in place but did not. The "New York Times" test is related to the "sin of omission" in that such an omission does not hold up to public inspection. The threat to secure operations is caused by errors and omissions by staff, which create professional liability. A company requires processes, training, product reviews, and quality control to demonstrate that it is performing due diligence and exercising responsibility. The first line of defense is to prevent unauthorized access to the system.

16.3.3 Physical security

Physical security is the first line of defense to prevent unauthorized entry to the factory plant shop floor or damage caused by an insider threat (e.g., inventory theft or data loss). A physical security and surveillance system is part of the overall security and safety solution. The company has published a security policy that includes physical controls to mitigate vulnerabilities that could be exploited by competitors, criminals, hackers, and malicious insiders. Some physical security measures that should be implemented to protect sensitive resources include:

1. Construction (e.g., walls, doors, fire rating and resistance to forcible entry, emergency marking, placements, locked or controlled entrances, alarms, secure hinges, electric power supplies, backup and alternate power supplies, water and gas lines, fire detection, and rating)
2. Sensitive asset protection (i.e., prevent theft, unauthorized use, damage/sabotage, service interruption, compromised integrity, inventory and periodically audit equipment items, apply technical controls, apply physical controls, furnish travelers with security cables, maintain robust and secure backups, install safes and other secure storage, limit off-premises transport, and use of sensitive items)
3. Personnel (i.e., a controlled list with facility access permissions enforced at entry points using badges, access control cards, or other means)
4. Access controls (e.g., fencing, perimeter intrusion detection and assessment system, lighting, guards and patrol force, dogs, locks, storage areas, badges, and biometrics)
5. Surveillance (i.e., intrusion detection, alarms, and visual recording system)

16.3.4 Security operations

Well-planned, properly staffed, and adequately resourced security operations are essential. Security operations include sensitive information and processes, including information systems (e.g., communication data, routing information, and operational information), human resources (e.g., engineers, support staff, and third-party vendors), operations, network services, IT services (e.g., operation management software, application software, development tools, and utilities), physical sites (e.g., heating, ventilation, and air conditioning (HVAC), and fire sprinkler system) and assets (e.g., switches, cables, and terminal equipment), and compliance. Key elements include personnel, third parties, and system and network management. Security operations include activities and resources to maintain the effectiveness of controls and the overall risk posture of the system. Primary elements of a secure operations strategy and procedures implemented for the factory include:

- Asset discovery automation and inventory processes at the Level 1 process control layer composed of programmable logic controllers (PLCs) and control devices that directly interact with physical processes such as sensors and actuators of the Purdue Enterprise Reference Architecture (PERA) model are facilitated by the framework's passive and safe active enumeration of network-connected assets, such as those enabled by using the MOSAICS framework.

- Limited remote access and monitoring compliance
- Automated patch and antivirus system updates
- Technical access controls (e.g., preventative, detective, mitigative, and deterrent controls)
- Change controls (e.g., governance and configuration management)
- Audits (i.e., the process of examining systems and processes for correct design and use to ensure compliance with required operational practices, audit trails, and logs)

Behavioral analytics and machine learning to alert on human and system errors or malicious security incidents. The power of using the MOSAICS framework, as described below, lies in the application of behavioral analytics. The MOSAICS framework enables the use of machine learning and analytics to identify malicious behaviors by analyzing differences and patterns discovered in behavioral data, trends, and anomalies from otherwise normal network activities in near real-time

16.3.5 Wide area network (WAN)-connected devices

A challenge for organizations, such as those in manufacturing, is the policy requirement for a secure wide area network (WAN) to interact with other company facilities (e.g., headquarters), materials suppliers, customers, and transportation providers. For example, WAN is used in an additive manufacturing (3D printing) facility, or any factory, to support large, complex, and expensive areas. Additive manufacturing includes programming various automated tools used in product fabrication, collecting and analyzing measured data quality, and supporting business functions. WAN is used at a lower speed and is commonly employed in asynchronous transfer mode (ATM) switching technology over synchronous optical network (SONET) fiber-optic links. WAN technologies include dedicated lines that can use either packet or circuit switching, or both. The asynchronous network category is an encapsulation of data bytes in processing bits that each have start, stop, and parity control. For the additive manufacturing solution, a software-defined approach using a centralized interface to a cloud-delivered software-defined wide area network

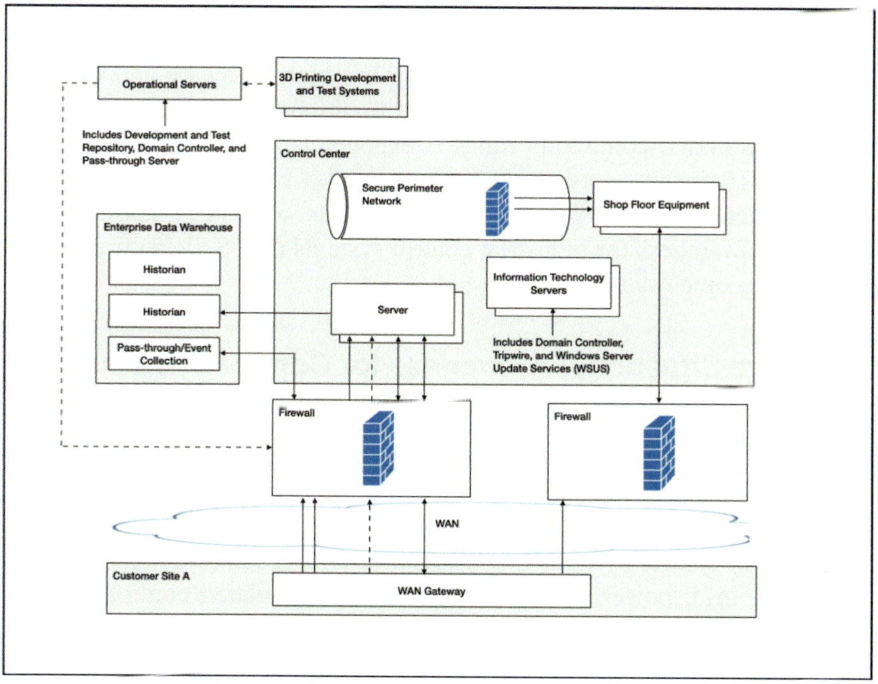

Figure 16.2 Additive manufacturing (3D) facility WAN and WAN gateway.

(SD-WAN) would simplify management and operations by decoupling the networking hardware from the control mechanisms. The WAN and customer WAN gateway are depicted in Figure 16.2. The elements and functions of a WAN to implement secure communications for an additive manufacturing facility are like any factory:

- Support for simultaneous connections and a variety of protocols
- Meet non-stop production demand with optimized performance (i.e., low downtime) and automatic failover for reliable and consistent connectivity
- Cost-effectiveness
- Used in business applications (e.g., email, content management systems, database applications, and web delivery)
- Support of data replication, backup, data migration, virtualization, and disaster recovery (DR)

16.3.6 Quantum computing

The emerging area of quantum computing has both hardware and software technology components. While similar to traditional computing, the speed of compilation and computing times is exponentially increased, and quantum physics is increasingly used to solve complex problems. Automation is needed to detect changes that may indicate threats, which is where the more situational awareness for industrial control systems (MOSAICS) framework approach becomes valuable.

16.4 More Situational Awareness for Control Systems (MOSAICS) Framework

More situational awareness for industrial control systems (MOSAICS) is the initial cyber defensive capability framework for critical infrastructure. It is based on the operational, functional, and technical engineering requirements for industrial control systems (ICS) found in critical infrastructure, such as the electric grid, power, water/wastewater, facility-related control systems (FRCS), and fuel. The MOSAICS framework is an unclassified, publicly releasable, non-prescriptive, vendor-agnostic framework defining a body of engineering requirements (operational, functional, and technical) for control system cyber threat defense. The cybersecurity capability is realized through a collection of integrated technologies that satisfy minimum viable specifications to meet specific MOSAICS Block configurations mapped to specific, traceable engineering requirements (e.g., functional, operational, and technical).

In 2019, when the DOD initiated the MOSAICS experimentation, it marked the first demonstration of a cyber-defensive operating capability for control systems (Aleksandra Scalco, 2019). Additionally, Navy system operators successfully demonstrated cyber defensive capabilities for industrial control systems (ICS) to mitigate cyber vulnerabilities during a military utility assessment (MUA) in August 2021. MOSAICS is an enabler of a Zero Trust strategy. Additionally, MOSAICS enables the asset owner and operator to utilize the advanced cyber industrial control system tactics, techniques, and procedures (ACI TTP) in an automated fashion (USCYBERCOM, 2016). Some of the Zero Trust pillars that MOSAICS was always designed to focus on, such as visibility and analytics, significantly allow an asset owner and operator to get toward Zero Trust.

16.4.1 Predecessor efforts

Understanding the more situational awareness for industrial control systems (MOSAICS) joint capability technology demonstration (JCTD) helps clarify the path to initial experimentation. First, we need to go back to 2008, when the Defense Science Board Task Force on DOD energy strategy issued its report, "More Fight – Less Fuel." A rallying cry of the Naval Information Warfare Center (NIWC) Atlantic MOSAICS JCTD Technical Manager, Salvatore Richard Scalco (former National Security Agency (NSA) Senior Executive Service (SES) engineer), would later be: "*No Fuel – No fight!*" The Defense Science Board's Task Force report was issued three years before "cyber" would be declared a warfighting domain in 2011. The Task Force identified two primary energy challenges to the DOD in their report: "Unnecessarily high and growing battlespace fuel demand" and "Military installations are almost completely dependent on a fragile and vulnerable commercial power grid, placing critical military and Homeland defense missions at unacceptable risk of extended outage" (DSB, 2008). Additional findings included: "Finding #3: The Department lacks the strategy, policies, metrics, information, and governance structure necessary to properly manage its energy risks," and "Finding #6: Operational risks from fuel disruption require demand-side remedies; mission risks from electricity disruption to installations require both demand- and supply-side remedies" (DSB, 2008).

The advanced cyber industrial control system tactics, techniques, and procedures (ACI TTP) discussed earlier in Chapter 7, "Threats to Information-intensive Systems," and other predecessor efforts, such as the DOD and Department of Energy's (DOE) smart power infrastructure demonstration for energy reliability and security (SPIDERS) joint effort, led to the successful MOSAICS enabler and the capability to realize Zero Trust principles for control systems in critical infrastructure in an automated fashion. Recall from Chapter 7, "Threats to Information-intensive Systems," that the ACI TTP was an output Joint Base Architecture for Secure Industrial Control Systems (J-BASICS), which was chartered in February 2014 to address the following problem statement: "Network managers supporting DOD industrial control systems (ICS) lack TTP to detect, mitigate, and recover from nation-state level cyber attacks"(Walsh, 2016). The output of J-BASICS was the manual ACI TTP. The objective of J-BASICS was to address concerns about cyber threats to the Department of Defense (DOD) industrial control systems (ICS). The output consisted of manual procedures in three main sections: detection, mitigation, and recovery, along with detailed baselining and routine monitoring sections. For example, the ACI TTP provided baseline development information, allowing operators to compare observed system behavior with a baseline. The overall results of the field tests demonstrated that operators could detect the presence of adversaries and mitigate and restore devices (Walsh, 2016). The fundamental goal of the detection procedures was to enable operators to identify malicious activity and anomalous behaviors in a system. The primary goal of the mitigation procedures was to enable an operator to maintain control of the system's functional capabilities in the event of a cyber-attack. The primary goal of recovery was to restore the system to a fully mission-capable (FMC) state (USCYBERCOM, 2016). The next planned steps after the field tests were to promote ICS as part of the cyberculture across the Department of Defense (DOD) and to inform all industrial control system (ICS) operators, including commercial partners, to prepare for a potential war-fighting engagement (Walsh, 2016).

While J-BASICS addressed TTPs, other experiments explored methods to protect critical infrastructure from physical disruptions to the electric grid resulting from cyber events. The JCTD SPIDERS conclusion followed in 2015. SPIDERS was a 4-year effort to demonstrate the value of utilizing a cyber-secured microgrid by integrating smart grid technologies to enhance mission assurance. Smart power infrastructure demonstration for energy reliability and security (SPIDERS) partners included the DOD Office of the Secretary of Defense (OSD), USINDOPACOM (then known as USPACOM), USNORTHCOM, DOE, DHS, US Army Corps of Engineers (USACE) and Engineer Research and Development Center (ERDC), US Marine Corps, US Navy, US Army, US Air Force, DOE National Laboratories – Sandia National Laboratories, Pacific Northwest National Laboratory (PNNL), National Renewable Energy Laboratory (NREL), Oak Ridge National Laboratory, and Idaho National Laboratory (INL) – US Army Tank-automotive and Armaments Command (TARDEC), the Naval Facilities Engineering Command (NAVFAC), local utility companies Colorado Springs Utilities and Hawaiian Electric Company, and the US States of Hawaii and Colorado (Johnson, 2015). The operational objectives of the SPIDERS JCTD were to demonstrate the technical capability to:

1. Protect critical defense infrastructure from power loss due to physical disruptions or cyberattacks on the bulk electric grid.
2. Sustain critical operations during prolonged utility power outages.

3. Integrate renewable energy sources, energy storage, and other distributed generation technologies to power critical defense infrastructure in times of emergency.
4. Manage DOD installation electrical power and consumption efficiently to reduce petroleum demand, carbon footprint, and cost (Command, 2015).

The final report was made public to provide timely information to federal and non-federal agencies, as well as the broader industry. Microgrids and smart microgrids, along with some commonly encountered systems engineering challenges, are detailed further in Chapter 20, "Foundations of Systems Engineering."

16.4.2 Critical infrastructure at risk

The non-proprietary, vendor-agnostic MOSAICS framework, which the US government developed in collaboration with national laboratory partners, was grounded in these successful predecessor experimentation investments and other predecessor experiments, all of which were heavily invested in by government departments and agencies to secure critical infrastructure from cyber vulnerabilities. A pivotal point in efforts to overcome cyber vulnerabilities in critical infrastructure is when the problem statement evolves from a Department of Defense (DOD) to a broader national interest. The problem statement for the MOSAICS JCTD was "US critical infrastructure is at risk" (Rich Scalco, 2018). The proposed solution was to prototype the integration of commercial off-the-shelf (COTS) and government off-the-shelf (GOTS) technologies to achieve enhanced situational awareness and the initial cyber defensive capability for industrial control systems associated with task-critical assets. GOTS was used only where no COTS were available. The technical team successfully demonstrated the capability in a military utility assessment (MUA) on a power site. As promised, the prototype successfully delivered the following capabilities to operators:

- Integrated, operational capability to enable the defense of control systems
- ICS baselining tools and programmable logic controller sensors
- Tailored visualizations, analytics, and automated cybersecurity orchestration (Rich Scalco, 2018).

Importantly, MOSAICS automated most of the ACI TTP. The ACI TTP is valuable to operators, but it is nearly impossible to use manually due to the high cyber speed. The principles of the integrated adaptive cyber defense (IACD) strategy and framework, which are designed for IT systems, were applied to operational technology (OT) (APL, 2019). IACD demonstrated the ability to automate risk determination and the decision to act on cyber anomalies found in a system. Like the ACI TTP, IACD was an output of a government-funded and led initiative. IACD, initially sponsored by the National Security Agency (NSA) and the Department of Homeland Security (DHS), introduced key technology areas, including orchestration, information sharing, and playbooks, into experimentation. The playbooks guide cybersecurity responses related to orchestration services. Essentially, the ACI TTP provided the playbooks, and the IACD strategy and framework provided the methodology for applying the TTP as automated response actions or courses of action (COAs). These automated COAs changed the delivery of actionable situational awareness and enterprise information sharing to near real-time. Perhaps most importantly, MOSAICS delivers what no asset owners had: the ability to inventory key system devices and components, map communication and data flows, and categorize system components based on criticality and vulnerability. This ability to view what is on the network with granularity, including devices, applications, and users, is essential for implementing any Zero Trust security strategy. Automated cyberattacks and automated cyber defenses are a game-changer in the cyber world. MOSAICS demonstrated the ability to automate sections of the ACI TTP.

16.4.3 MOSAICS as part of a Zero Trust strategy

So, going back to the earlier statement, some of the Zero Trust pillars (Figure 16.3) that MOSAICS was always designed to focus on, such as the Zero Trust pillar for visibility and analytics, significantly enable an asset owner or operator to meet Zero Trust requirements. Many of the Zero Trust pillars are already reflected in the principles found or "baked into" the ACI TTP. Because the initial MOSAICS prototype was a JCTD demonstration, typical program artifacts for the acquisition and sustainment of a system were yet to be developed. For example, a JCTD does not require an initial capabilities document (ICD), which describes the capacity gap, associated tasks, conditions, and operational performance standards and metrics. The MUA completed at the end of a JCTD demonstration is augmented by a final report that qualifies the results as equivalent to an ICD. The report supports the development of a subsequent capability development document (CDD) or capability production document (CPD) (Chairman of the Joint Chiefs of

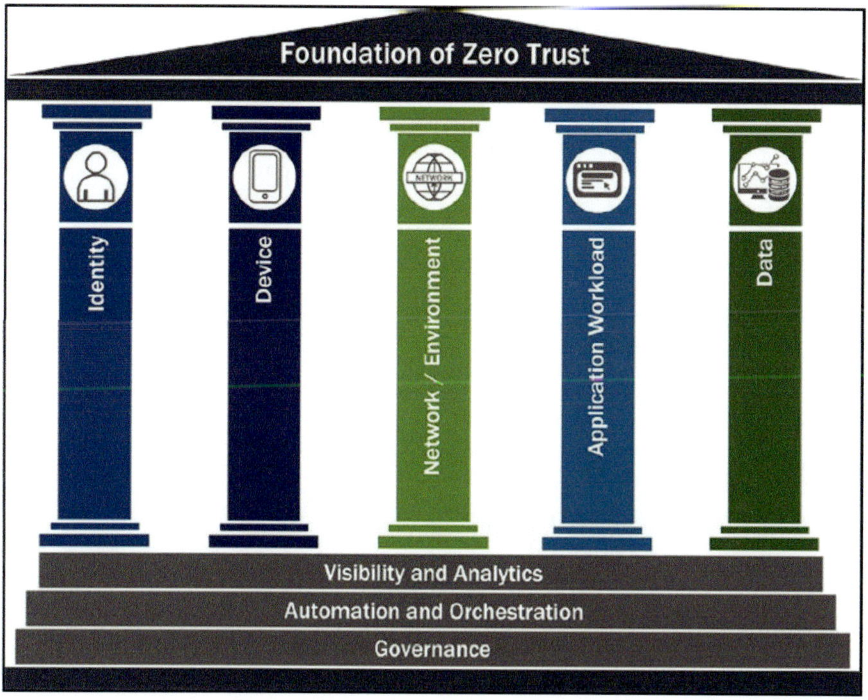

Figure 16.3 Zero Trust pillars (CISA, 2023)

Staff, 2012). But, as the adage goes, "*How do you eat an elephant*?" The response – "*one bite at a time*." The MOSAICS prototype was decomposed into two acquisition blocks, which were then further decomposed into smaller "bites": MOSAICS Block I A and Block I B, and MOSAICS Block II A, Block II B, and Block II C. The later blocks introduce additional innovative technology and become more complex to implement, requiring more research and testing before they can be deployed in the field. The Block II future state involves continuous monitoring and near real-time mitigation for the operational defense of critical infrastructure, aiming to overcome cyber vulnerabilities. This approach utilizes advanced technologies, including software-defined networking (SDN), software-defined perimeter (SDP), virtual local area networks (VLAN), machine learning (ML), and eventually, advanced artificial intelligence (AI) tools.

For example, MOSAICS Block I capabilities cover many of the capabilities under each of the seven Zero Trust pillars:

- Visibility and analytics pillar, MOSAICS Block I capabilities address
  - 7.1 Log all traffic

 - ○ 7.3 Common security and risk analytics
 - ○ 7.4 User and entity behavior analytics (UEBA), and
 - ○ 7.2 Security information and event management (SIEM)

- Automation and orchestration pillar, MOSAICS Block I capabilities address
  - ○ 6.3 Machine learning (ML) and
  - ○ 6.7 Security operations center (SOC) and incident response
- Network and environment pillar, MOSAICS Block I address
  - ○ 5.1 Data flow mapping
- Data pillar, MOSAICS address
  - ○ 4.1 Data catalog risk alignment and
  - ○ 4.4 Data monitoring and sensing
- Application and workload pillar, MOSAICS Block I address
  - ○ 3.1 Application inventory
- Device pillar, MOSAICS Block I, enables advanced Zero Trust capabilities
  - ○ 2.1 Device inventory
  - ○ 2.3 Device authorization with real-time inspection
  - ○ 2.4 Remote access, and
  - ○ 2.7 Endpoint focuses on computers, laptops, and other devices and extended detection and response focuses beyond endpoints to include networks, cloud applications, and email (EDR and XDR)
- User pillar, MOSAICS Block I address
  - ○ 1.4 Privileged access management

MOSAICS Block II addresses additional Zero Trust pillars and capabilities. MOSAICS Block II adds:

- Visibility and analytics
  - ○ 7.5 Threat intelligence integration; automation and orchestration
  - ○ 6.5 Security orchestration, automation, and response (SOAR)

- 6.1 Policy decision points (PDP) and orchestration
- 6.2 Critical process automation

- Network and environment
  - 5.4 Micro-segmentation
  - 5.2 Software-defined networking (SDN); and
- User pillar
  - 1.1. User inventory

16.5 Summary

In basketball, with the three-point perimeter, a player no longer needs to be the highest jumper or the tallest participant. The three-point shooter is difficult to guard because its perimeter is much more extensive than the "paint" closer to the basket. The three-point addition changed the game. A challenger does not have to be a phenomenal basketball shooter – the player must be versatile and adept at making open shots. Taking outside shots at the basket opens the court for other players to drive and creates opportunities, forcing the defense back enough to open the net for a jump-shot attack. The bottom line is that different strategies and skills are needed to guard a network when modern technologies are introduced. When a promising perimeter player is given space, it will be almost impossible to stop them from scoring. When a centralized controller is compromised in SDN, for example, it presents a single point of vulnerability for an attack, following an intention to reduce the otherwise greatly increased perimeter surface. The technology might make it look easy, but, as Steph Curry puts it, "the behind-the-scenes stuff is the challenge."

These new cyber capabilities require the organized application of systems engineering principles and practices to be successfully integrated into operations. What is astonishing is the advanced capabilities that can be incorporated into critical infrastructure sectors today to mitigate cyber vulnerabilities. You can start applying the same level of cyber-defensive operating capability to your control systems today. You can start now. This section, "Systems Engineering Methods and Practice," tells you how to do it, and Part III of this book, "Stakeholder Alignment," tells you how to get everyone in agreement to want to do it and work together to do it. Basically, how to get the team together in what feels like chaos and to perform like champions.

References

(DSB), D. S. B. 2008. More Fight – Less Fuel: Report of the Defense Science Board Task Force on DoD Energy Strategy. *In:* (DOD), D. O. D. (ed.). Washington, D.C. 20301–3140: Office of the Under Secretary of Defense for Acquisition, Technology, and Logistics.

(USCYBERCOM), U.S.C.C. 2016. Advanced Cyber Industrial Control System Tactics, Techniques, and Procedures (ACI TTP) for Department of Defense (DoD) Industrial Control Systems (ICS). *In:* (DOD), D. O. D. (ed.). United States Cyber Command.

Aleksandra Scalco, M. J., Steve Simske 2019. More Situational Awareness for Industrial Control Systems (MOSAICS) Joint Capability Technology Demonstration (JCTD): A Concept Development for the Defense of Mission Critical Infrastructure. *Homeland Defense & Security Information Analysis Center.*

APL), J. H. U. A. P. L. J. 2019. *Integrated Adaptive Cyber Defense (IACD)* [Online]. https://www.iacdautomate.org: Johns Hopkins University Applied Physics Laboratory (JHU APL). [Accessed December 9 2019, 2019].

Chairman of the Joint Chiefs of Staff, J.-. 2012. Instruction CJCSI 3170.01H Joint Capabilities Integration and Development System. *In:* Defense, D. O. (ed.). Washington, DC.

Command, N. F. E. 2015. Smart Power Infrastructure Demonstration for Energy Reliability and Security (SPIDERS) Joint Capability Technology Demonstration (JCTD). *In:* (DOD), D. O. D. (ed.) *Technology Transition Final Public Report.*

(CISA), C. A. I. S. A. 2023. Zero Trust Maturity Model Version 2.0. *In:* Division, C. (ed.). Cybersecurity and Infrastructure Security Agency (CISA).

(DISA), D. I. S. A. & TEAM, N. S. A. N. Z. T. E. 2022. Department of Defense (DOD) Zero Trust Reference Architecture Version 2.0. *In:* (DOD), D. O. D. (ed.). Robert Freter, Defense Information Systems Agency (DISA) Zero Trust Program Lead (ID2).

Dransfield, M. 2022. Mission-Critical Control System (MCCS) Cybersecurity. *In:* National Security Agency (NSA), C. S. C. (ed.). Cybersecurity & Information Systems Information Analysis Center (CSIAC): National Security Agency (NSA).

John Kindervag, S. B., Kelley Mak 2012. Build Security into Your Network's DNA: The Zero Trust Network Architecture. Forrester: Forrester.

John Kindervag, S. B., Kelley Mak, And Josh Blackborow 2016. No More Chewy Centers: The Zero Trust Model Of Information Security. *Vision: The Security Architecture And Operations Playbook.* Forrester: Forrester.

Johnson, M. 2015. JCTD SPIDERS Technical Overview Phases 1 3. Federal Utility Partnership Working Group: Engineer Research and Development Center.

Larry Bird, B. R. 1990. *Drive: The Story of My Life*, Bantam.

Rich Scalco, D. B. W., Jorge Lacoste, John Andrews, Bill Beary, Ross Roley, 2018. *More Situational Awareness for Industrial Control Systems (MOSAICS) Joint Capability Technology Demonstration (JCTD)* [Online]. IACD Automate: Johns Hopkins University Applied Physics Laboratory (JHU APL). Available: https://www.iacdautomate.org/may-2018-integrated-cyber [Accessed October 2019 2019].

Walsh, G. 2016. Joint Base Architecture for Secure Industrial Control Systems (J-BASICS). ITEA, 16 March 2016, 2016. International Test and Evaluation Association (ITEA): The International Test and Evaluation Association (ITEA).

17

Architecture Approaches

"The mother art is architecture. Without an architecture of our own, we have no soul of our own civilization."

– Frank Lloyd Wright (1867–1959),
American architect

Abstract

Throughout history, architects have shaped our world by interpreting the aesthetics of functional elements to create building structures. Architecture is a skill that is unique to designing a building, and it is different from the construction itself. The architect serves both the aesthetic soul of the building and the practical functional requirements of the construction. Working with the systems architect is an essential aspect of systems engineering for critical infrastructure. Both the systems engineer and the systems architect are fundamental to the premise of model-based systems engineering (MBSE). MBSE formalizes aspects of systems engineering. Model-based systems architecture practice (MBSAP) further formalizes and allows for the reuse of design patterns. MBSE is discussed in more detail in Chapter 24, "Creating a Model-based Organization." Cybersecurity is addressed throughout the entire lifecycle stages of a system, from conception to operational use and maintenance, by both the systems engineer and systems architect, who guide all the component elements to realize the required functionality and capability, formalizing all aspects and allowing for the reuse of design patterns. Designing and developing applications for mitigating cyber vulnerabilities in critical infrastructure requires unambiguous traceability to systems engineering requirements and clear language for describing the system. Unambiguous language for describing system architecture is essential to contribute to the effective development and use of a system, as is an open, extensible architecture, and the eventual realization of a service-oriented architecture (SOA) that allows for bi-directional information sharing (or shared situational awareness (SA) of threats).

DOI: 10.1201/9788743809128-17

17.1 Introduction

While the systems engineers are generally viewed as the central coordinators of system development, the system architect uses a formal representation of an entity (e.g., structure, behavior, and rules) to support system development. The architect remains involved throughout the design and building of a system, utilizing an architecture model to address overarching issues such as the diverse concerns of a system's stakeholders, stated and derived system requirements, planned system evolution, policies, and other constraints, as well as the operational suitability of the delivered system. The well-known American modern architect Louis H. Sullivan influenced modern building architecture with his "form follows function" philosophy. Systems engineers can use the same elegant idea to architect critical infrastructure systems in a cyber world. The water system must function as intended. The form of connection to the Internet must follow that function. Ultimately, the objective of a water system is to deliver water. Cyber is the form that follows that function, not the other way around. The addition of cyber functionality to critical infrastructure should not interfere with the purpose of the construction but instead should appreciate it, respect it, and allow it. The addition of cyber functionality is why a SECOPS (security concept of operations) is created during the system architecture phase. The system architect brings together stakeholder concerns at various levels of abstraction. The system architect promotes modular, technology-independent designs to manage complex, information-intensive systems, thereby reducing risk, facilitating stakeholder dialog, and enabling the incremental development of capabilities (Herber, 2020).

The system architecture provides a structured, unambiguous system description to address complexity. Unambiguous language empowers the approach to complexity management, including decomposition, composition, modeling, reuse, and metrics. "Architecting involves the search for optimum solutions in problem spaces with many dimensions, and the architect must balance competing interests while keeping at the forefront the needs of the ultimate customer. Clear and consistent terminology helps in meeting that challenge" (Borky and Bradley, 2019a). The Department of Defense (DOD) and the National Aeronautics and Space Administration (NASA) utilize model-based systems engineering (MBSE) to more effectively trace operational needs to technical requirements in a structured and unambiguous manner, thereby enabling the reuse and interoperability of models (Herber, 2020).

17.2 Model-based Systems Engineering Architecture (MBSE)

The fundamental premise of model-based systems engineering (MBSE) and the basis for model-based systems is that the discipline formalizes systems engineering using models and accomplishes systems engineering tasks and processes upon a foundation of a formal architecture model, thereby transitioning from a "document-centric" to an "architecture-centric" systems engineering approach. The goal of MBSE is to capture the system design from the enterprise level (e.g., IT business and cloud applications) down to the component level (e.g., operational technology (OT) and programmable logic controllers (PLCs)). MBSE can address the perennial difficulties with complex systems (Ibid.).

Model-based systems engineering (MBSE) is a tool for creating a complete, mathematically based, standards-compliant architecture that complies with SysML, a universal standard medium for capturing a model. Using MBSE means that systems engineering is a metrics-managed process. The fundamental principle is to move away from legacy text-based documents, graphical diagrams, and spreadsheets and replace them with mathematical models. Mathematical models can be used to "perform sanity checks on results of complex analysis and simulations" (Alexander Kossiakoff, 2011). The model itself essentially becomes the framework with rigorous metric control.

Architecture is the core of MBSE. It is a model-based paradigm anchored in models. "MBSE formalizes [systems engineering] using models to ensure rigor, repeatability, and producibility in [systems engineering] processes; promote quality, completeness, and correctness in system designs; reduce risk in requirements analysis, design, integration and test, and other activities, enhance communication and synchronization of activities across organizations and disciplines" (Borky and Bradley, 2019a).

The system architecture definition used in Borky and Bradley's book, the model-based system architecture process (MBSAP), is closely tied to the systems engineering processes that occur throughout a system's life cycle, providing a foundational, structured approach and methodology using model-based systems engineering (MBSE). "MBSE formalizes SE using models to ensure rigor, repeatability, and producibility in SE processes, promote quality, completeness, and correctness in system designs, reduce risk in requirements analysis, design, integration, and test, and other activities, and enhance communication and synchronization of activities across organizations and disciplines" (Borky and Bradley, 2019a). The architecture is the core of model-based systems engineering (MBSE), offering traceability throughout the entire architectural model. The systems engineering flow is anchored in the architecture model (Herber, 2020).

Enterprise architecture applies to the top level of an ensemble of nodes, systems, elements, or other resources that collaborate to fulfill the functions of an overall organization or business process. System architecture applies to an ensemble of elements (ultimately, hardware and software components) that collaborate to meet defined requirements allocated to a node or system, implying that a clear system boundary and user interfaces are defined. Service-oriented architecture (SOA) describes interactions among elements and with the external environment in terms of capabilities exposed as services that are invoked by users or other elements, typically in a request-response model using messaging. Executable architecture represents a design in models that can be run as simulations to quantify and analyze behaviors.

Designing and developing applications for mitigating cyber vulnerabilities in critical infrastructure requires unambiguous traceability to systems engineering requirements and clear language for describing the system. Unambiguous language for describing system architecture is essential for contributing to the effective development and use of a system, as is an open and extendable architecture. An open system is characterized by and contributes to the development and use of a system by the ability to:

- Select and assemble "best of breed" products from various sources without expensive modifications and integration efforts
- Selectively replace, modify, or upgrade at the level of relatively fine-grained components, such as a processor or software application, with minimal (ideally, no) impact on the remainder of the system, other than utilizing the new or improved functionality
- Scale system capabilities by the addition of modular units of hardware and software, achieving performance commensurate with the installed assets; similarly, create new system components that can be reused for equivalent functions in other systems, and reuse such components from other deployed systems
- Interoperate with networks and other systems with minimal (ideally, no) added effort to provide adapters or other special-purpose components
- Make extensive use of commercial-off-the-shelf (COTS) products to minimize development effort and enjoy mass-market pricing" (Borky and Bradley, 2019a)

17.3 Open, Extensible Architecture

The more situational awareness for industrial control systems (MOSAICS) framework, as described in previous chapters, is an example of an open and extensible system. The approach was to utilize "best-of-breed," vendor-agnostic, non-proprietary commercial off-the-shelf (COTS) components from various sources to integrate into a system's critical infrastructure site. This approach assures minimal development effort, avoids vendor lock, and encourages competition, resulting in better prices. Existing components that meet the requirements are used, and only selective, fine-grained components are replaced to achieve the desired capability, as realized through the functionality that meets the engineering requirements. Each site may have a slightly different solution based on its existing components and desired functionality. The US Navy implementation, for example, integrated MOSAICS into the existing Navy Control System Platform Enclave (CSPE) as part of an enterprise strategy. The US Air Force is employing a similar approach to the Air Force Community of Interest Network (COIN), selectively replacing, modifying, or upgrading components with minimal impact on the remainder of the system, except for the introduction of new functionality. Using this approach enables entities to scale system capabilities through the incremental, modular integration of hardware and software, such as software-defined networking (SDN), in subsequent increments to the foundational implementation, allowing for future enhancements. The scalability of the MOSAICS framework is a crucial system engineering tenet that enables the reuse of equivalent functions in other sites and systems, as well as the reuse of components. This approach allows for interoperability with other DOD service component networks and systems, eliminating the need for specialized add-ons to these systems.

The MOSAICS components include the serial tap, baselining tool, host-based agents, intrusion detection system (IDS), active directory certificate services, alert architecture, firewall, firewall net flow, security information event management (SIEM), security orchestration, automation and response (SOAR), serial analytics, and a visualization interface. The serial tap passively collects serial communications and translates the data for ingestion into IT tools (Rich Scalco, 2018; Haegley, 2020; Rich Scalco, 2021). The baselining tool creates a snapshot of IT and control system devices for future comparison and reference. The host-based agent component collects information about activity on servers and workstations. The intrusion detection system (IDS) monitors and analyzes network traffic within the network environment for malicious activity, sending an alert when unauthorized activity is

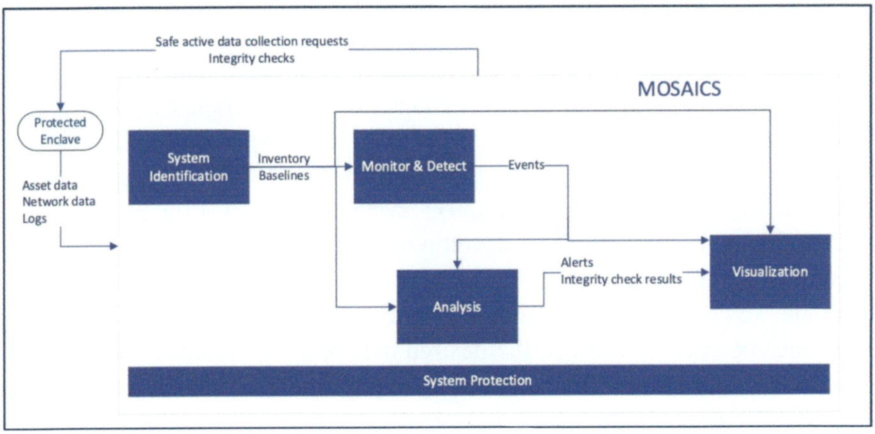

Figure 17.1 MOSAICS functional architecture (Munoz, 2025).

detected. The active directory certificate services manage digital certificates and enable access control. The alert architecture enables an analysis of chains of alerts across different devices. The firewall allows for the segregation of traffic and devices to enforce access controls. The firewall's netflow feature enables traffic analysis to identify deviations from normal communication patterns. The security information and event management (SIEM) solution collects logs, aggregates, analyzes security events, and stores cybersecurity data collected by MOSAICS. Security orchestration automates collection, analysis, response, recovery, and sharing actions. Serial analytics analyzes serial communication within the control systems, and the visualization interface provides a graphical representation of the data and actions taken. The selected commercial technologies are tailored to the specific site needs and the asset owner's strategy.

The MOSAICS functional architecture consists of data analytics at the Level 0 physical process layer, network and serial detection at the Level 1 intelligent device layer (such as relays and remote terminal units (RTUs)) to realize visualization and baseline capability, and end-point detection at Level 2 and Level 3 operations control with security orchestration and a data repository, as shown in Figure 17.1 (Munoz, 2025).

Each site may have a unique approach to architecture, tailored to the existing site components, site-specific requirements, and the asset owner's objectives. The basic functional architecture remains consistent and traceable to both engineering requirements and the behavioral tactics, techniques, and procedures (TTPs) outlined in the advanced cyber industrial control systems tactics, techniques, and procedures (ACI TTPs)

Table 17.1 MOSAICS components (Rich Scalco, 2018) (Haegley, 2020), (Rich Scalco, 2021).

Component	Capability
Serial tap	Passively collects serial communication and translates it to IT tools
Baseline tool	Creates a snapshot of IT and control system devices for future comparison
Host-based agents	Collects information about activity on servers and workstations
Intrusion detection system	Monitors and analyzes IT network traffic
Active directory certificate services	Manages digital certificates and enables access control
Alert architecture	Enables analysis of chains of alerts across various devices
Firewall	Enables segregation of traffic and devices to enforce access controls
Firewall netflow	Enables traffic analysis to identify deviations from normal communication
Security information event management (SIEM)	Aggregates and stores cybersecurity data collected by the MOSAICS integrated, composed solution
Security orchestration and automation (and response (SOAR))	Automates collection, analysis, response, recovery, and sharing actions
Serial analytics	Analyzes serial communication within the control systems
Visualization interface	Provides a graphical interface for understanding and taking action.

Reference: MOSAICS Industry Days (Rich Scalco, 2018), (Haegley, 2020), (Rich Scalco, 2021).

described in Chapter 16, “Modernization Strategies.” However, the architecture taxonomy will vary based on context. Three examples of systems engineering are provided in Figures 17.2–17.4 to illustrate how sites and systems can differ significantly in their taxonomy. The axis of the organization represents the organizational and structural hierarchy, as well as the basis for decomposition to implement modularity. The axis of categories identifies groups of systems that share common characteristics. There are many possible groupings. The axis of abstraction models the progression of architecture from abstract to concrete, providing the basis for levels of modeling and simulation.

Architecture can support many valuable roles in system development. The roles architecture can play in a program and how they can contribute to

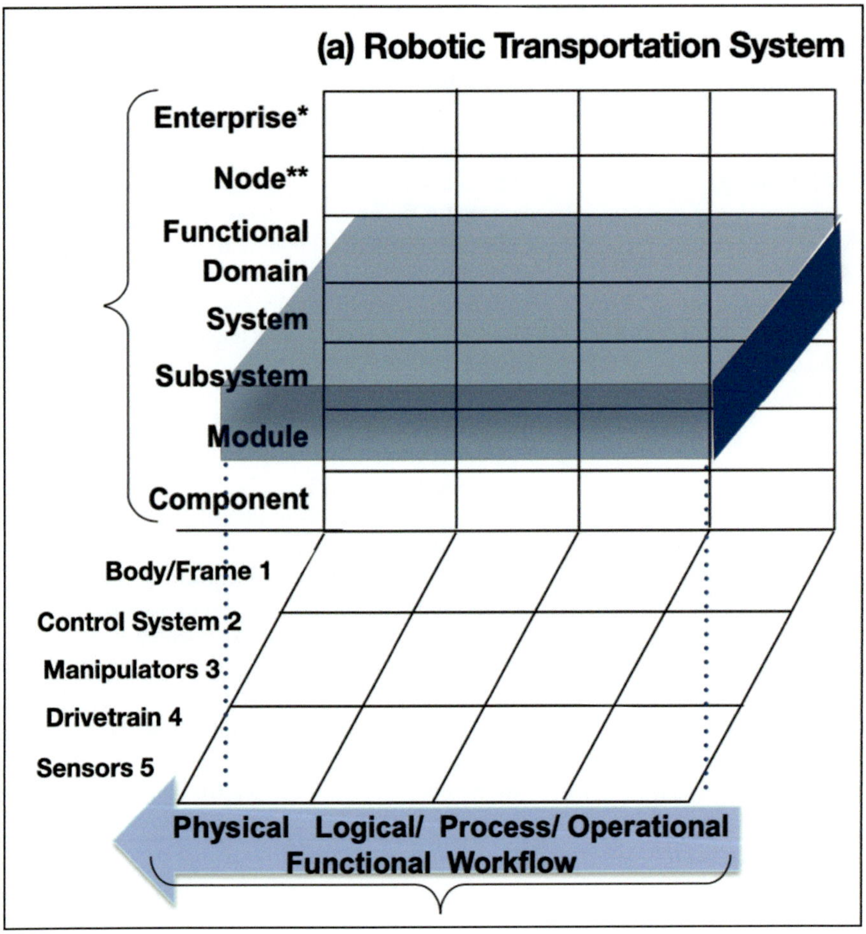

Figure 17.2 Robotic transportation system.

successful system development include complexity management, customer dialogue, technical execution, authoritative design capture and configuration management, and team and subcontractor management. Complexity management is valuable for dealing with complex entities and their elements, maintaining traceability from requirements to physical components, and ensuring that all system behaviors are captured and mapped to solution elements. Customer dialog helps to bring the need and the solution to life clearly and intuitively. It supports rapid prototyping, "what-ifs," human-in-the-loop simulations, technology demos, requirements refinement and prioritization, and design choices. Technical execution is the technical conscience

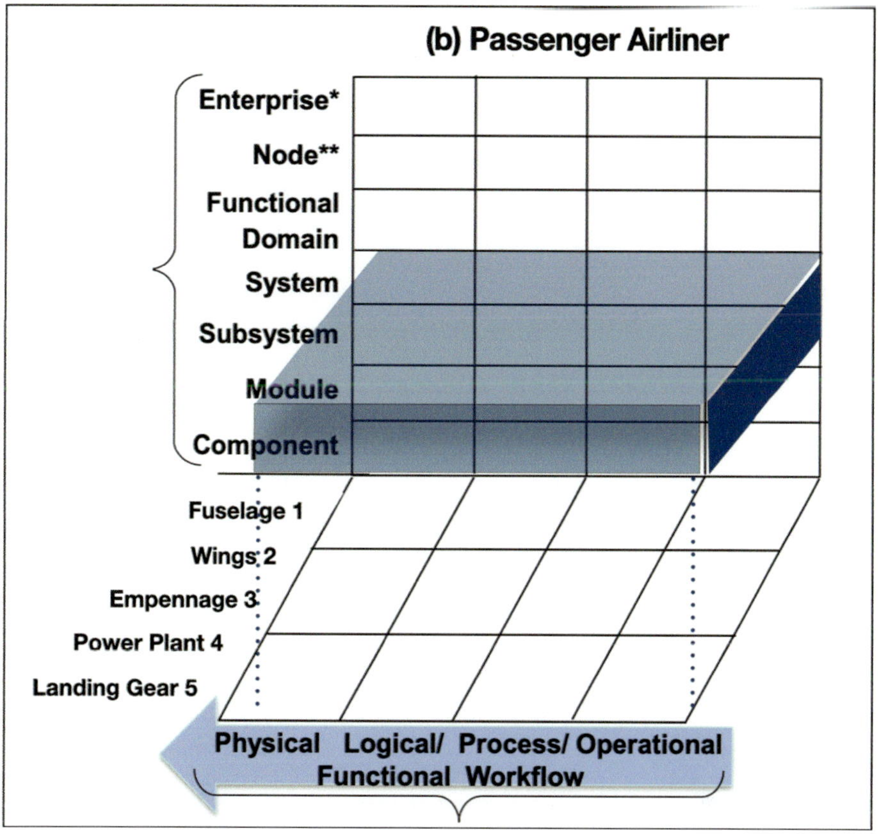

Figure 17.3 Passenger airliner.

(or "soul") of the program, and program management has the best insight into status and potential problems. Authoritative design capture and configuration management help guide trade studies, performance assessments, risk evaluations, and product evaluations. Team and subcontractor management create a shared, configuration-controlled design repository and normalize key processes (e.g., models, metrics, versions, and reports) while allowing local options (e.g., choice of compiler).

Some additional specific examples of ways in which the major systems engineering activities shown in the process flow could be supported by architecture include linkage of the evolving architecture to both physical and virtual (simulation) prototypes, rigorous traceability through the flow of architecture development from customer requirements to the elements of a delivered solution, support for implementation of service-oriented architecture (SOA), a

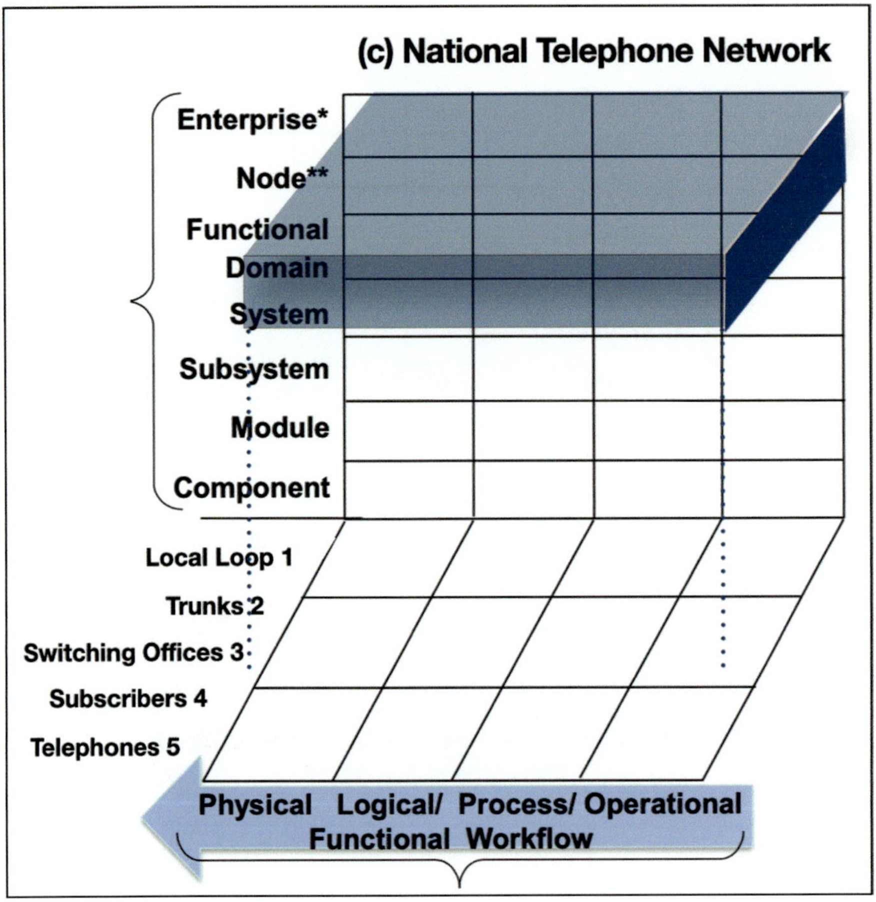

Figure 17.4 National telephone network.

good architecture model facilitates understanding and communication by and among its users (Borky and Bradley, 2019b).

If model-based systems engineering (MBSE) formalizes aspects of systems engineering, the architecture practice takes it a step further. Model-based systems architecture practice (MBSAP) further formalizes and allows for the reuse of design patterns such as the operational view (OV), which is the conceptual data model (i.e., data discovery), the logical view (LV), which is the logical data model (i.e., data management), and physical view (PV), which is the physical data model (i.e., data management). These viewpoints and activities are associated with the traditional systems engineering "Vee" diagram mapped to the left side of the "Vee." The nature of the transitions is decomposed flow-down from the top-level partitioning content, primary system functions, overall information categories, domain services, and supporting information to the logical view (LV) content of functional components and interface specifications, class and class collaboration functions, systems information entities, specific functional services, and supporting information to the physical view (PV) hardware and software and component specifications initial standards provide and forecast, database tables, physical service definitions, and support information (Herber, 2020).

17.4 Service Oriented Architecture (SOA)

The introduction of service-oriented architecture (SOA) to a legacy business process at the IT enterprise level introduces several issues, including changes to the process, as well as cultural and technical challenges. Critical infrastructure traverses both the IT enterprise-level and operational technology (OT) lower-level components. The transition methodology involves planning, testing, deploying, and sustaining each unique implementation site with trained experts to support operators and to enable standardized integration and maintenance efforts (Kilcoyne, 2021) The SOA approach (as shown in Figure 17.5 for the MOSAICS SOA) to overcoming cyber vulnerabilities in critical infrastructure requires visibility from the cloud to the programmable logic controller (PLC). This approach is dramatically different from the more traditional management of disparate systems. The system owner and operator processes and systems require significant reorganization to support the introduction of new architectural concepts. Steps to help prepare an organization's personnel include ensuring that the affected personnel understand the goals and benefits of service-oriented architecture (SOA). Specifically, providing "top-down support, empowered decision-making with involvement of all stakeholders, continuing focus on alignment with policy and strategic objectives, and clear direction to the organizations responsible for implementing such projects are universal" (Borky and Bradley, 2019a).

Strategies such as software as a service (SaaS) and infrastructure as a service (IaaS) are considerations in a modernization strategy. The primary difference between SaaS and IaaS is that the IaaS platform is more closely aligned with being fully on-premises. IaaS clients are responsible for man aging applications, middleware, and data. IaaS providers manage the servers, hard drives, networking, visualization, and storage. SaaS is furthest from on-premises responsibility and is primarily run by a third party, which poses a greater overall security threat. A SaaS implementation may require exchanging data with the backend SaaS data center to perform necessary software functionality. This data exchange means that business enterprise information may be exposed to additional security vulnerabilities. Essentially, when

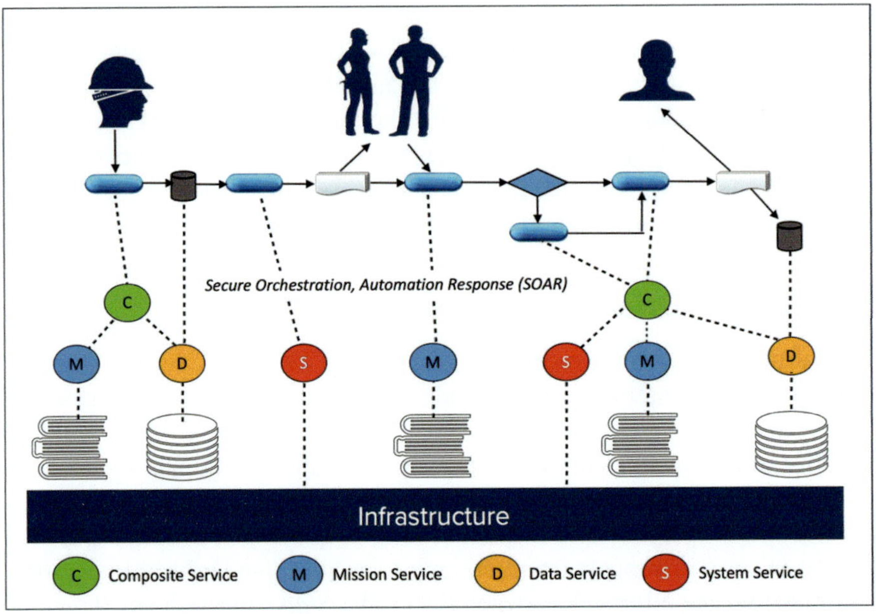

Figure 17.5 MOSAICS service-oriented architecture (SOA).

using a SaaS solution, the client hands over control to a third-party provider for software versions, updates, data, and governance (e.g., applications, data, runtime, Middleware, and operating system), which increases the overall security threat. IaaS also has security limitations. The client still needs to understand its organizational data security policy and governance. While the client controls the applications, data, runtime, middleware, and operating system, insider threats or system vulnerabilities sourced from the host or virtual machines (VMs) will pose a threat vector. Both SaaS and IaaS present risks that need to be managed and continuously evaluated against evolving threats. Unless an organization is growing its cryptographic security device (e.g., Type 1 security) and performing a security evaluation of the code, then the vulnerabilities are truly unknown. Cyber defense requires constant and consistent assessment of security decisions; however, it should not divert all resources to assessments and overlook the implementation of actual mitigations for cyber vulnerabilities.

17.5 Summary

Architecture approaches serve both practical and cyber purposes. One cannot separate the two in the cyber world. What distinguishes architecture from construction is its stability and permanence, as well as the experience of the ideas that flow through its form. The architect Frank Lloyd Wright's Prairie style of architecture is an enduring influence that is recognized and admired today. His buildings still work as intended, both functionally and spiritually. Cybersecurity must be addressed throughout the entire lifecycle stages of a system, from conception to operational use and maintenance, by both the systems engineer and systems architect. The form follows the function, guided by all the component elements to realize the required functionality and capability, formalizing all aspects and allowing for the reuse of design patterns. Designing and developing applications for mitigating cyber vulnerabilities in critical infrastructure requires unambiguous traceability to systems engineering requirements and clear language for describing the system. The architecture approach selected is core to the design.

References

Alexander Kossiakoff, W. N. S., Samuel J. Seymour, and Steven M. Biemer 2011. *Systems Engineering Principles and Practice*, John Wiley & Sons, Inc. Publication.

Borky, J. M. & Bradley, T. H. 2019a. *Effective Model-Based Systems Engineering*, Springer.

Borky, J. M. & Bradley, T. H. 2019b. Protecting Information with Cybersecurity. *Effective Model-Based Systems Engineering.* Springer International Publishing.

Haegley, D. 2020. DOD Cyber Strategy Implementation. *MOSAICS Industry Day.* MOSAICS Industry Day: OUSD.

Herber, D. R. 2020. Systems Engineering Architecture Lectures, SYSE 567. *System Architecture and Model-Based Systems Engineering.* Colorado State University: Colorado State University.

Haegley, D. 2020. DOD Cyber Strategy Implementation. *MOSAICS Industry Day.* MOSAICS Industry Day: OUSD.

Kilcoyne, M. 2021. MOSAICS Transition Strategy. MOSAICS Industry Day #3/TechConnect, 2021 Washington, D.C.: MOSAICS JCTD.

Munoz, C. 2025. More Situational Awareness for Industrial Control Systems (MOSAICS) Block I Integration Specification. *In:* (NAVFAC), N. F. E. S. C. (ed.). Naval Facilities Engineering Systems Command (NAVFAC) CIO Website: Naval Facilities Engineering Systems Command (NAVFAC).

Rich Scalco, D. B. W. 2021. More Situational Awareness for Industrial Control Systems (MOSAICS) Joint Capability Technology Demonstration

(JCTD) Overview Briefing. MOSAICS Industry Day #3/TechConnect, October 19, 2021 2021 Washington, D.C. DTIC, RDP21: MOSAICS JCTD, 12.

Rich Scalco, D. B. W., Jorge Lacoste, John Andrews, Bill Beary, Ross Roley, 2018. *More Situational Awareness for Industrial Control Systems (MOSAICS) Joint Capability Technology Demonstration (JCTD)* [Online]. IACD Automate: Johns Hopkins University Applied Physics Laboratory (JHU APL). Available: https://www.iacdautomate.org/may-2018-integrated-cyber [Accessed October 2019 2019].

18

Software Approaches

"The Internet+ is coming. With little forethought, or architecting, or planning. It's coming. It's going to change everything in ways we can only imagine and in ways we can't yet imagine"

– Bruce Schneier (1963–),
American cryptographer and computer scientist
(Schneier, 2018)

Abstract

If you want to raise eyebrows while reading in public, then read Bruce Schneier's book "Click Here to Kill Everybody: Security and Survival in a Hyper-connected World" (Schneier, 2018). An essential takeaway from Schneier's book is that while we might be able to disconnect from the internet individually, collectively, it is much more difficult. Software approaches in computing connect us all. Bruce Schneier said this before the massive wave of artificial intelligence (AI) that hit the planet from 2022 to the present, and so it is even more provocative at the moment. The term "Internet+" refers to software applications that utilize the internet and associated technologies throughout critical infrastructure sectors, including the economy, manufacturing, power generation, water facilities, healthcare, agriculture, and other related fields. Internet+ is also China's concept of "Internet Plus," unveiled in 2015 by the State Council of the People's Republic of China to deepen the integration of the internet with all sectors of the economy, including infrastructure (China, 2015). Connecting to the internet requires integrating software applications into the system. As described in earlier chapters, this can introduce unintentional vulnerabilities to the system. However, the objective should not be to stop a few malware threats. Our goal should be to design systems that are less susceptible to attack in the first place (Schneier, 2018). Incorporating secure software development into software engineering activities is a vital and continuous process. The entire team must embrace core cybersecurity elements. Purchased and open-source

DOI: 10.1201/9788743809128-18

software requires evaluation for vulnerabilities. Dynamic, measurable testing needs to be employed to provide a realistic understanding of software faults. Technologies such as artificial intelligence (AI), machine learning (ML), and software-defined networking (SDN) are valuable software approaches to focus on as key enablers of systems engineering for critical infrastructure in the cyber world.

18.1 Introduction

Incorporating security into software engineering activities involves determining information protection needs, defining security requirements, embedding security elements into the design, testing software during development, assessing the effectiveness of protection, and continuous development. Secure software development must be adopted by the entire team, including customers, designers, developers, testers, system administrators, and all other team members with relevant roles. Six core cybersecurity elements are:

1. Confidentiality – to protect the information and only allow authorized persons to access the information
2. Integrity – to protect the operational capacity of the system
3. Availability – to protect the availability of the system so that production can continue under adverse conditions
4. Authentication – to guarantee that the authenticating party is who they claim to be
5. Authorization – to control the access to information managed by the system; and
6. Access control – to ensure the authenticated person has the need-to-know. The requirements must be aligned with the security goals. Security has to be factored into the software early and cannot be an afterthought

Purchased and open-source software are common sources of vulnerability and require evaluation. Commercial software components purchased or licensed by vendors may be deployed and configured for the facility's processes as part of the system. The following are elements of a software strategy to ensure the code used is secure:

- Establish the functional and non-functional requirements to be verified, including policy and mandatory security controls.

- Ensure rigorous analysis of results, statistically significant data collection, use of appropriate tools, techniques, facilities, and expertise, as well as high-fidelity operational and support environments and configuration data.
- Perform internal testing and mitigate all discovered vulnerabilities. For example, bug detection is a method of deliberately injecting known dangerous, illegal, or semi-random inputs into a system to detect bugs and vulnerabilities.
- Use a third-party service for ethical hacking and penetration testing (PenTest) of the applications – the code should not be tested by the developers.
- Keep code secure by scheduling and performing periodic penetration tests (PenTests), automated software fuzz tests, and code scans.
- Using a third party frequently enough to validate that organic testing is adequate.

The randomness of fuzz testing can provide probabilistic reliability against software application faults. Fuzzing benefits from randomness, as it enables testing to uncover bugs that would otherwise be impossible to find using traditional testing methods. The benefits include simplicity, speed, and improved coverage of testing security attack entry points. Fuzzing detects exploitable issues without assumptions made prior to starting the test. Fuzz testing is an automated software testing technique that provides invalid, unexpected, or random data and records how the software behaves, monitoring for exceptions (e.g., crashes). The fuzz testing program is monitored by deliberately injecting malformed, illegal, semi-random, or known dangerous inputs.

It is important to use both use and abuse case scenarios in defining the secure concept of operations (SECOPS) to model and understand forms of defense, as well as forms of abuse that we want to prevent. The attack and defend standpoints are two main approaches used in writing cyber strategies. "When written from the attack perspective, cyber strategies focus on the security testing techniques that will be used to find and fix security vulnerabilities" (Diogenes and Ozkaya, 2019). A use case is an abstract scenario that describes the interaction between an application and its users or stakeholders. It helps the design development team deliver code and applications that meet the requested capabilities of a need statement. Use cases are, essentially, the "path of least resistance" for the delivery of what the client wants the application to do. An abuse case is a type of interaction where the outcome is harmful to the system, one of the actors, or the stakeholder's mission. The abuse case is an approach that involves thinking like a hacker to check each function and see how an unethical user or hacker could potentially break it. The abuse case describes the minimal abuse of privilege necessary to cause harm. Essentially, it describes how not to use the system.

18.2 Artificial Intelligence and Machine Learning (AI/ML)-driven Supply-chain Security

Artificial intelligence and machine learning (AI/ML)-driven supply chains are those in which intelligent analytics, including machine learning and artificial intelligence, are used to define the manner in which the supply chains work. Supply chains, which connect providers, goods, services, distributors, warehouses, consumers, and investigative personnel, by definition, comprise a vast and often vulnerable threat surface, making them obvious targets for cyber and cyber-physical threat agents. However, due to the long-standing track-and-trace model implemented in compliance with GS1 and other standards, the generalized supply chain also offers the opportunity for bi-directional application of AI/ML; that is, for both deep learning and deep unlearning. GS1 is an international, non-profit standard for maintaining barcodes and corresponding prefixes for companies that issue barcodes. GS1 provides a common digital language that enables companies to find, track, and fulfill products easily (GS1, 2025). By incorporating simultaneously "forward-looking" (i.e., predictive) and "backward-looking" (i.e., remediating) AI/ML, supply chain security can be ensured at both the structural and algorithm levels simultaneously. This insight is used to structurally and procedurally develop more secure and more responsive supply chains through five primary tasks:

1. Models for hybrid deep learning
2. Models for hybrid deep unlearning
3. Models for deprecation, ablation, and "ballot stuffing"
4. ML/AI and the identification of threats, including poisoning of federated networks; and
5. ML/AI and the remediation of threats. Through these approaches, we will be able to recommend a secure concept of operations (SECOPS) that builds upon the systems engineering concept of operations (CONOPS)

and provides traceability and provenance from the ground up, even in complex supply chains.

A key aspect of utilizing ML/AI in supply chains to enhance their security is, of course, having access to relevant data. Relevant data includes both de-identified supply chain data, such as that used to assess illicit supply chains for several industries (González Ordiano, 2020b; González Ordiano, 2020a), and simulated supply chains to which various analyses (Simske, 2023b) and incentivization approaches (Simske, 2022) (Simske, 2023a) can be applied. Our approach will be to further the work initiated in these earlier studies and augment them by simultaneously analyzing multiple supply chain networks. Independent discrepancies, of course, can be correlated to provide better localization of where supply chain irregularities are occurring. Our approach is to take multiple independent network analyses to the next level by training ML/AI algorithms to look for cross-network anomalies, that is, anomalies that have some presence on two networks at once. In this type of analysis, the term multi-network authentication (MNA) is analogous to multi-factor authentication (MFA). MNA is envisioned as a powerful means to prevent the more insidious types of attacks that are yet to come, such as ML/AI-based, statistically based, and dynamic/learning attacks.

18.2.1 Task 1: Models for hybrid deep learning

Our first task consists of defining models for hybrid deep learning. Model definitions will encompass both structural and spatiotemporal aspects. By "structural," this means that the multiple networks being evaluated will be analyzed by the algorithm most appropriate to their own nature. For example, a network that is largely intended for encrypted transactions (e.g., a financial or micro-service-based network) may be readily analyzed with expert system or simple machine learning approaches, including classification based on entropy and other network traffic descriptors; while a network aggregating sales, incentives, rebates, refunds, and other supply chain events is a clear candidate for advanced AI technologies such as convolutional neural networks (CNNs) and generative adversarial networks (GANs) where sufficient training is possible. By spatiotemporal, this means that the networks will be architected such that data deprecation and augmentation are both readily accomplished, and that consensus mechanisms, such as distributed ledgers, can be readily employed to prevent various types of attacks, including distributed denial of service (DDoS) and statistical attacks, such as poisoning of federated networks. We anticipated that "federated models" would emerge to accommodate federated networks, meaning hybrid models comprising two or more network-specific deep learning systems, as a result of this task.

18.2.2 Task 2: Models for hybrid deep unlearning

Once threat scenarios are identified, models for remediating them are appropriately investigated. One of the more promising models that have been designed using systems engineering principles (project definition, test and measurement, verification and validation (V&V), and sensitivity analysis steps performed as part of the concept of operations (CONOPS) is the hybrid deep unlearning approach, wherein elements of the training data and specific algorithms involved in a hybrid scheme such as the multi-network authentication (MNA) scheme described above are partially or fully retrained, or even deprecated, based on the changing needs of the overall network. These "macro" level approaches to deep unlearning will be applied after each hybrid deep learning module is posted. These macro-level approaches will be guided by generative adversarial network (GAN) approaches, specifically in the case of convolutional neural networks (CNNs) (Mansourifar, 2022), as well as by systems engineering test and measurement, verification, and validation (V&V), and sensitivity analysis approaches.

18.2.3 Task 3: Models for deprecation, ablation, and "ballot stuffing"

In addition to the "macro" level deep unlearning of Task 2, "micro" level adjustments to the hybrid ML/AI learning models will be performed, both separately and in conjunction with the "macro" level approaches described in Task 2. As such, Task 3 will be offset from Task 2. Work to date has indicated that deprecation of specific data, ablation of large, undertrained networks (to prevent them from hiding malware within network connections), and removal of "ballot stuffing" (where training data, usually synthetic, overestimates a particular type or distribution of data) are necessary. Ablation, in order to more precisely match the size of the network to the size of the intelligence problem posited by the training data, is of particular importance here.

18.2.4 Task 4: ML/AI and the identification of threats

Machine learning is an excellent starting point for temporally misaligned networks, which are themselves important categories of MNA. Past work in supply chain authentication, for example, has shown that certain events (e.g., the mining of legitimate unique identifiers from legitimate shipments in one location) can anticipate later illicit events (e.g., the re-printing of these unique identifiers on counterfeit products). The worldwide illicit economy is estimated to be in the trillions of dollars annually (World Economic Forum (WEF) Council on Illicit Economy) and is very dynamic. Hence, the threat identification will, by necessity, include the identification of poisoning attacks on federated networks and obscured malware in undertrained networks, matching the anticipated "macro" and "micro" attacks listed above. Where possible, a red-team/blue-team approach will be employed, with Task 4 serving as the "red team" and Task 5 serving as the "blue team." Threat identification will be led by a team distinct from the threat remediation team. This approach is aided by the fact that the work will be carried out by students and faculty in three separate Departments (see "Facilities and Resources" below).

18.2.5 Task 5: ML/AI and the remediation of threats

Threat remediation is the final task to initiate, as it will benefit from the insights gained from the other four tasks. Task remediation is viewed as a combination of structural and spatiotemporal approaches, which is in line with our overall hybrid view of the supply chain network. Structural remediation encompasses changes in multi-factor authentication (MFA) and other identifier-based approaches, while spatiotemporal remediation involves the addition of new anomaly detection and network size optimization algorithms. Here, recent red team activity has identified the most significant emerging threat to ML/AI-based systems as that of network-resident latent malware (NRLM). With NRLM, rogue agents place malware within the deep learning networks. To remediate NRLM, AI-based "fuzzing" (utilizing approaches such as long short-term memory (LSTM) and evolutionary approaches like V-fuzz, or vulnerability-oriented evolutionary fuzzing) is employed, and the output from the network is monitored. Anomalies in network configuration (indicating infiltration and potential in-place data operations), network output (indicative of exfiltration and other unintended communication), and overall network behavior (e.g., measurable changes in response to the same inputs) are examples of signals that indicate the need for remediation.

Remediation then proceeds along the lines of Task 3 but is guided by the AI to perform "friendly adversarial" remediation of the latent malware in the network. Generative AI has been shown to provide improved AI defense mechanisms in imaging applications (Rejeb, 2021; Rejeb, 2020), including those in which time series data is wrapped into neural networks to benefit from their huge number of features (Anderson, 2023).

AI/ML-driven supply-chain security is an important project to offset the increasingly sophisticated AI-based attacks on networks, such as the poisoning and hidden malware attacks described herein. Using ML, AI, and generative artificial intelligence (GAI) tools ensures a healthy red-team/blue-team situation in which the "ethical hackers" help train the system protectors to both understand and incorporate cutting-edge ML/AI tools in their network defense.

18.3 Software-defined Networking (SDN)

Another type of software-driven approach is software-defined networking (SDN), discussed earlier in Chapter 16, "Modernization Strategies." In Chapter 16, we discussed how SDN can be examined as part of an overarching cyber strategy. Here we discuss this software approach in more detail. Software-defined networking (SDN) is a software technology of interest in the energy sector, which originated from the need for an agile data center network fabric that dynamically connects users to servers and storage in the cloud computing era. SDN is an evolution of the classic WAN model in which the control and data planes, rather than being integrated into a network element, are now physically separated. In SDN, the control plane function resides in a centralized platform named the SDN controller, which is also equipped with an open application programming interface (API). In this way, the network becomes programmable, allowing other applications to control and monitor the network using the controller programmatically.

As an example, analysis of alternatives (AOA), we examine software-defined networking (SDN), which is located in software and is referred to as a flow controller, against the traditional network control plane situated in network appliance hardware (e.g., switches). SDN technology offers greater capabilities. However, while it is fast emerging in use in enterprise networks, it is still in development. It is expensive and may pose security risks that are less prevalent in hardware with built-in firewalls. The comparison is further examined in the economic analysis and decision-making, which may counter initial objections based solely on price and stability attributes.

To ensure that the ultimate requirement will be met at (or before) the critical design review, it is essential to assess the lifecycle cost of production of any unique products, which may have high acquisition, sustainment, and maintenance costs to support the systems in varying configurations; assess the critical useful-life of the items(s) for maintenance and component replacement plans; maintain configuration control throughout the process, and effectively implementing the design process effectively. Design review, evaluation, and feedback are effective steps and represent systems engineering best practices for a continuous, ongoing, and interactive day-to-day process review and evaluation. Formal design reviews at specific gateways for design and system acquisition decisions are necessary to realize the applicable technical performance measure (TPM) of life-cycle cost. For example, considerations before the critical design review for systems developed for the DOD typically include government-off-the-shelf (GOTS), systems developed exclusively for the purpose, or commercial-off-the-shelf (COTS) products, which are commercially developed and produced. GOTS is custom-developed software that can add to the total lifecycle cost of the system. A better option is to look for available, high technology readiness level (TRL) commercial off-the-shelf (COTS) products and, after performing an analysis of alternatives (AoA), then down-select an appropriate commercially available product to meet the requirement specifications. Another step is to "maintain some form of configuration control (i.e., "change" control) in the process" (Fabrycky, 2011).

Adherence to the baseline management can aid in maintaining the system design and acquisition process. Stringent configuration management (CM) contributes to the integrity of the system design by maintaining the vital system development baselines, which include the functional baseline, the allocated baseline, and the product baseline. These baselines are essential elements throughout the system lifecycle. Recall in Chapter 14, "Detailed Design," that design criteria are applied at each system level throughout the system design, preliminary design (subsystem level), detailed design and development, and system integration and testing. The CM must be maintained throughout, capturing each change and justification. "Formal change control of system-level changes is usually exercised by a designated group composed of senior engineers with recognized technical and management expertise capable of making judgments among performance, cost, and schedule" (Alexander Kossiakoff, 2011). The goal of integration is to engineer the new system into a compelling operating whole.

The software-defined networking (SDN) network architecture is a key benefit of the technology. The power of SDN lies in the application programming interface (API), which enables rapid changes to network connectivity and services to be synchronized and automated through applications. In effect, it converts the network from a static to a dynamic state of operation. "Shortening the acquisition process (cycle) continues to be a desired objective" (Fabrycky, 2011). Promises of SDN are found in the dynamic ability to control operations. Modern grid operations are becoming increasingly dynamic with the move to a smart grid.

Additionally, dynamic network services are necessary to fully leverage the benefits of cloud computing, as they enable more responsive bandwidth utilization. In addition to centralized network intelligence and optimization, SDN offers several other advantages to utilities. These include integrating multiple technologies (e.g., microwave, IP/MPLS, or optical) and technologies from different vendors, as well as enabling a multi-tenant shared network through network slicing. The controller thus also plays a unifying role. In addition, it offers both enhanced network security and agile service automation, and, importantly, it can be cloud-enabled in the current utility IT environment.

Depending on the use-case requirements, software-defined networking (SDN) can be implemented as an overlay to or an evolution of an existing wide area network (WAN). Three example use cases include extending the network reach to microgrids, as demonstrated in the Duke Energy OpenFMB microgrid, and IT/OT integration through network slicing and real-time multi-layer network optimization [Engerati-Duke Energy Microgrid Proves Interoperability]. As energy resources become more decentralized, microgrid deployment is becoming increasingly prevalent, both in large facilities such as corporate campuses, college campuses, data centers, and industrial plants, as well as in small residential communities.

Utilities will need to expand their operational networks to reach all microgrids, which will require rapid provisioning and management of a large number of gateway routers. An SDN operating over a private WAN network, service provider VPN, or the internet would give the utility complete flexibility to reach the deployed microgrids while retaining standard provisioning and operational procedures. Recall earlier efforts mentioned in Chapter 16, "Modernization Strategies," such as the DOD and Department of Energy's (DOE) Smart Power Infrastructure Demonstration for Energy Reliability and Security (SPIDERS) research that explored microgrids. The SDN automated provisioning and configuration management capability also allows the microgrid site connectivity to be provisioned readily without requiring a network specialist.

IT and OT organizational integration in critical infrastructure is essential. Up to now, IT and OT applications have generally been managed separately under different organizations and run over separate communication networks. A key initial step in transitioning to IT and OT convergence is to integrate the respective networks into a common, shared network while preserving each organization's current operating practices. While some advanced network management systems enable multiple organizations to share a common network through the use of virtual private network (VPN) technology and span of control capabilities, they fall short of empowering each organization to administer its network resources with full autonomy.

A software-defined networking (SDN) capability enables operators to virtually segment the integrated network into "slices" for multiple organizations or "tenants." Each organization owns a set of pre-assigned network assets, such as an equipment port complete with full administrator privileges, thereby enabling them to retain their operational processes.

Network infrastructure optimization is another benefit of SDN. Applications in a modernized grid are becoming more bandwidth-intensive and dynamic. For instance, operation crews are equipped with tablets running applications such as geo-databases and video conferencing. During power restoration after storms, as crews access and exchange information, networks in the affected areas can experience a surge in bandwidth. SDN offers utilities the opportunity to dynamically direct traffic across the network to avoid congestion by utilizing the path computation engine (PCE) within the SDN controller. An SDN controller can monitor key network performance indicators, such as packet loss and delay when provided with full network service and network-wide information. When congestion is indicated, the SDN controller can compute new optimal paths for the affected services. By providing smarter end-to-end network control, automation, and service agility, SDN rises to new operational challenges. SDN is a suitable strategy for utilities to follow on their modernization journey and is likely to flourish in the electricity market.

However, as with any other new technology, there are challenges associated with it. Test and evaluation (T&E) is required to introduce any new technology into a system, regardless of the functionality's technology readiness level (TRL). For example, SDN must be tested in the appropriate operational environment. Of particular interest is the potential security vector of attacks using timing to disrupt OT found in critical infrastructure. OT-SDN must be tested for resource efficiency, scalability, and centralized traffic burden (i.e., OT-SDN may not be capable of supporting network automation on large-scale networks, such as in an enterprise strategy) to inform the implementation strategy. There needs to be security testing for single points of failure, such as denial-of-service (DOS) attempts to control the control plane and manipulation of forwarding rules, as well as testing for user and server synchronization. Key questions are whether single points of failure are known, and for large-scale networks, who is responsible for industrial control systems (ICS) and supervisory control and data acquisition (SCADA) used to control geographically dispersed assets.

Therefore, additional consideration should be given to the system's concept of operations (CONOPS). For example, does the asset owner follow a centralization or decentralization operational strategy? An additional challenge is found in the data strategies that separate monitoring and controlling data, as well as understanding the flow of information, such as the inventory of incoming data. Still, the workforce must be trained to meet the expectations outlined in any critical infrastructure concept of operations (CONOPS). Singapore published career pathways for IT and OT professionals in 2021, providing performance expectations for various regulated occupations within the "Operational Technology Cybersecurity Competency Framework" (Singapore, 2021). The DOD published a similar guide, the "DOD Cyber Workforce Framework (DCWF)," in 2023 (CIO, 2023), which was referenced in earlier chapters. The DOD included workforce identification and coding guides in the framework for government and civilian work roles working on government contracts but has not gone as far as Singapore in regulating occupations. The workforce is discussed in more detail in Chapter 21, "Workforce."

18.3.1 Arc flash protection

Other potential uses of SDN have yet to be realized. Co-invention with technologies such as software-defined networking (SDN) can have a tremendous positive impact on critical infrastructure in applications like cyber vulnerability mitigation and preventing personal injury from equipment hazards, such as arc faults. Arc faults are causes of personal injury hazards, collateral damage to equipment, and interruption of processes in the power sector of critical infrastructure. Early detection of an arc is crucial for rapidly clearing the fault and protecting the switchgear or control gear associated with power systems. Arc faults release copious amounts of unpredictable electric energy in the form of flashes. Arc-resistant equipment is designed to provide a protection barrier by redirecting an arc flash hazard away from an operator. A technology used to limit arc flashes is currently being implemented to reduce and extinguish arc energy when detected. Reducing incident energy levels and decreasing arc flash time are key components of the protection relay system, which helps minimize hazards and reduce collateral damage. The systems engineer understands the network design, interfaces, and thresholds, and the power engineer understands the voltage and voltage systems.

Co-invention with power engineering could lead to more effective experimentation and collaboration for the development and design of optical sensor types that can be used to enhance microprocessor-based relay protection capabilities for arc flash protection relays utilizing light detection. For example, research and development in energy projects are shifting toward cyber-controlled physical systems driven by technologies such as software-defined networking (SDN). Power systems today are controlled by serial relays. Within 2 to 3 years, power engineering operations of energy systems or subsystems will no longer be able to purchase serial protection relays. The commercial market will be IP-based relays.

Today, it is possible to respond to an arc flash within 15 milliseconds, while SDN can react to an arc flash within less than seven milliseconds. Co-invention between systems engineering of the architecture of systems with power engineering becomes interesting in combining technologies to "detect and interrogate both arc flash and the associated arc current signatures in combination with arc-resistant switchgear or controlgear [that] can provide a coordinated solution for controlling the level of incident energy at various points within the distribution network of a forest product facility" (Zhao et al., 2019). It would be interesting to apply systems engineering principles to the development of a multiplexer (GHz to THz converter) utilizing super electrically conductive materials, such as graphene, to retrofit protection relays for compatibility with software-defined networking (SDN) networks. This approach would enable the higher spectrum THz wavelength and frequency to be utilized for arc flash detection.

Early detection of an arc flash is necessary to extend beyond existing safety standards, with a focus on the speed and reduction of incident energy. Ultra-fast arc flash protection can be achieved using "light and current based arc flash protection, the arcing time consists of the operation time of the protection relay (2- 15ms) and the operation time of a circuit breaker or contactor (typically 50-80ms)" (Zhao et al., 2019). The invention could "step up" a cyber-physical system protection relay to the speeds of IP switches to detect electric arc faults for safety and collateral equipment protection. This approach would bring together the strengths of digital logic systems engineering and power engineering disciplines to more effectively address a problem statement by co-invention of a solution.

18.4 Encryption and Entropy

Two methods of design for system hardening use multi-factor authentication (MFA) or one that uses two inputs (two-factor authentication (2FA)) for the hash (e.g., PKI public key and the media access control (MAC) address of the device assigned to each device connected to the network), particularly when combined with a trusted timestamp approach; and secondly, one that accelerates authentication upon suspicious behavior that would require re-authentication. Both system-hardening approaches reduce the risk of unauthorized access to protected information. Multi-factor Authentication (MFA) requires more than one credential to verify a user's identity. MFA immediately neutralizes risks associated with compromised passwords by adding a layer of security to protect personal information (University, 2019). When MFA is combined with a trusted timestamp approach, synchronization of communication is part of the data transmission (i.e., party A sends the email to party B at this specific time, rather than party A's data). While re-authentication is an essential method for reducing the risk of unauthorized access, it comes with some usability issues, as it burdens the user who needs to re-authenticate if incongruent behavior is detected in a specific context.

18.4.1 Encryption

Encryption is a key security strategy for both data at rest and in motion. A security strategy must consider both states to ensure that information is protected when stored (e.g., servers, databases, local hard drives, and cloud storage) and when it is in motion (e.g., file sharing, web uploads and downloads, WiFi network transmission, and cloud data sync). The security approach also needs to take into account the data in use, which activity logs and behavior analysis can track. Search analytics and machine learning can be used to automate the detection of threats to the data.

The encryption strategy is part of a broader security defense-in-depth (DiD) and Zero Trust strategy. The encryption strategy encompasses the governance and guidelines that start with an effective data classification program. Data classification policies can be utilized to enhance the management of sensitive data and promote a culture of security within the facility. Data encryption in the facility's information system should be implemented for both data at rest and data in motion. Data-in-use is a challenge that may be addressed using a distributed solution in deployed cloud services. Data-at-rest encryption encompasses encrypting file systems (EFS), virtual encryption for storage protection, database encryption for protecting structured data, and full disk encryption (FDE) for endpoint protection. Data-in-motion should be used for virtual private network (VPN) remote access, secure sockets layer (SSL) for web browser-to-server communications, and Wi-Fi-protected access (WPA) for wireless access points. Technologies such as SSL VPN are crucial in defending against man-in-the-middle attacks. The preferred use of private (symmetric) and public (asymmetric) keys or public key infrastructure (PKI) can be deployed for handling key distribution and management, which consists of a certificate authority (CA) that issues and verifies digital certificates to prove ownership, a registration authority (RA) that verifies the CA prior to the issuance of a digital certificate, a directory where the CA is held, and a certificate management system. A certificate management system for the facility should encompass the entire lifecycle, from initial issuance to formal retirement. The system includes trusted administrative oversight, inventory (i.e., certificate type, deployment, expiration date, the responsible entity for certificate management, monitoring compliance and inventory, tracking expiration, certificate installation support, and revoking or retiring certificates as required by policy).

Public key infrastructure (PKI) is a solution for non-repudiation of message transmissions through digital signatures. Non-repudiation is achieved by having a unique public key, which means that it cannot be refuted that the member transmitted a message as the author once it has been sent digitally, linking the sender to the sent message. Non-repudiation means proof of the integrity and origin of data. Authentication is the verification of the identity of an access request to a system. For example, a digital signature may be considered to offer non-repudiation. The concepts are similar in that they provide a high level of assurance that a message was transmitted by the sender as identified. Non-repudiation is the assurance that the information was not forged or corrupted in any way.

However, no regulation in the United States legally backs non-repudiation, so there may still be a potential dispute regarding the level of certainty of authentication. Generally, public key infrastructure (PKI) comprises the certificate authority (CA) and the registered certificates issued to CA users. Every member has a public key and private key pair. The single public key infrastructure (PKI) embedded in a certificate serves as a unique identifier for the key holder. The public key can be shared with any entity and is used to encode a message sent to a member. The private key is kept secret and used to decrypt the message when received. Essentially, every member has a unique public key, which allows a communicator to verify that the person with whom they are communicating is who they claim to be. A security strategy must consider both states to ensure that information is protected when stored (e.g., servers, databases, local hard drives, and cloud storage) and when it is in motion (e.g., file sharing, web uploads and downloads, WiFi network transmission, and cloud data sync). Encryption is a crucial security strategy for both data at rest and in motion. A security strategy must consider both states to ensure that information is protected when stored (e.g., servers, databases, local hard drives, and cloud storage) and when it is in motion (e.g., file sharing, web uploads and downloads, WiFi network transmission, and cloud data sync). The security approach also needs to consider data in use, which activity logs and behavior analytics can track. Search analytics and machine learning can be used to automate the detection of threats to the data. The data at rest in the historian and the transition of data in motion are both encrypted.

18.4.2 Timestamp

A timestamp guarantees uniqueness. A timestamp and record of all online transactions validate all data entries, detect and block any malicious inputs, and ensure the integrity of all data. A hash function is used as a hash (message and time stamp) and a hash (message). The hash of the message and the message timestamp are compared upon receipt to ensure secure delivery to the correct recipients, who sign the receipts. The recipient identification (ID) is required in the headers. Key questions are whether timestamp data is configured accurately, synchronized, and managed across the system, and whether timestamp data is used to reference persistent time-based trends (Scalco, 2022).

18.4.3 Public and private keys

Public key infrastructure (PKI) is a solution for ensuring the non-repudiation of message transmissions through digital signatures. Non-repudiation is achieved by having a unique public key, which means that it cannot be refuted that the member transmitted a message as the author once it has been sent digitally, linking the sender to the sent message. Generally, PKI comprises the certification authority (CA) and the registered certificates issued by the CA to its users. Every member has their own public-key and private-key pair. The single PKI embedded in a certificate is a unique keyholder identifier. The public key can be shared with any entity and is used to encode a message sent to a member. The private key is kept secret and used to decrypt the received message. Every member has a unique public key, which ensures that the person you are communicating with is who they claim to be.

18.5 Summary

Many potential software approaches for security have yet to be realized. However, like any other innovative technology, it comes with its challenges. Incorporating security into software engineering activities involves determining information protection needs, defining security requirements, embedding security elements into the design, testing software during development, assessing the effectiveness of protection, and continuous development. Secure software development must be adopted by the entire team, including customers, designers, developers, testers, system administrators, and all other team members with relevant roles. Team engagement is especially important in the era of AI since more and more software will be produced by AI, with the first "human" looking at the software akin to the role played by a pair programmer or a software tester. The framework for software testing is increasingly incorporating the role of code walkthroughs, dating back to the object-oriented (OO) and even earlier days of software development. Fortunately, lessons learned from those days can be borrowed forward for the testing of AI-generated code. For example, OO designs significantly increased software modularity by providing encapsulation (usually with well-designed and well-defined classes), reusability (classes could be imported into other software environments), scalability (class definitions can be used to effectively partition systems in a manner like that of network segmentation used in cybersecurity), and maintainability (inheritance, abstraction, and encapsulation primary among the maintenance advances). AI code can be guided to follow the best aspects of OO design and development, and in doing so, provide code security akin to that of the best OO systems.

In addition to testing, methods of system hardening design must be applied consistently to create systems that are less likely to be successfully attacked. To be more secure, software approaches require rigorous systems engineering, not just application development focused on functionality. Software approaches can also be leveraged to achieve safety and security. Other technological innovations, such as biometrics, can be used to secure application and system access. We will explore biometrics next in Chapter 19, "Biometrics."

References

Alexander Kossiakoff, W. N. S., Samuel J. Seymour, and Steven M. Biemer 2011. *Systems Engineering Principles and Practice*, John Wiley & Sons, Inc. Publication.

Anderson, W. J., and Steven J. Simske 2023. Forensic Identification of Environmental Sensors Through Challenge-Based Biometrics. *IEEE Sensors Journal,* 23, 15721–15731.

China, T. S. C. T. P. S. R. O. 2015. China Unveils Internet Plus Action Plan to Fuel Growth. *English.gov.cn.*

CIO, D. 2023. DOD Cyber Workforce Framework (DCWF) Military & Civilian Workforce Identification & Coding Guide. *In:* (DOD), D.

O. D. (ed.). Office of the DOD Deputy Chief Information Officer for Resources & Analysis, Workforce Innovation Directorate.

Diogenes, Y. & Ozkaya, E. 2019. *Cybersecurity — Attack and Defense Strategies,* Birmingham, UK, Packt Publishing Ltd.

Fabrycky, B. S. B. A. W. J. 2011. *Systems Engineering and Analysis*, Prentice Hall.

González Ordiano, J. Á., Lisa Finn, Anthony Winterlich, Gary Moloney, and Steven Simske 2020a. A method for estimating driving factors of illicit trade using node embeddings and clustering. *Mexican Conference on Pattern Recognition.* Springer International Publishing.

González Ordiano, J. Á., Lisa Finn, Anthony Winterlich, Gary Moloney, and Steven Simske 2020b. On the analysis of illicit supply networks using variable state resolution-Markov chains. *International Conference on Information Processing and Management of Uncertainty in Knowledge-Based Systems.* Springer International Publishing.

GS1. 2025. *The GS1 US Solution* [Online]. GS1 Website: GS1. Available: https://www.gs1us.org [Accessed 2025].

Mansourifar, H., Alexander Moskovitz, Ben Klingensmith, Dino Mintas, and Steven J. Simske 2022. Gan-based satellite imaging: A survey on techniques and applications. *IEEE Access,* 10, 118123–118140.

Rejeb, A., Karim Rejeb, Steve Simske, And Horst Treiblmaier 2021. Blockchain technologies in logistics and supply chain management: a bibliometric review. *Logistics,* 5, 72.

Rejeb, A., Steve Simske, Karim Rejeb, Horst Treiblmaier, and Suhaiza Zailani 2020. Internet of Things research in supply chain management and logistics: A bibliometric analysis. *Internet of Things,* 12, 100318.

Scalco, A. 2022. *Measuring Disagreement in Segments of the Cybersecurity Profession as a Means of Identifying Vulnerabilities,* Colorado State University.

Schneier, B. 2018. *Click Here to Kill Everybody: Security and Survival in a Hyper-Connected World,* New York, N.Y., W. W. Norton & Company, Inc.

Simske, H. K. A. S. 2022. Encouraging reuse in the corrugated packaging industry using persuasion and operant conditioning. *Sustainability,* 14.

Simske, H. K. A. S. 2023a. Demographic Considerations in Incenting Reuse of Corrugated Cardboard Boxes. *Sustainability,* 15.

Simske, H. K. A. S. 2023b. A Life Cycle Analysis and Economic Cost Analysis of Corrugated Cardboard Box Reuse and Recycling in the United States. *Resources,* 12, 22.

Singapore, C. S. A. O. 2021. Operational Technology Cybersecurity Competency Framework. Cyber Security Agency of Singapore.

University, C. M. 2019. *Multi-Factor Authentication: What It Is and Why You Need It* [Online]. Information Security Office Computing Services: Carnegie Mellon University. Available: https://www.cmu.edu/iso/news/2019/mfa-article.html [Accessed 2019].

Zhao, L., Zhou, Y., Hsieh, T. & Lee, W. 2019. Arcing Fault Type Identification with Light Spectrum. *IEEE IAS Electrical Safety Workshop (ESW)*. Jacksonville, FL, USA.

19

Biometrics

"Be yourself; everyone else is already taken"
– Oscar Wilde (1854–1900),
Irish author, poet, and playwright.

Abstract

Biometrics are cyber-physical processes that use individuals' physical characteristics to validate identity, such as fingerprint scanners or facial or voice recognition technology. The security risk of using biometrics is that, eventually, any data can be hacked or stolen. Since pieces of an individual's physical identity are generally irreplaceable, this means that once the biometric data is stolen, it may no longer be in the individual's control and thus largely loses value for their personal identity. Good hash mechanisms for biometrics are unavailable, as hashes do not work well with biometric data. There is no recourse if biometric data is lost. Keeping biometric information limited or opting out of providing it reduces the security risk. To paraphrase Oscar Wilde, "Be yourself." To be yourself in the cyber world requires authentication and an expectation that everyone has their own unique, immutable digital qualities that can be used to authenticate their identity. Biometrics can be used to enhance risk management, authentication, and user access for critical infrastructure systems.

19.1 Introduction

Several of the ideas and broad elements of this chapter were originally presented in a 2009 IEEE conference [Sims09], and the concerns expressed there have only been magnified in the years since. One of the main concerns about the field of biometrics in cyber-physical security (biometrics are all about the physical person) is that the need for real-time multi-factor authentication (MFA) has never been higher and that authenticating the person involved in a service, application, or transaction is a way of providing this. One reason multi-factor authentication (MFA)is so important in today's security

DOI: 10.1201/9788743809128-19

processes is that the "traditional" security authorization triangle (possession, knowledge, identity) has, in some cases, been reduced to a single point (knowledge) because of the limitations to possession attributable to virtualization, and because of the limitations to identity attributable to the use of static biometrics. This triangle – privacy, security, and access control – may be revisited in some cases using dynamic biometrics, and it is our intention in this chapter to help point the way for this to happen.

19.2 Types of Biometrics

There are three main types of biometrics: physical biometrics, behavioral biometrics, and innate biometrics [Nana02]. Often, physical and innate biometrics are combined and referred to as biological biometrics [Shar19] since the differences between them are large differences in overtness. While each of these can be used for assigning identity, the processes, costs, accuracies, and security considerations vary considerably between them and within each of these three categories. Assigning identity through the measurement of physical attributes or behavior is increasingly using more sophisticated approaches, including artificial intelligence and dual-factor authentication (DFA) or, more generally, multi-factor authentication (MFA). Still, given the nature of the current system and network vulnerabilities, it is almost certainly merited.

19.2.1 Physical biometrics

A non-exhaustive but representative set of human biometrics includes physical, behavioral, and innate – or chemical – biometrics [Nana02]. Physical biometrics include (1) face recognition; (2) fingerprint representation; (3) hand geometry (outline, shape, and/or size of the hand); (4) iris recognition (the peri-pupillary portion of the front of the eye); (5) retinal recognition (back of the eye, accessed through the pupil); and (6) vein recognition (hand, earlobes, etc.)

These physical biometrics are the least dynamic of all biometrics, meaning that if a rogue agent captures them, they are effectively compromised by another authentication factor, such as video or human validation. If you are relying on physical biometrics for authentication, the authentication protocol is only as strong as the weakest link in the whole system around the biometrics. This weakness could be exploited to access the measurement itself (e.g., collecting your fingerprint from a surface) or the sensor registers, transmission, or backend database. However, physical biometrics can be an excellent factor in an MFA process.

19.2.2 Behavioral biometrics

Behavioral biometrics include two biometrics often categorized as "physical" biometrics: (1) heartbeat and associated measurement (the electrocardiogram, the vectorcardiogram, blood pressure, cardio phonics, etc.) and (2) voice recognition. These biometrics are based on continuous signals and sometimes multiple signals, and as such, have a broader temporal and behavioral component than the physical biometrics described above.

Other behavioral biometrics include (3) arm sweep action (movement of an arm through 3D space); (4) finger writing (e.g., analysis of touchpad usage); (5) gesture analysis (like arm sweep but here with the hands); (6) handwriting/signature analysis; (7) keystroke dynamics (latencies, pressures, etc.); and (8) gait (walking) analysis. The characteristics of these motion-based biometrics are often relatively static in the way they are employed but can be incorporated into more dynamic biometrics, as described below.

19.2.3 Innate biometrics

Innate or chemical biometrics are those not usually associated with physical or behavioral/continuous biometrics, and as such, typically involve more sophisticated equipment for the analysis. These include genetic analysis (DNA, RNA, mitochondrial RNA, etc.); (2) tissue assaying (chromatography, ligand binding, immunoassay [ELISA, etc.]); and (3) mass spectroscopy. Since these are typically more difficult to obtain than physical biometrics (being "covert" rather than "overt"), they are more expensive to analyze/assay and access. Still, otherwise, they are no more suitable (being "static") than physical biometrics for the task of dynamic biometrics. Since they are harder to access and more expensive to assay, they are, in general, uninteresting to us in biometrics for authentication.

19.3 Biometrics and the Security Triangle

As mentioned above, the security authorization triangle (possession, knowledge, identity) has, in some instantiations, boiled down to a single point of knowledge. A robust security triangle consists of possession (the assurance that you have an object required for authorization in your possession), knowledge (having the information required for authorization), and identity (being who you claim to be). One long-familiar device that illustrates this triangle is the ATM (automated teller machine): to use it, you have the card (possession), you know the personal identity number, or PIN (knowledge), and are typically being surveilled by a camera. A fourth element for security – location – is also known for the ATM and other fixed secure locations, although the location may be regarded as a form of knowledge. While there are, of course, problems with the ATM (perhaps the most obvious being the addition of a fraudulent interface or skimmer), other devices have even less security built in. Consider, for example, logging onto a computer. If you log in remotely, you can claim to be any user whose password you have discovered (through stealth, intelligent guessing, keyboard entry trapping, coercion, bribery, etc.). Logging in all boils down to knowledge. A similar "boiling down" is the use of personal questions ("What is your mother's maiden name?" "What was the name of your first pet?" etc.) as an additional layer of security.

As noted above, the most commonly used biometrics, such as face and fingerprint authentication, are static in nature. Once known, they do not (appreciably) change. This static nature means that a randomly selected person is unlikely to have a static biometric that matches yours. However, it also means that once information on one of your static biometrics is known, you can no longer use the biometric (unless a secondary factor is also employed). Also, simply using a different finger each time a previously used finger is compromised is unsustainable.

If the triangle of possession, knowledge, and identity is to collapse into a single point, it is probably better that it collapses into identity. Not identity-as-knowledge, for example, a fingerprint, but an identity as in passing-the-Turing-test identity. Dynamic, session-based, interactive identity, consistent in spirit with the way we know if we are talking to our friend or her doppelganger. For this purpose, a security rights layer is appropriate. The tradeoff here is between the security policy imposed by the multi-user system on its users and the privacy policy that the users wish to impose on the system. Many security policies impose a high upfront "cost" or annoyance on the user – logging in, providing a password, using an approved card or device, and submitting to surveillance, among other requirements. Users who prefer a high level of privacy may require additional security measures to ensure their data remains secure. Playing off these two policies – security and privacy – are the access control privileges granted. If access control is variable – that is, dynamic throughout the session – then customized access rights are required.

19.4 Dynamic and Challenge-based Biometrics

Biometrics can be used to provide dynamic session-based security. By "dynamic," we mean that the level of security can change during the session, and the level of authorization (e.g., access privilege) granted is adjusted according to the current session confidence. In general, dynamic biometrics are based on two or more behavioral biometrics. This confidence can either mean: (1) the presence of two or more sensors for behavioral biometrics, or (2) the capture of two or more independent, streamable behavioral biometrics. An example of (1) is a keystroke dynamics recorder, which captures both latencies and pressures during keyboard entry. Another example is a touchpad that simultaneously records finger-writing dynamics and pressure (e.g., a six-axis accelerometer-affixed touchpad). An example of (2) is an inertial device (e.g., a 6-axis accelerometer) affixed to a person's hand, which records both macro-motion (arm motion/gesture) and finger motion/gesture.

A dynamic biometrics system (DBS) describes how the available biometrics are combined. If, for example, three biometric recorders are available (speech recording, touchpad recording, and inertia recording), then the DBS will indicate which attributes of these three biometrics, when combined, are best for validating the user's identity. Thus, the relationship between the multiple biometrics can be personalized to the aspects of the user that stand out the most from their peers. Using a DBS, the key for combining the individual biometric data streams is the *session key* for possession, and the dynamic biometrics linked to or specified by the key validate the user's identity. The user's privacy policy is also provided; otherwise, the default privacy policy for the session is used. The service then requests specific information necessary to authorize access with the desired level of privacy. The service, meanwhile, assesses what combination of these resources (biometrics, possession, knowledge, location) is available. From this, a session "key" tuned to the available dynamic biometrics is conveyed to the user. User access to the service is thereafter maintained throughout the session by a high-confidence match between the user's biometrics and the identity biometrics available to the service. In this way, a "biometric VPN" (virtual private network) is maintained, and the DBS can be considered a secure link between the individual and the privileged information provided by the service.

19.4.1 Biometrics and multifactor authentication (MFA)

First off, we introduce the nomenclature here. For each person, a signature set of signals, $\hat{S}$, will be collected from the overall biometric content, $\hat{C}$, which comprises the set of all possible elements of the signature set, $\hat{s}$. The current active biometric fingerprint, or signature, $\hat{S}$, serves as the session key and is the current set $\hat{s}$ of signature signals. With the advent of advanced AI, this signature can be optimized in a plethora of ways. Once computed, it can be provided by an authentication service itself (historical signature). Otherwise, it can be updated through the interaction between the client and the service (interactive or learning signature). The second approach is preferred because it allows the signature $\hat{S}$ to be dynamic. A set of signals is relevant to a given portion of a session and, as such, can be time-averaged to give the salient set of elements in a general biometric content set, $\hat{C}$, that together comprise the session key (or possession key). Within this set of potentially employable biometrics, a subset can be automatically selected for the purposes of the identity key (described next).

For a given task, generating a biometric signature, $\hat{S}$, is a classification problem that requires training the DBS, determining the optimal $\hat{S}$ for the task, and using this $\hat{S}$ thereafter for identity determination. More accurate classification for specific tasks can be garnered using classification techniques. Many traditional types of classification can be useful for challenge-based and/or cognition-based biometrics. Biometric fusion – combining two or more biometric signals into a single identity classification – may require advanced classification approaches. Increasingly, combined or hybridized classification based on the combination of nonparametric and parametric classifiers [Chau09] and the combination of disparate classifiers using meta-algorithmic patterns [Sims06] are being considered to improve the accuracy of various image, text, and other classification tasks. These classification techniques will, in general, provide higher accuracy for specific biometric identity tasks by allowing the overall classification approach to be selected from a wide variety of possibilities. The best choice of classifier will depend on the specifics of the dynamic biometric system, as described in the next section.

19.4.2 The dynamic biometric system (DBS) and identity

The dynamic biometric system (DBS) is concerned with two different session needs. These are (1) proving that you are the person you purport to be, which is required for access privileges; and (2) proving that you are not you, which, once established, signals an end to the session. Errors against (1), termed "false positives" for identity, are more serious when the data or resources accessed are of higher security. Errors against (2), termed "false negatives" for identity, on the other hand, are more serious when disconnection is disruptive. From a security perspective, the DBS is usually designed to prevent false positives. However, this is, in general, a more difficult classification task.

These needs align with the distinction between identity (what name best matches this biometric data?) and determination (does my biometric data match the database data?) [Nana02]. As an example, for identity, suppose for a given dynamic biometric signal in a DBS, the odds of any two users "matching" are 1 in 10^4. Suppose the requirement is that false positive errors only have a 1 in 10^{12} chance of occurring. In that case, three independent signals with odds of false identity of 1 in 10^4 must be used to establish identity.

After establishing identity and authorizing access, the purpose of the DBS is to restrict access privileges if/when the system's confidence in the currently authorized identity drops. This type of analysis is based on the detection of rare events. For example, an ASR engine can look for evidence of outside-accent formants [Mok23]. A heartbeat monitor can look for a non-characteristic electrocardiogram (ECG) or vectorcardiogram (VCG). A challenge-based system, as described below, can request that the user provide feedback for which the user has the lowest historical variance (the optimal biometric "challenge response") for any of the measured signals.

19.4.3 Auditing, denial of service, identity, and evidence

The signature set, $\hat{S}$, can be used in combination with a unique "Session Key", or workflow key, $\hat{W}$, stored with the Service and not transmitted or shared with the User. This key, $\hat{W}$, can be determined in multiple ways, depending on the needs of the Service, including but not limited to:

1. Historical basis: Past user interactions in any or all of the biometric modalities are used to determine the set of salient coefficients.
2. Confidence basis: signatures with the highest classification confidence (usually specificity) are chosen.
3. "Randomized" selection of the signatures to prevent reverse engineering.
4. Challenges that are linked to the set $\hat{S}$ through some algorithm.

Proof of identity is required in at least three situations: (1) establishing the session, (2) increasing access rights, and (3) reasserting identity when the confidence of the DBS is insufficient to allow current access privileges. From a security standpoint, it is important to prove with an accepted level of tolerance that the identity of the person using the DBS is correct. Ideally, the confidence in identity matches or exceeds the confidence level required to access the rights (data, services, etc.) currently authorized by the DBS. If not, more frequent determination of identity – for example, through the addition of more biometrically challenging and response-analyzing features – is required.

General Bayesian probability is used to determine the confidence in identity, as shown here:

Eqn (19.1). Bayesian probability used to determine confidence in identity.

$$p(A \mid B) = \frac{p(B \mid A) * p(A)}{p(B)} = \frac{p(B \mid A) * p(A)}{p(B \mid A) * p(A) + p(B \mid !A) * p(!A)} \quad (19.1)$$

Here $p(A)$ is the probability it is you, $p(A|B)$ is the probability of it being you if event B occurs, $p(B)$ is the probability of event B occurring, $p(B|A)$ is the probability of event B occurring given it is actually you, $p(B|!A)$ is the probability of B occurring given it is not you and $p(!A)$ is the probability of it being someone other than you, or $1 - p(A)$. The overall probability $p(A|\beta)$ may comprise a vector β of many events B1, B2, …, BN, as shown here:

Eqn (19.2). Overall probability $p(A|\beta)$.

$$p(A \mid B) = \frac{\sum_{i=1...N} p(B_i \mid A) * p(A)}{\sum_{i=1...N} [p(B_i \mid A) * p(A) + p(B_i \mid !A) * p(!A)} \quad (19.2)$$

In this equation, $p(A|\beta)$ provides the means for combining multiple biometric signals to determine statistical confidence for identity. A different set of probabilities is stored for each biometric signal under a variety of conditions – for example, under different emotional states for voice, different bodily postures for heartbeat, etc. This equation can be used to compute, in addition, the probability that it is not you or $p(!A|\beta)$ (simply switch "$!A$" and "A" in the equation). While setting $p(A) \geq (1 - 10^{-9})$ is a reasonable rule (one in a billion chance of it not being you), it is not as obvious what setting should be used for $p(!A)$ to discontinue access privileges. This setting depends on several factors, including (1) the signal-to-noise ratio (SNR) in the signal; (2) the relative stability of the signal for identifying the individual; and (3) the odds of the metric identifying someone else with higher odds (herein termed the person-to-population ratio, or PPR).

This list of factors identifies interesting research opportunities in biometrics and, more generally, in analytics. One research thread for dynamic biometrics that is likely to continue is how to dynamically (and reliably) assign a probability value for $p(A)$ and, with most likely more difficulty, $p(!A)$. In addition, hybrid patterns [Sims13], capable of intelligently combining the output of multiple biometric systems, will be crucial for these determinations. How are multiple biometrics combined to give the best overall confidence that a person is who she claims to be, or that someone else is now posing as her? The use of a current biometric signature, Ŝ, allows for different biometrics to contribute differentially to the assessment of identity over time, making the user's identity much more difficult to steal or spoof.

Once the previous and current "fingerprints" Ŝ are compared, the value $p(!A)$ is computed. If it is statistically certain (with the system's desired level of confidence) that the person is not the person claimed, then services (augmented access control, etc.) are denied. If $p(!A)$ is not sufficient to disprove the purported identity, then $p(A)$ is computed and logged. This approach is important for several reasons:

1. Auditing. This is the archiving of the $p(A)$ and $p(!A)$ at appropriate sampling times.
2. Denial of service. When there is a large change in $p(!A)$ – independent of change in Ŝ, for example – then the access privileges currently authorized can be denied. Regaining these access rights requires re-authorization, for example, through behavioral biometrics, multiple biometrics, or challenge-based biometrics as described in the next section.
3. Biometric (identity) descriptor. If the data collected is associated with the correct identity throughout the interaction, it can augment the individual's biometric descriptor – the historical biometric data associated with the user. This information can be rolled back if needed. However, augmenting this allows for the adaptation of the user's behavior over time.
4. Evidence. If allowed in the jurisdiction, the information collected by the system can be encrypted, stamped, and retained for later evidentiary purposes.

Next, the privacy and security policies are applied, with direct implications for the required $p(A)$ and $p(!A)$ values. With the given set of {security policy, privacy policy, $p(A)$, $p(!A)$}, the individual can be allowed different levels of authorization. For example, if $p(!A) < 0.001$ after having collecting the original biometric signature, then generic (read-only) access may continue to be granted while $p(!A) < 0.1$, advanced (read-append) access while $p(!A) < 0.01$, and full access (read-write-append) while $p(!A) < 0.001$. More generally, the following policies may be applied:

1. While $p(!A) < T_{ro}$, read-only access is granted
2. While $p(!A) < T_{ra}$, read-append access is granted

3. While $p(!A) < T_{rwa}$, read-write-append access is granted

where T_{ar}=threshold probability for the specific access rights (ar) and $T_{rwa} < T_{ra} < T_{ro}$. Various access right privilege strategies can be adopted, although the graceful change in access rights incumbent with this approach makes it relatively easy to maintain baseline access rights (that is, continue a task at hand) without the continual need for ongoing biometric input. Then, additional access rights can be attained through the successful input of identifying biometric input.

19.5 System Considerations

An important system consideration is the choice of biometric signals. Due to the high bandwidth requirements and difficulties with illumination, among other factors, visual biometrics are not always suitable. Additionally, mobile devices are helping to make auditory and touch biometrics increasingly important in this area. Inertial (acceleratory) and chemical biometrics are also increasingly important and benefit from the increasing maturity of nanotechnology-based sensors.

Speech has been useful as a biometric at least since the invention of the telephone. Humans are very adept at voice recognition, and language interpretation includes the ability to differentiate between, for example, declaration and interrogation solely from the inflection (rising or falling pitch). Behavioral speech biometrics encompass changes in inflection, pitch, loudness, timbre, rhythm, and other factors that are influenced by emotions, health status, and other variables. Obviously, identity "passwords" can be created that more accurately identify the purported speaker's voice characteristics – choosing from the historical biometric data the word or words which provide the minimum $p(!A)$ – if a higher level of access privilege is needed or if identity confidence for the current level of access privilege is not being attained.

Biometric signals can be collected continuously, as is the case for face recognition from video, speech recognition from voice data, and fingerprint or gesture recognition through a touchscreen. Additionally, when current identity confidence is statistically insufficient, more interruptive methods of obtaining biometric data are required. Two such methods are (1) challenge-based biometrics and (2) cognition-based biometrics. A challenge-based biometric is a prompt for user input that is used to re-establish identity. Existing challenge-based prompts include requesting the re-entry of a password or the re-swiping of a fingerprint, which constitutes a biometric challenge. Cognition-based biometrics are concerned with the person's ability to perform a given task. The cognitive aspect can be based on user history; that is, the user may have rehearsed the response or provided it in the past, such as a signature or a vocal response to a query. Although disruptive to the user's task, the user already understands that occasional re-assertion of identity is needed to ensure security. Importantly, challenge- and cognition-based biometric tasks may be requested sequentially until either identity is established with sufficient statistical confidence or until the access control rights are revoked.

For behavioral, challenge-based, and cognition-based biometrics, latency (timing considerations) in the user's response is important. Identity counterfeiting is made more difficult when differences in an individual's adeptness to respond are measurable and unaffected by connection. This adeptness to respond is consistent with where a DBS is best deployed – when there is a fast and reliable internet connection.

The DBS can be used for virtually any system requiring security, privacy, and access control. Centering next-generation security systems around biometrics provides additional advantages. The sensors used to provide access control can also be used to provide environmental or personal monitoring. In medical applications, biometrics can provide both security through identification and clinical information suitable for analysis and interpretation by medical professionals. One example is the electrocardiogram (ECG). Information from the ECG and its related vectorcardiogram (VCG), blood pressure profile, and heart sounds can provide a "signature" for the individual under resting, sitting, standing, and other conditions, with values for $p(A)$ and $p(!A)$ that, while modest, are suitable for patient identification in a clinical setting. The clinical value of these sensor-input biometrics is obvious.

The DBS can be employed in financial, political, and corporate security systems – collectively designated as "enterprise" security systems. Here, the auditing and evidence aspects of the DBS can be deployed consistently with the system's governance. As such, compliance with the DBS can be largely determined by the service, with obvious incumbent advantages to user policy adherence.

The development and deployment of a DBS offer several further threads of research interest. Massive amounts of ground-truthed (training) data have been accumulated since the biometric data can be directly associated with a given user. The DBS, therefore, can provide a central repository of biometric data that is useful for analytics and data mining purposes. This central repository of data is also an absolute requirement for more advanced artificial intelligence algorithms, which notoriously require huge training data sets [Samb21]. This data may be helpful in determining better usability for applications, more facile means of gathering user biometric data to ensure security, privacy, and access right policies, and more generally for the community concerned with the connection between biometrics and health, safety, and environmental monitoring.

Ultimately, the goal of the DBS is to disappear into the background – or at least become as non-disruptive as possible. There are compelling current trends that support this goal. The traditional WIMP (window, icon, menu, pointing device) paradigm faces increasing competition from new human-machine interaction models, such as touch and voice control. Gesture and facial expression control – if not direct brain control – models may also augment the experience. Clearly, the human-computer interface is no longer a generic mouse and keyboard. The new modes of interaction are largely based on, not coincidentally, biometrics. New modes should make collecting dynamic biometric information suitable for the maintenance of a "biometric VPN" easier and less invasive.

19.6 Summary

The dynamic biometric system (DBS) outlined in this chapter considers current trends in sensing, identity research, mobile devices, cloud services, and advanced biometrics. It is meant to provide a fix for the collapse of identity and possession into knowledge and call for action on new research in security, privacy, biometrics, and biometric fusion.

Many aspects of this system pose interesting research challenges. Specifically, it corrects for differences in location, connection quality, device quality, and model, as well as the user's health and condition, and the effective selection of dynamic biometric signals to collect. Location and connection quality will impact the integrity and signal-to-noise ratio (SNR) of the signals collected. Research is often necessary to correct or calibrate for these differences. Device quality and model will affect the gain, phase, signal-to-noise ratio (SNR), and frequency response, among other factors, of the sampled signals. Again, device qualification and periodic calibration will be necessary to ensure the highest accuracy of the DBS. User health and condition, including emotional state and point in the diurnal cycle, must also be considered to optimize the accuracy of the DBS. Fortunately, these needs align with the requirements for data collection in advanced AI technologies, which are becoming increasingly important in today's biometrics.

It is worth noting that security is more of a net than a hook. Broad, interdisciplinary approaches that take advantage of multiple technology threads are generally more robust, scalable, and adaptable to the changing needs of the security community. The DBS is designed to provide both security and privacy in mind. In the Internet of Things (IoT) sensor-rich, mobile, distributed, and biometric world of today, a DBS makes more sense than ever before. A DBS certainly will coexist well in a world increasingly augmented by IoT, AI, and machine learning.

References

[Chau09] Chaudhuri P, Ghosh AK, and Oja H, "Classification based on hybridization of parametric and nonparametric classifiers," IEEE Trans Pattern Analysis Machine Intell, 31(7), 2009, pp. 1153–1164.

[Mok23] Mok PPK, Fung HSH, Cao GWL, and Leung CW, "How similar are the formants in the speech of bilingual speakers?" *International Journal of Speech, Language & the Law* 30, no. 1 (2023).

[Nana02] Nanavati S, Thieme M, and Nanavati R, "Biometrics: Identity Verification in a Networked World," New York: John Wiley &Sons, 2002.

[Samb21] Sambasivan N, Kapania S, Highfill H, Akrong D, Paritosh P, and Aroyo LM, ""Everyone wants to do the model work, not the data work": Data Cascades in High-Stakes AI," In *Proceedings of the 2021 CHI Conference on Human Factors in Computing Systems*, pp. 1–15. 2021.

[Shar19] Sharif M, Raza M, Shah JH, Yasmin M, and Fernandes SL, "An overview of biometrics methods," in *Handbook of multimedia information security: techniques and applications* (2019): 15–35.

[Sims06] Simske SJ, Wright DW, and Sturgill M, "Meta-algorithmic systems for document classification," Proc DocEng 2006, New York, NY: ACM, 2006, pp. 98–106.

[Sims09] Simske SJ, "Dynamic biometrics: The case for a real-time solution to the problem of access control, privacy and security," In 2009 First IEEE International Conference on Biometrics, Identity and Security (BIdS), pp. 1–10. IEEE, 2009.

[Sims13] Simske SJ, "Meta-algorithmics: patterns for robust, low cost, high quality systems," John Wiley & Sons, 2013.

20

Foundations of Systems Engineering

"Champions keep playing until they get it right"

– Billie Jean King (1943–),
American tennis player ranked number
one in the world from 1966 to 1968.

Abstract

Yet another metaphor that speaks to cyber – in this case, the need for modularity in cybersecurity planning, testing, and evaluation – can be drawn from the game of tennis. A great tennis player possesses a solid foundation in their technical skills, as well as their ability to respond accurately to a tennis ball, and their agility to move quickly on the tennis court. Mentally, a skillful player understands court positioning to maximize opportunities and can anticipate an opponent's shots, skillfully putting themselves where the tennis ball will be while maintaining confidence. Tennis requires the ability to adapt one's game to an opponent's strengths and weaknesses in each match situation. Everything about the game is winning without conceding any points. The tennis player must decide in a split second what tactic to use for every shot played. The best tennis player has a strong work ethic. Their work ethic helps them master the various skills needed to excel on the court, just as a "best" system is one that is engineered to satisfy a specific need and integrates diverse components to achieve the most balanced solution. This solution must consider and incorporate into its design the critical attributes of cost, schedule, and performance of the development, all while utilizing advanced technology effectively. A "best" system employs a systematic, multidisciplinary approach to integrate elements that work together to deliver the necessary capabilities. Top tennis players dedicate a substantial amount of their time to strength training and conditioning. In addition to its physical aspect, the sport of tennis is also a technical and mental challenge. Top players train all modalities to be world-class champions. Modularity in systems engineering involves decomposing complex systems into smaller, distinct

DOI: 10.1201/9788743809128-20

components or modules, each with a well-defined functionality. Similarly, the great tennis player practices in each of the functional areas. Modularity, operational test and evaluation (OT&E), prototyping, and various types of testing are all foundational to systems engineering. The development of advanced tennis players involves several years of developmental effort, just as advanced development for systems containing unproven concepts takes years to develop.

20.1 Introduction

A "best" system is one that is engineered to satisfy a specific need and integrates diverse components to achieve the most balanced solution of critical attributes, including cost, schedule, and performance, all the while incorporating appropriately advanced technology. An example of a "best" system employs a systematic, multidisciplinary approach to combining elements that function together to deliver the necessary capabilities in existing or proposed systems. An analysis of alternatives (AoA) study can be used to select the best design approach. Existing or proposed systems for critical infrastructure may use various modernization strategies to achieve cyber vulnerability mitigations as described in earlier chapters, such as the Zero Trust strategy, the More Situational Awareness for Industrial Control Systems (MOSAICS) framework, micro- or service mesh architecture, software approaches such as machine learning (ML) and artificial intelligence (AI), software-defined network (SDN), software-defined perimeter (SDP), virtual area network (vLAN), and/or hardware approaches such as quantum computing and biometrics.

The role of a systems engineer is critical, as the functional components and analysis of alternatives (AOA) can be performed by various independent engineering team members at different locations. For example, the initial cyber defensive capability for industrial control systems, developed through the MOSAICS framework, was realized by the Department of Defense (DOD), and Department of Energy (DOE) national laboratories (e.g., Sandia National Laboratory, Pacific Northwest National Laboratory (PNNL), and Idaho National Laboratory (INL)), as well as the Johns Hopkins University Applied Physics Laboratory (JHU APL). Each lab contributed advanced technology research on a subcomponent of the MOSAICS framework. The role of the systems engineer is to lead the diverse team of technical members needed to design, develop, integrate, test, and evaluate the system, ultimately transitioning it for field use while balancing desired performance against cost and minimum acceptable performance. The systems engineer takes a holistic view of subcomponents, system interfaces, and system interactions. Systems engineering principles encompass firsthand knowledge of the environment in which the system will be used, the operators who will utilize the system, technical domain knowledge of cybersecurity and mission assurance, as well as technical domain knowledge of industrial control systems (ICS) and supervisory control and data acquisition (SCADA) systems.

In addition to multidisciplinary technical knowledge, the systems engineer's leadership skills and their ability to link members of disparate organizational entities into a cohesive team are critical to achieving a common objective. The principle of a systems engineering "sanity check" helps to ensure that experimental results can be used to derive the magnitude of test results and overcome any skepticism of the design approach with a healthy dose of a skeptical yet positive attitude (Alexander Kossiakoff, 2011).

20.2 Modularity

The term "modularity" refers to the measure of individual component independence from its interfaces with other system components in functional allocation at the subsystem level baseline. The concept of modularity is essential to systems engineers. Characteristics of a modular system include the functional allocation into subsystems. The modular system is well-defined at the subcomponent level, with the most basic interfaces and component boundary interactions (Alexander Kossiakoff, 2011). It is analogous to the best object-oriented software system design. MOSAICS is an example of a system designed for modularity. A systems engineer reuses modules, such as security orchestration and automation playbooks, as well as high-definition modeling and simulation code, found in IT systems. These models and simulation codes will be reused across commonalities in critical infrastructure sectors (e.g., power, water, wastewater, heating, ventilation, and air conditioning (HVAC), and facility control systems). For example, reuse bridges complex signal and data systems with material and energy systems in oil refineries and electric power plants (Alexander Kossiakoff, 2011). Recall that operational technology found in critical infrastructure encompasses a broad spectrum of task-specific systems that are highly customized for specific industry applications and are physics-based, resulting in a physical transformation of a product or service (Engineers, 2017).

In contrast, traditional enterprise IT systems support business functions that support supply chain management and corporate operations. The team must utilize expertise from engineering and specialized technician competencies. Risks associated with applying generalized IT solutions to OT systems can include compromised operational systems that are less efficient and more vulnerable to hacking, as well as risks to life and safety, and decreased system fault tolerance, leading to unacceptable system downtime for critical infrastructure OT, such as power utilities (Engineers, 2017). Countries such as the United States, Canada, China, Australia, and the European Union (EU) are all researching smart grid applications and technologies and actively integrating advanced metering and two-way communication features into the power utility's advanced metering infrastructure (AMI) or back-haul system. These advanced technologies and smart applications will generate vast amounts of output data that need to be analyzed and handled securely, whether transmitted over wired or wireless communication infrastructure from the smart meter to the utility data center. Technology suited for one environment may not be suitable for another or abnormal situations such as those presented by a storm. Wireless technologies can be constrained by bandwidth and security limitations; however, their lower installation costs can make them a more cost-effective choice for large-scale deployments. Wired technologies offer greater capacity, reliability, and security, but can be costly (Gungor et al., 2011). Engineering helps make these types of trade-off decisions. As smart grids are integrated into power utilities, cybersecurity from cyber threats is of critical concern in maintaining grid integrity and reliability (D et al., 2025). Artificial intelligence (AI) and machine learning (ML) enhance predictive analysis and automated decision-making capability for grid operations such as load forecasting and energy management. However, critical near-real-time monitoring and control for dynamic management require the advanced use of digital twin technology and knowledge of power utility operations to provide virtual replicas for simulation and testing (D et al., 2025). The increased use of smart grid applications and technologies introduces risks that must be considered in the risk assessment, which for the DOD is the Risk Management Framework.

Additionally, the Risk Management Framework (RMF) controls are baselined for each independent entity, such as the Navy or Air Force. The RMF controls used by the Navy are mapped against the Air Force controls for reuse, with the gaps addressed independently by each service for authorization to operate (ATO) by the respective designated authority (DA). Eventually, the RMF will give way to near real-time system monitoring, which will enable continuous ATO driven by artificial intelligence (AI) analytics. The initial demonstration in the power sector is modeled and is reusable for integrating, testing, and evaluating the design solution in other sector systems, making this a highly modular system. "Risk assessment is a basic systems engineering tool, which is used throughout development, but especially during advanced development" (Alexander Kossiakoff, 2011).

Schedule and performance are competing characteristics to which the law of diminishing returns may apply in developing a system. System performance might reach a more desirable threshold if more time for development is available in the schedule. Agile development practices can also enable the faster delivery of capabilities at any level (initial, mostly functional, and fully functional) to the field. A trade-off analysis might demonstrate that a solution can provide increased performance while also enhancing personnel protection. The timely delivery of a rapid acquisition program within an agreed-upon schedule, versus the opposing situation wherein an organization can achieve higher performance with a longer development time, is an example of the type of trade-off expected when optimizing one aspect of the system competes with another aspect of the system. Systems engineering focuses on meeting technical performance requirements, cost, and schedule. Maintaining focus on the requirements in relation to the interdependence with the schedule means that trade-offs need to be made, consistent with the law of diminishing returns. The principles of systems engineering, such as life cycle analysis (LCA), help achieve the minimum acceptable system performance to meet the desired schedule. In the example of a rapid acquisition system that provides personnel protection, a lead systems engineer may decide to rapidly field a solution using an agile development process in the first spiral, achieving a minimum acceptable performance, and then introduce additional capabilities not met within the original schedule later.

The systems engineer must maintain a balance between performance and schedule, and the term "good enough" should be used to meet minimally acceptable performance standards. "Virtually all critical attributes are interdependent; a proper balance must be struck in essentially all system design decisions" (Alexander Kossiakoff, 2011).

Systems engineers use different processes and methodologies in the development of a system. Examples of systems engineering process models are the linear, "Vee," spiral, and waterfall models for system development (Ibid.). The use of the model may be better suited to a project with well-defined requirements. The "Vee" model focuses on early lifecycle verification and validation to ensure the project requirements are met. In the waterfall model, testing activities are conducted continuously after development activities have been completed. The spiral model offers greater design flexibility, which is advantageous for design changes implemented at various project phases. The spiral model is also more challenging to use. It can be better suited to complex systems engineering projects. For example, it is significantly better in the MOSAICS framework as it is dynamic by design and syncs with the various laboratories' research development processes and learning curves. It poses minimal risk to the overall project, as it enables the adoption of new technologies as features. The spiral model is iterative, with incremental releases throughout the project that may be used to refine and improve the system's performance and development. The model that will be used must be identified so that members understand the development approach taken by the lead system engineer to meet the system requirements. Usually, the systems engineering process model is identified in the project systems engineering plan (SEP). Systems engineering incorporates elements from several system domains, including traditional engineering disciplines and disciplines from related fields such as management (e.g., project management, human resources (HR), and research operations), modeling and simulation, quality, architectures, and control systems. It also incorporates political and legal, human and social, and other engineering field domains. These domains impact the systems' development to varying degrees, mainly depending on the scale of the environment. "Scales at the extremes are of particular interest due to their complexity" (Alexander Kossiakoff, 2011). For example, engineers working on systems for space operations outside of the earth's environment have considerations that require additional attention (e.g., the effects of space radiation on the systems, interactions between the space environment and the system, and structural integrity of the system in space). Systems engineering encompasses exposure to the broader picture and a comprehensive perspective, combined with in-depth knowledge in the technical domain, to address complex problems.

20.3 System Design Hierarchy: Electrical Power Plant Example

Using the power plant system as an illustration (Figure 20.1 and Table 20.1), the systems engineer would possess a broad knowledge of the properties of the entire engineered system, interfaces, subsystems, and critical technical components from the power plant to the customer.

20.3.1 Smart grid

Electric power plant systems are part of a larger electricity distribution system that ties together electric power plants, known as the "smart grid." The smart grid supports interoperability between the power plant system, power generation, and distribution. Systems engineers support smart grid infrastructures across a hierarchical three-tier architecture. The access tier supports near real-time information exchange between the user's home area

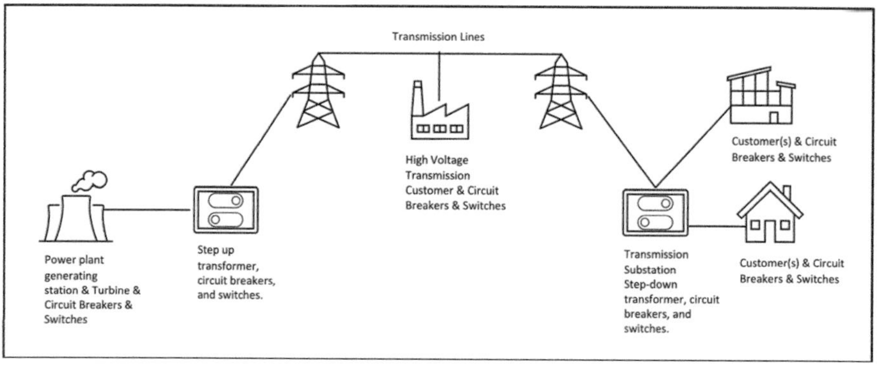

Figure 20.1 System design hierarchy for an electrical power plant.

network (HAN) or building area network (BAN) monitoring and control devices and the near real-time distribution substation control management field area network (FAN). The distribution tier connects the HAN or BAN with the smart grid FAN to provide reliable communications to sensors and actuators that control the power system assets. The core tier is a wide area network (WAN) that establishes a high-capacity, low-latency communication backbone spanning large distances, offering network protection, monitoring, and control.

20.3.2 Smart microgrid

Microgrids are a grouping of connected, distributed energy resources that operate as a single, controllable system, allowing them to connect or disconnect from the smart grid. Microgrids are localized, small community options that provide efficient, low-cost energy and enhance the reliability and resiliency of regional electric power systems in the smart grid. Microgrids can operate in various modes as a component system to a broader energy distribution system or as a local source of power generation for a specific location, such as a military installation or a geographically dispersed rural community.

There is increasing interest in smart microgrids, as described earlier in this book in the MOSIACS Framework section of Chapter 16: "Modernization Strategies." An example given was the joint experimentation effort between the DOD and the Department of Energy's (DOE) Smart Power Infrastructure Demonstration for Energy Reliability and Security (SPIDERS). Smart microgrids utilize two-way digital technology to manage the power supply to larger communities. An advantage of smart microgrids is the ability to

Table 20.1 System design hierarchy for an electrical power plant.

Systems	Function	Design specialist	Systems engineer
Power plant generating station	Generates power for electrical grids, which vary in size from single facilities to national grids (countries) and transnational grids	Electromechanical engineering/transport of energy/power from generation through various substations and conversion of mechanical energy into electrical energy, operational maintenance	Signals, data, material, energy, and activities, boundaries, and interface (e.g., primary substation flow toward load, consumers served), maintenance and lifecycle support, regulation, and policy
Subsystems			
Transmission substation	Carries overhead lines that transfer generated electrical energy from generation to distribution substations	Electromechanical engineering/transfer of generated electrical power, stepping electrical voltage up for transmission or down for distribution, operational maintenance	Signals, data, material, energy, and activities, boundaries, and interface (e.g., primary substation flow toward load, consumers served), maintenance, and lifecycle support
Components			
Transformer	Carries power from distant sources to demand centers	Electromechanical engineering/high voltage electrical engineering conversion of mechanical energy into electrical energy, transfer of power from one level to another, operational maintenance	Signals, data, material, energy, and activities, external environment, operator interfaces, system boundaries and interfaces, maintenance, and lifecycle support
Subcomponents			
Turbine	Drives the generator to convert energy into electricity	Electromechanical engineering/mechanical engineering conversion of mechanical energy into electrical energy, operational maintenance	Signals, data, material, energy, and activities, external environment, operator interfaces, system boundaries and interfaces, maintenance, and lifecycle support
Parts			
Circuit breakers, switches	Interrupts the current flow after a fault is detected	Electromechanical engineering/mechanical, operational maintenance	Signals, data, material, maintenance, and lifecycle support

operate autonomously and integrate various energy sources. Disadvantages include the costs and time-consuming effort of implementing and managing the systems. Microgrids can also help in the ability to restart parts of a power system to recover from a blackout. These microgrids can be used to restart the system from the bottom up and can help insulate the system from widespread disruptions that cyberattacks could cause. Resilient when functioning, the smart microgrid can also quickly be disconnected from the user. Engineering a smart microgrid also has its challenges.

Some commonly encountered systems engineering challenges that would be likely to arise with a smart microgrid include:

- Heterogeneity – The primary and smart microgrid systems may be built on varying technology baselines and may respond to dissimilar priorities.
- Complexity – The functions, interfaces, constraints, and technologies may exceed the capabilities of any individual system, particularly as new functionality and capabilities are added to the smart microgrid system over time, such as SOAR.
- Extending systems engineering models – A generally accepted inter-systems engineering strategy and methodology is needed. For example, the Navy's power utility system may have the authority to operate (ATO) based on the Navy's Control Systems Platform Enclave (CSPE) security. Smart microgrid systems may not have the inherent protection of the CSPE; therefore, a strategy for obtaining an ATO for these systems needs to be developed.
- Requirements definition and validation – The primary grid and the smart microgrid system may have competing priorities and needs. For example, what services are prioritized when there is an unscheduled power outage or intentional brownouts are used for load reduction in an emergency? Typically, the primary grid requirements will give priority to medical and emergency services.
- Interface definition and management – The primary and smart microgrid systems may have technical and functional differences. For example, the primary grid may utilize SDN capability, which features interfaces within the SoS that differ from traditional routing and cabling.
- Enterprise evolvability – Changes to an enterprise, such as the introduction of 5G at a smart microgrid system level (e.g., SmartBase/SmartCity), may not be utilized for the management of the primary grid

at Purdue Enterprise Reference Architecture (PERA) model levels 6, 5, and 4 due to safety and security concerns. Open architectures and loose coupling are crucial to ensure that the system of systems (SoS) can evolve in response to these changes.

- Governance – The service strategy and service operations design become increasingly complex for additional systems that utilize separate control configuration boards (CCBs) and governance models.

20.4 System Design Hierarchy: Passenger Vehicle Example

Automotive manufacturers have utilized motorsports events, such as the 24 Hours of Le Mans, as testing grounds for innovation and to demonstrate the effectiveness of innovative technologies. Recall from Chapter 11, "Lifecycle Phases," that the value of this 24-hour endurance race is a test of both driver and machine and a testament to the interdisciplinary systems engineering field in managing complex systems throughout their lifecycle. The racing events serve as field test environments and significant development proving grounds that inform the design and development of production vehicles. Systems engineering rigor is crucial for understanding how the integration of new technology affects vehicle components, such as the chassis and powertrain, as well as performance aspects, including fuel consumption and overall vehicle performance. From a systems engineering perspective, this concept is very similar to the smart grid. All of these parts come together, considering different requirements to meet the performance expectations of the driver and the team. The smart grid's three-tier architecture, home and building area network (HAN/BAN), field and neighborhood area network (FAN/NAN), and wide area network (WAN), can be likened to a Le Mans factory production team, with HAN/BAN serving as the driver, FAN as the race team, and WAN as the manufacturer's design team. The essentials of the systems engineering principles and practice are based on the same foundations. The three-tier architecture enables near real-time control over the distribution grid.

Principal parts, subsystems, and components for a passenger vehicle block diagram relating the subsystems and components to the system as illustrated in Figure 20.2. The systems engineer must possess a broad knowledge of various disciplines to develop complex systems effectively. The domain of the systems engineer encompasses several categories, extending from the system component level, and working in collaboration with the design

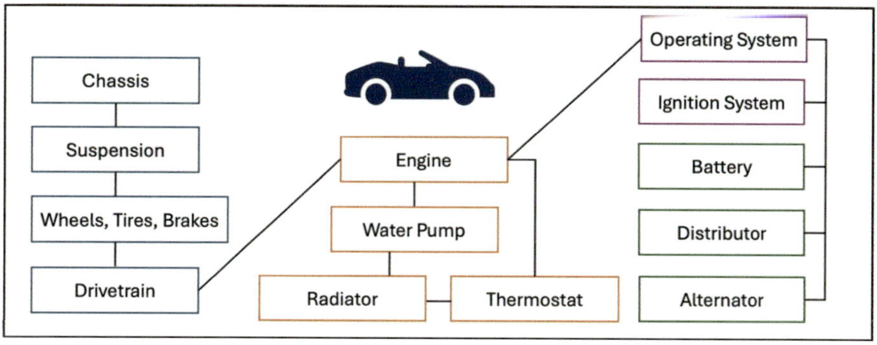

Figure 20.2 System design hierarchy for a passenger vehicle.

specialist, whose expertise and experience typically lie within a single discipline. Interfaces are a critical engineering concern that involves including subsystems and components to achieve the desired operational outcomes. Key activities for a system engineer that require technical knowledge at the component level include developing specifications, requirements, cost estimates, and analysis of alternatives (AoA) or trade-off studies. The systems engineer needs to probe into sub-component levels to effectively identify and discuss technical problems and component material solutions that meet the functional and physical specifications of the system design, as well as technology readiness levels (TRL), to achieve the best performance versus cost. The principal parts, subsystems, and components of a passenger vehicle are shown in Table 20.2.

20.4.1 Security engineering – The collaboration between security operations (SecOps), development operations (DevOps)

At the start of an initiative, describe who in the organization is responsible for the systems engineering and the systems engineering plan (SEP), and how system cybersecurity design considerations are addressed. Risk assessment frameworks are based on best estimates. Using a weighted methodology to normalize both quantitative and qualitative data is valuable in the decision matrix. Stakeholders determine the best site to implement a first cybersecurity defensive capability using a use case decision matrix. An example of determining the best initial site at a facility is provided in Table 20.3. Input from all stakeholders with system knowledge is needed.

Table 20.2 System design hierarchy for a passenger vehicle.

Category	Component	Significance	Functional element
Mechanical			
	Chassis	Singularity	Provides the basic frame to which the parts are attached
	Drive train	Important function	Transfers power from the engine
	Suspension	Singularity or Important function (depending on make and model)	Increases friction between the tire and the road to provide stability to ride and handling
	Wheels/tires	Important function	Transfers power to the ground
Thermomechanical			
	Engine	Important function	Generates motion
	Radiator	Commonality	Dissipates excess heat from the coolant
	Thermostat	Commonality	Works with the radiator to dissipate excess heat from the coolant
	Water pump	Commonality	Works with the radiator to dissipate excess heat from the coolant
Software			
	Operating system	Singularity or Important function (depending on make and model)	Control system
	Ignition system	Singularity	Causes combustion in fuel to make the car run (components are spark plugs, distributor, ignition wires, and coils)
Electromechanical			
	Battery	Commonality	Delivers the electricity to the vehicle's electric components
	Distributor	Commonality	Connects the ignition system components (i.e., coil and spark plugs) to the engine's cylinders in the correct firing order.
	Alternator	Commonality	Provides electricity to run headlights, brake lights, and internal electrical components

Table 20.3 Use case decision matrix.

Decision Factors		Facility 1	Facility 2	Facility 3	Facility 4	Facility 5
Criteria	Wt.	1	2	3	4	5
Strategic Objective						
Leverage Existing Tools						
Actionable (Location)						
Cost Affordable						
Time						
Support						
Weighted Scores		0.0	0.0	0.0	0.0	0.0

Criteria	Definition
Strategic Objective	How well does the site meet the target objective and benefit
Leverage Existing Tools	How much flexibility and support is available at the site
Actionable (Location)	How quickly and easily can the site design be customized
Cost Affordable	High Cost = Low Point; Low Cost = High Point)
Time	Meets Milestone Objectives
Support	How good is the sustainment and maintainabuility support

20.5 Unproven Components

Unproven components that utilize new technology can be attractive options for consideration in new system development. New technology can be an attractive option to replace an obsolescent system because a new system in development needs to meet performance requirements for many years beyond the original design, thereby reducing update costs throughout the system's life cycle. The maturity level of technology requires assessment to determine readiness for complex system development. The technology readiness level (TRL) is a standard for evaluating the maturity of a technology to determine if it is a usable choice for complex system development. TRL 1 is the lowest level of maturity, and TRL 9 is the highest. Many new complex system developments incur significant risks when choosing immature technology, as these components have not yet been integrated into representative systems for integration, testing, and evaluation (IT&E). Therefore, additional laboratory tests are necessary to measure performance parameters and make analytical predictions for subsystems. Reasons the complex system developments incur substantial risks include incomplete specifications until late in the system development life cycle, poor requirements definitions, unaddressed risks, lack of required expertise or inadequate expertise in the new technology chosen, and extreme weather conditions. New technologies often lack the rigorous test data that mature technologies possess.

Reliable performance requires years of development and testing. An example of when immature technology choices paid off is Apple's SMART Watch. Microsoft's similar earlier effort, the smart personal object technology (SPOT), is an example of a failure. Both companies developed smartwatches using new wireless technology to enable small-size-factor, integrated radio receiver solutions for personal messages, calendar updates, and other data. Microsoft's smartwatch relied upon one-way FM radio as cellular broadband technology emerged. Once cellular broadband became the norm, the data limitations via FM radio were considered outdated, and the project was retired. Apple and Google developed wearable smartwatches leveraging the disruptive technology of mature cellular broadband.

An example of model-based systems engineering (MBSE) that best illustrates the systems engineering principles discussed is the use of an initial architecture for a cyber defense system that utilizes the MOSAICS framework to modernize a critical infrastructure system. Risk assessment is used throughout the development process, especially during the advanced stages of developing a complex, real-world example. Technology is evolving at such a rapid rate that it can become obsolete before it is even applied. MBSE enables stakeholders to digitize the risk management plan, thereby reducing the significance of each risk area. "[T]he objective of the [risk management] plan is to minimize the total cost of managing each significant risk area" (Alexander Kossiakoff, 2011). The objective of managing costs associated with risk-related requirements makes model-based systems engineering (MBSE) a powerful tool. Basic requirements include coverage and capacity, scalability, robustness against failures, security (including both cyber and physical security), throughput, reliability and availability, costs, control, and other relevant factors. Continuous risk assessment using model-based systems engineering (MBSE) is a fundamental tool in systems engineering.

20.6 Operational Test and Evaluation (OT&E)

Operational test and evaluation (OT&E) is a foundational aspect of systems engineering. It is iterative throughout a system's incremental build approach. It will occur iteratively against the requirements stated in the operational requirements document (ORD), which outlines the system requirements, including its interface with other systems and the environment. For example, a complex hardware system used in an extreme environment is a vessel traffic system (VTS), which is involved in shore-based marine environments for port operations. It includes a system control unit with sensors at remote, all-weather condition sites in the port vicinity. The VTS operates in different temperature and humidity conditions. The system's operational performance requirements include monitoring, tracking, and surveillance, as well as automated and manual data processing, voice and non-voice communications, recording, and reporting. Operational performance requirements are identified in the operational requirements document (ORD) and the system specification. The ORD and system specification identify critical technical parameters. These parameters need to be evaluated, including accuracy, prediction capability, availability, environmental, data access, and data archive.

20.6.1 Management

The management identifies the entity and who will review the system's major test and evaluation (T&E) reports. A test management oversight team (TMOT) is established and serves as the primary point of contact with the local maritime and port representatives. It assists in preparing the design test and evaluation (DT&E) plan, executing test procedures, providing technical expertise, resolving test and evaluation (T&E) problems, reviewing test activities, and providing oversight during the system's acceptance.

20.6.2 Design test and evaluation (DT&E) overview

Design test and evaluation (DT&E) is iterative throughout the incremental build approach. The initial installation will be a commercially available system, rather than a custom-developed one. The test plan and procedures will focus on the system's technical performance acceptance at the first site. DT&E is discussed in detail in the DT&E Plan, including reliability, maintainability, availability, system safety, staffing and training, survivability, logistics supportability, environmental concerns, electromagnetic and environmental effects, interface testing, and evaluation. Test support equipment requirements and test site specifications for the test targets need to be identified.

20.7 Prototyping

The primary motivation for prototyping as part of a system development effort is to verify problem fixes, verify trade studies, and support ongoing engineering activities in smaller developmental increments. Such an approach provides for experiments from smaller builds with few system elements "all the way up to the full system" (Borky and Bradley, 2019). These development builds are small in scope and less expensive to execute. An example of such an approach was used in the initial development of the MOSAICS prototype. The technical team tested the prototype in small increments throughout the development phases, culminating in the final test increment. This approach enabled system corrections and test data verification, informing ongoing engineering development efforts.

20.7.1 Verification methods

A comprehensive approach to verification and validation (V&V) involves building test cases against the requirements in a simulation environment using virtual prototyping. "[T]he use of virtual prototyping involves its use to develop, analyze, refine, and communicate the smart microgrid (SMG) architecture" (Borky and Bradley, 2019). As mentioned earlier, virtual prototyping was the approach used by the initial MOSAICS prototype for power utility development. The approaches to using spiral development successively integrate new capabilities to meet requirements in each spiral prototype, which is tested against the requirements in a virtual environment and then in a physical environment with both hardware and software in the loop. Due to the complexity of the power utility system and its cyber defense requirements, this approach enables incremental verification and validation (V&V) and productive feedback from the engineering team. Effective feedback is used to correct problems.

Additionally, a well-formed model of the target system is developed in a simulation environment that enables software testing within the system model. Trade studies are used to augment the various levels of abstraction to quantify component performance in the system. Finally, the system is tested with the operator in a human-in-the-loop configuration as part of the final military utility assessment (MUA) demonstration. At this point, the prototype is determined to have utility or not for integration into production operations. A test and evaluation master plan (TEMP) scopes and documents the test activities for a composed solution capability such as MOSAICS.

20.8 Test and Evaluation Master Plan (TEMP)

The test and evaluation master plan (TEMP) documents the scope, content, methodology, sequence, management, and roles and responsibilities for tasks and activities needed to perform test activities. This test and evaluation master plan (TEMP) serves as the basis for verifying test results and validating the composed commercial solutions fielded using the proposed methodology. A test analysis report is created at each level of testing. The TEMP is prepared when developing the composed solution requirements. Problems identified during testing shall be tracked under configuration control as described in the configuration management (CM) plan. Changes and improvements based on the test results shall be incorporated into the composed solution capability document and maintained in the composed commercial solution library, such as for MOSAICS. The TEMP guides the management of test activities, encompassing organization, roles, and responsibilities. The TEMP also provides guidance on the extent of necessary testing, which serves as the basis for verifying test results and validating the system. The validation ensures that the functional requirements are met and that the subsystems are not adversely affected.

The purpose of the TEMP is to provide users with vendor-agnostic configurations that meet the demand for commercial off-the-shelf (COTS) products in composed solutions. The TEMP provides guidance and lessons learned from proof-of-concept demonstrations that can be reused, thereby shortening the time it takes to deliver timely, reliable, secure, and composed solutions to users. The TEMP provides the detailed data that is used in the capability package reference architecture and configuration information. A TEMP is developed for each composed solution capability package. The capability package enables users to implement secure solutions independently by utilizing composed COTS products (Scalco and Wince, 2012).

A TEMP is developed for each composed solution, such as for a MOSAICS implementation at a specific site in a particular sector. The system description provides an outline of the system's purpose and objectives. All subsequent features and requirements are based on this original description. The system description of the composed solution describes the system- or subsystem-wide design and the architectural design of a system or subsystem. The system description is captured in the capability package, along with its associated interface design descriptions (IDDs), and serves as the basis for further system development by each user case. It establishes the general data and content requirements for each composed solution. The composed commercial solution is described in terms of its general nature. The history of the system's development, operation, and maintenance is documented, along with a list of all relevant documents. The system's design is detailed from a user's point of view (Scalco and Wince, 2012).

The systems engineer ensures that previous testing is documented within the TEMP to illuminate when the testing was conducted and what the results were. Any lessons learned can be valuable in future testing. The system's key performance parameters (KPPs) and key system attributes (KSAs) need to be identified in the TEMP. Each parameter should include both a threshold and an objective value from the capability development document (CDD) and the capability production document (CPD). The CDD specifies the capability requirements in terms of key performance principles (KPPs), while the CPD captures all the necessary information to support production and deployment. The key interfaces segment of the TEMP should identify the interfaces with existing or future planned systems that are essential for completing the mission. Interoperability and necessary modifications should also be addressed. Finally, the TEMP summarizes systems engineering-based information that drives the technology development phase and the development of a prototype (Department of Defense, 2012).

20.8.1 Roles and responsibilities

The program manager (PM) is responsible for confirming test and evaluation results and shall control the integration testing. The test manager (TM) is responsible for overseeing the test and evaluation process, including the development of test plans, execution, review, and coordination of test and evaluation results acceptance. The test team is responsible for testing or assisting with the testing of the composed solution. The risk manager is responsible for conducting risk assessments on the proposed solutions, developing risk-handling options and mitigation plans, and identifying the required resources. The risk manager monitors the effectiveness of risk-handling options. The risk manager updates the risk assessments in the composed solution library to ensure that the information in the library is current and accurate with respect to the risk assessment. The risk manager reviews and recommends changes to the project manager (PM) regarding the overall risk management approach, based on lessons learned (DOD, 2012). Quality management is a critical job. The primary job function of the quality manager is to establish, plan, develop, and implement techniques, processes, and procedures for ensuring the desired level of quality in the proposed methodology and work products. The quality manager develops performance metrics and implementation procedures to track and control quality, utilizing systems that detect and resolve problems (Scalco and Wince, 2012).

20.8.2 Independent verification and validation (IV&V)

The independent verification and validation (IV&V) role is a rigorous, independent process that evaluates the correctness and quality of composed commercial solution products to ensure that customer requirements are met and that the products are well-engineered to mitigate risk factors. The IV&V purpose is to perform assessments by an independent third party, verify that the solution is well-engineered, and validate that the solution meets user requirements. An independent team performs IV&V to assess the proposed solution against reviews, analysis, evaluation, inspection, and test of the solution in the operational environment. The independent team provides an unbiased perspective and promotes the early detection of test variances, enabling the prompt implementation of corrective actions to bring the product up to user expectations and validate the solution, thereby ensuring compliance with defined requirements. The IV&V also serves to provide insight into potential solution risks and supports the idea that the solution complies with regulatory, performance, schedule, and budget requirements (INCOSE, 2015).

Composed solutions are evaluated to verify that the requirements align with the defined specifications, the design meets the defined standards, and the composed solution adheres to these standards, as well as other factors such as policy and legal statutes. Verification techniques include static test approaches, code reviews, and inspections of work products, such as documentation of the source code (Gemmill, 2011; Scalco and Wince, 2012).

20.8.3 Test strategy and method

A test summary report is developed after each test stage (unit, functional, integration, system, interface, performance, regression, user acceptance, and pilot testing) to record the results and obtain certification for the next test or the final implementation of the composed solution. The test process schedule for each composed solution variant includes (Scalco and Wince, 2012):

- Preparing the test plan
- Preparing the test specification
- Approving the test specification
- Establishing the test environment
- Preparing the test cases
- Reviewing the test plan
- Approving the test plan
- Reviewing the test cases
- Accepting test releases
- Test execution
- Updating of test reports
- Delivering final test reports

20.8.4 Unit testing

The objective of unit testing is to test the smallest units of a composed solution. Unit testing can be time-consuming and demands rigorous documentation. The benefit of the shared, composed solution capability package is that once all units of the proposed solution have been tested to work, larger components of the solution can be evaluated in integration testing. The composed solution unit tests are shared among users. It is not possible to test a solution for every possible operational scenario when the solution is used in a real-world environment. Ultimately, shared test data is why the user's integration testing is so important. The capability packages enable users to reach a fielded solution faster and more cost-effectively, without compromising sound system engineering best practices (Scalco and Wince, 2012).

20.8.5 Functional testing

Functional testing measures the quality of the functional components of the composed solution against the specified requirements. The functional test is used to determine if each component part performs in accordance with the specifications under the conditions under which it may be used, based on the requirements that the composed solution must deliver. As in all test phases, any defects detected are recorded and managed for risks. Each defect discovered during the functional testing is traced to a specific requirement or requirements that are not performing according to the specifications (Scalco and Wince, 2012).

20.8.6 Integration testing

The objective of integration testing is to combine units that have already been tested into a use case component and test the interface between them. Integration testing ensures that users receive suitable and effective solutions. Integrated tests can shorten the development lifecycle by establishing a test process that aligns with and synergizes all stakeholder objectives. Many units are combined into the components for testing, which are aggregated into a larger, composed solution. The concept is to test variations of the unit combinations and expand the process to evaluate the composed solutions alongside those of other component groups within the system. Eventually, the components making up the composed solution in the user product are tested together. Integration testing identifies problems that occur when multiple units are combined. Various approaches to integration testing can be applied to the composed commercial solutions. These approaches include the top-down approach, which requires that the highest-level units be tested and integrated first; the bottom-up approach, which requires the lowest-level units to be tested and integrated first; and the umbrella approach, which tests along functional data and control-flow paths (International Council on Systems Engineering), (Scalco and Wince, 2012). From the perspective of object-oriented system design, this is the stage of testing when interactions between multiple objects (e.g., classes) are thoroughly tested.

20.8.7 System testing

The objective of system testing is to verify and validate the functionality of the composed solution. Issues and discrepancies discovered during the system test are identified in the system test report, which is maintained in the composed solution CM database for each composed solution capability package variant. The technical team should demonstrate a minimum of three proof-of-concept systems to satisfy the composed solution capability package before publishing it as an option for use. System testing shall be performed on each use case before submission for approval by the designated approval authority (DAA) (Scalco and Wince, 2012).

20.8.8 Interface testing

The objective of interface testing is to identify problems, gain insight into the system's performance, and fix issues. The interface characteristics of the system components are verified to ensure they meet the information contained in the IDDs, verifying both interfaces among the components and the interfaces with external items, such as other systems, configuration items, and users. Test cases are developed to ensure that the components satisfy the requirements when two or more components are put together (Scalco and Wince, 2012). In object-oriented design, this is where the most complex interactions between multiple objects are put to the test.

20.8.9 Performance testing

The objective of performance testing is to determine how a composed solution performs in response to workloads. Quality attributes, such as scalability and reliability, are measured and validated as commercial off-the-shelf (COTS) products are integrated into composed solutions. The basic system assumption in this methodology is that the system will use COTS products that offer feature-rich capabilities that users demand. The performance testing types of the composed solution, as specified by the user for the specific use case, include load, stress, configuration, and endurance testing (Scalco and Wince, 2012).

20.8.10 Regression testing

The objective of regression testing is to verify that design errors have been resolved. Discrepancies are entered into an error-tracking system and reported back to the project manager (PM). The builds are submitted to system testing to verify and validate the functionality of the composed solution. This procedure is repeated until the exit criterion is satisfied (Scalco and Wince, 2012).

20.8.11 User acceptance testing

The objective of user acceptance testing is to ensure that the composed solution meets the user's expectations. The user acceptance test is focused on a set of established requirements. The user acceptance test occurs after a capability package has been prepared, and all changes to the baseline composed solution are identified and available to the test team. The requirements specification is baselined prior to user acceptance testing and is under CM control. The user acceptance test environment is baselined with the release baseline of the composed solution used. The integration test defects are fixed, validated, and documented prior to user acceptance testing (Scalco and Wince, 2012).

20.8.12 Pilot testing

The primary objectives of pilot testing are to determine the usability of the composed solution and identify additional functional requirements of the system (Scalco and Wince, 2012). In object-oriented designs, this is when all objects are at least stubbed out, and the complete system with at least simulated functionality in all objects is tested.

20.8.13 Test plan elements

Test resources are limited. Developing executable test plans is challenging. It includes the following elements:

- Evaluation criteria
- Incident management
- Requirements traceability matrix
- Review and approval process
- Test resources
- Test schedule
- Test case report
- Test log
- Incident tracking log
- Corrective action plan
- Test summary report

20.9 Summary

Modularity, operational test and evaluation (OT&E), prototyping, and various types of testing are all foundational to systems engineering. The development of advanced tennis players involves several years of developmental effort, just as advanced development for systems containing unproven concepts takes years to develop. A truly proficient systems engineer also takes years to build. "Advanced development is especially critical for systems containing extensive advanced development or unproven concepts that may involve several years of development effort" (Alexander Kossiakoff, 2011). The systems engineer possesses technical proficiency developed through engineering education and experience, as well as knowledge from various technical domains, and an understanding of software and hardware in both information technology (IT) and operational technology (OT) system domains. All of this is developed through years of foundational skill-building and continuous practice and development. Systems engineers must be experts who lead the development of the best system. A systems engineer, whether naturally or intentionally, often balances the workforce, which we discuss in Chapter 21, "Workforce."

References

(DOD), D. O. D. 2012. *Defense Acquisition Guidebook* [Online]. https://acc.dau.mil: Defense Acquisition University (DAU). Available: https://acc.dau.mil [Accessed].

Alexander Kossiakoff, W. N. S., Samuel J. Seymour, and Steven M. Biemer 2011. *Systems Engineering Principles and Practice*, John Wiley & Sons, Inc. Publication.

Borky, J. M. & Bradley, T. H. 2019. Protecting Information with Cybersecurity. *Effective Model-Based Systems Engineering.* Springer International Publishing.

D, S., K.bashkaran, Manivel, M., T.jayakumar, R.sivakumar & C, A. 2025. An extensive critique on advances in digital control techniques for smart grid and micro grid operation. *IEEE International Students' Conference on Electrical, Electronics, and Computer Science.* IEEE.

Engineers, U. S. A.C.O. 2017. Energy Division Operation Technology Foundation and Process. *In:* Center, U. S. A. E. A. S. (ed.). Huntsville, Alabama: U.S. Army Engineering and Support Center.

Gemmill, D. 2011. Lectures. *AerE/EE/IE 565 Systems Engineering and Analysis/Fall 2011.* Iowa State University: Douglas Gemmill.

Gungor, V. C., Sahin, D., Kocak, T., Ergut, S., Buccella, C. & Cecati, C. 2011. Smart Grid Technologies: Communication Technologies and Standards. *IEEE Transactions on Industrial Informatics,* 7, 529-539.

Incose. 2015. A Complexity Primer for Systems Engineers. Available: https://www.incose.org/docs/default-source/ProductsPublications/a-complexity-primer-for-systems-engineers.pdf [Accessed October 11, 2019].

Scalco, A. W. & Wince, K. 2012. A Proposed Methodology to Achieve Balanced Resilience and Risk. IE 566 Applied Systems Engineering: Iowa State University.

21

Workforce

"Humanity is not a spectator sport. It is not baseball, but rugby. It is a contact sport – if we don't get bruised sometimes, we are sitting on the bench."

– Thomas Frisby,

"Ain't All Good. Ain't All Bad: That's The Way It Was"

(Frisby, 2015)

Abstract

The rugby game does not stop for player tackles. The constant moving play only stops once a team has scored, if a ball is out of bounds, or if a penalty is called. The game is distinguished by continuous play, rather than time-outs, as seen in other similar games such as American football. Players also do not wear the same protective helmets, shoulder pads, or any other protective gear. It is a full-contact sport. All players play both offensive and defensive roles during the game. The team scoring the most points, called a "try" in rugby, at the end of two 40-minute halves wins the match. The principles of the game are to move forward, support the play, maintain the play, and exert pressure. The sport of rugby is like working in cyber. Offense going forward means moving the ball toward the opponent's end zone goal line, or "try line." Every "try" when a player touches the ball on the ground in the opponent's end zone is worth five points. The game can be intimidating. The rugby team embodies a powerful sense of competition and camaraderie on the field. It does not hurt to have the same intense sense of competition and camaraderie on a systems engineering team when facing a cyber game. The workforce is always a critical key to systems engineering. The systems engineer has to be knowledgeable about how to bring various component knowledge, skills, and abilities (KSAs) to the game, how to design and use an Integrated Master Plan (IMP) and work breakdown structure (WBS), how to lead integrated product teams (IPTs), effectively frame problems, be decisive, and most of all lead a team.

DOI: 10.1201/9788743809128-21

21.1 Introduction

The job market for control systems positions is changing fast. The expectations of workforce roles, as well as knowledge, skills, and abilities, are changing just as rapidly. Like the game of rugby, the job market for control systems does not stop for player tackles. The constant moving only stops once a team has scored, something about the system is found to be out of bounds, or a penalty is called (i.e., an advanced persistent threat (APT) is found in the system). Technological changes, including digitalization and automation, drive a revolution. The onset of the global coronavirus crisis in 2020 highlighted that telework, electromobility, social systems engineering, and climate goals can generate significant tensions as drivers of transformation. More changes are still ahead. These drivers will lead to changes in job roles and create new demands for digital skills and solutions tailored to the digital environment. Systems engineering methodology and practice can offer equilibrium to complex, context-sensitive, dynamic classes of control systems for critical infrastructure.

21.2 Component Level Knowledge, Skills, and Abilities (KSAs)

Three key activities for a system engineer that require technical knowledge at the component level are specification requirements development, cost estimates, analysis of alternatives (AoA), and trade-off studies. The systems engineer needs to probe into sub-component levels to effectively identify and discuss technical problems and component material solutions that meet the functional and physical specifications of the system design and technology readiness levels (TRL), thereby achieving the best performance versus cost. The middle-level systems building blocks that can require the systems engineer to have an interest at the component level are "namely, (1) advancing technology, (2) competition, and (3) specialization" (Alexander Kossiakoff, 2011). The systems engineer also needs to probe into the sub-component level to select "the most promising technology approaches, assessing the associated risks, rejecting those for which the risks outweigh the potential payoff, planning critical experiments, and deciding on potential fallbacks" (Alexander Kossiakoff, 2011).

21.3 Integrated Master Plan (IMP)

The purpose of the integrated master plan (IMP) is to identify the design development timeline for the complete design management and acquisition process of the composed commercial suite of solutions. The integrated master plan (IMP) is a product under the management control of the project engineer. The IMP serves as the baseline planning tool for implementing the composed commercial suite of solutions. The integrated master plan (IMP) is used to develop the integrated master schedule, which is also referred to as the "Level 2" schedule. The IMP is used to create the detailed schedule, also referred to as the "Level 3" schedule. The Level 3 schedule is used to establish the work breakdown structure (WBS) and to develop cost estimates. The WBS is cross-referenced to the integrated master plan (IMP) (Scalco and Wince, 2012).

Because traditional government cybersecurity solutions can take as long as three to five years to develop, build, certify, and field, establishing a baseline of sound systems engineering methodology is necessary to support decisions on where the schedule may be crashed in the process. This process addresses the problem statement that, in many cases, business enterprise cyber solutions do not meet the requirements of high-grade government solutions, which are already outdated by the time these security solutions are fielded. They are obsolete technology before they are even made operational (Scalco and Wince, 2012).

The integrated master plan (IMP) complements the requirements and the statement of work (SOW) in a work breakdown structure (WBS) format and is linked to the risk mitigation plan for each composed solution capability package. The IMP is used to integrate functional activities and field a comprehensive solution, providing insight into the overall effort required for integration into a user product. This also demonstrates interrelationships and critical paths in quickly fielding a solution that utilizes a commercial off-the-shelf (COTS) item. A single numbering system is used to tie the IMP to the WBS and SOW, ensuring consistency in work products and that each tool references the same work task. Consistency enables the project team to establish metrics that can be used to monitor the fielded use cases for changes that may negatively impact the solution capability. The IMS becomes a valuable management tool that maximizes government/industry teaming to help manage risk and provide inputs for risk assessments (Scalco and Wince, 2012).

21.4 Work Breakdown Structure (WBS)

Developing a detailed work breakdown structure (WBS) for a system development project is a crucial aspect of project management. However, the systems engineer also makes a necessary contribution to the successful project management of a systems development effort. "It is essential that the technical concepts and implied design and interfaces are feasible" (Alexander

Kossiakoff, 2011). There are several roles that the systems engineer plays in defining the work breakdown structure (WBS), in addition to the "systems engineering" task activities. The integration of systems engineers with project management is evident throughout the lifecycle, most apparent in the early phases of a project. For example, the role of the systems engineer is to share three critical task functions, namely, project planning and control. These include task definitions (e.g., task allocation and project reviews), risk management (e.g., risk assessment and risk mitigation), and customer integration (e.g., technical management). The task functions that are primarily those of a systems engineer are as follows: systems architecture (e.g., concept design and functional allocation), technical coordination (e.g., technical disciplines and subcontractors), and systems integration (e.g., interface management, verification, and validation) (Alexander Kossiakoff, 2011). The systems engineer collaborates with project management to address risk management throughout the lifecycle and in the development of the systems engineering management plan (SEMP). The SEMP is the implementation plan for all systems engineering tasks of a project, defining the roles and responsibilities of all participants. The two things that are particularly insightful in the article "Applying the work breakdown structure to the project management lifecycle" by Shelly Brotherton, Robert Fried, and Eric Norman about effective WBS characteristics include that (1) "It is created by those doing the work;" and (2) it "communicates project scope to all stakeholders" (Brotherton et al., 2008). These two points are insightful, as the WBS may sometimes be viewed as a project manager's tool for managing the project. Both examples indicate that the WBS, when effectively utilized, has a significantly greater impact on the entire project team, particularly the systems engineer.

Building collaboration in a complex system requires considerable integration skills. The role of the systems engineer is to integrate multiple individuals who are performing various tasks within an exceptionally complex system. Effective technical communication is highlighted as "the single most important task of the project systems engineer" to maintain both internal and external relationships within and outside the company (Alexander Kossiakoff, 2011). There are often many concurrent work activities to adjust to the integration of new engineering requirements for cyber capabilities into an original system that may not have included operational and functional requirements for operating in the cyber domain. Managing requirements, the systems engineering process, and the systems engineering skill sets for managing change are essential. "The systems engineering organization spans disciplines and participating organizations but also adapts to the company organizational structure" (Alexander Kossiakoff, 2011). When a contract is rebid, a company can potentially lose valuable sets of professionals who are familiar with a system, and the organization may have to rebuild its staff. It is essential to be methodical in the design and consider future customer requirements. Documenting the rational need for allocated values enabled agility. Requirement iterations enabled an organization to keep the technical baseline in synchronization. Systems engineering, a hierarchy of complex systems, is a "mindset of problem-solving." A systems engineer approaches it by looking at the whole as a problem solver and leader in the program, addressing many of the technical issues.

The work breakdown structure (WBS) reflects the work elements to be accomplished and identifies the associated functions, tasks, work packages, and budget required to complete the composed solution project. The project schedules and cost estimates are prepared in accordance with the work breakdown structure (WBS). Cost estimates and task responsibilities are assigned to project elements using the work breakdown structure (WBS) (Scalco and Wince, 2012).

21.4.1 Example work breakdown structure (WBS)

The work breakdown structure (WBS) maps to the task deliverable summary and funding.

1. Task 1: Engineering
   - 1.1. Perform requirements analysis
   - 1.2. Conduct a feasibility study
   - 1.3. Define system operational requirements and maintenance concepts, and identify and prioritize the technical performance measures.
   - 1.4. Develop system-level functional analysis.
   - 1.5. Allocate system lower-level requirements.
   - 1.6. Prepare requirements and planning documentation.
     - 1.6.1. Shall include a system specification, Type A
     - 1.6.2. Shall consist of a test and evaluation master plan (TEMP)
     - 1.6.3. Shall include a systems engineering management plan (SEMP)
       - 1.6.3.1. Technical program tasks shall be identified in the systems engineering management plan (SEMP)

1.6.3.2. Activities and milestones necessary to accomplish the systems engineering process shall be identified and defined in the SEMP

1.6.3.3. Engineering integration requirements shall be identified in the system engineering management plan (SEMP)

1.7. Provide the required resources, such as a functional area subject matter expert (SME), to accomplish the engineering activities

2. Task 2: Composed commercial solutions

2.1. Comply with the International Society of Automation (ISA)/ International Electrotechnical Commission (IEC) 62443 Standard.

2.2. For DOD critical infrastructure, comply with

2.2.1. The National Institute of Standards and Technology (NIST) standards

2.2.2. And comply with the fundamental principles and procedures described in the Department of Defense Acquisition Guidebook (DAG) (DOD Directive 5000.01 and DOD Instruction 5000.2) (Department of Defense) to ensure investments in technologies and product support meet National Security Strategic standards that satisfy user needs

2.3. Identify all governance documentation and relational dependencies

2.4. Define variants of demand signals

2.5. Identify, define, describe, and document the functional flow process of the demand signal to the product fielding. This includes all sub-processes, demand signal types, demand signal entry points, stakeholders, and customers.

2.6. Identify risk-critical components of composed solutions.

2.7. Create a commercial off-the-shelf (COTS) composed solution library

3. Task 3: Management and accounting process reporting

3.1. Requirements development and analysis

3.1.1. Provide synthesis, analysis, and evaluation of the product and processes

3.2. Program management, control, and evaluation

   3.2.1. Plan, coordinate, and conduct formal design review meetings

   3.2.2. Monitor and review system test and evaluation activities

   3.2.3. Coordinate and review all formal design changes and modifications to ensure improvements

3.3. Initiate and establish necessary ongoing liaison activities throughout the product life cycle from design, production, integration, utilization, logistics, and retirement

21.4.2 Example project schedule

The project schedule is developed based on the work breakdown structure (WBS), as shown in Table 21.1(a–c).

21.4.3 Example task deliverables summary

The task deliverables should ideally align with the work breakdown structure (WBS). The performance standard and acceptable quality level are identified, along with the monitoring methods, including the method of surveillance and the surveillance plan. Additionally, any remedies or incentives are specified, as well as the expected work product. The performance standard and acceptable quality level for the requirements analysis deliverable are specified in the work breakdown structure (WBS). An unacceptable product is a product that is not completed, has flawed requirements analyses, assumptions, or conclusions, or has numerous grammatical problems (i.e., no more than two formatting or typographical errors per page of documentation). The requirements analysis must be delivered on time. The monitoring methods for the requirements analysis include a 100% surveillance review. The surveillance plan stipulates that the requirements subject matter expert or their designee will review the work within five working days of receipt. The remedy or incentive is rework (at no change in target cost or fixed price) for accuracy. Consistent problems or delays will result in negative ratings. The work product is in a specified report format.

21.5 Integrated Product Teams (IPT)

Systems engineers are valuable partners in integrated product teams (IPTs). IPTs bring four main benefits to engineering systems: involving senior

Table 21.1(a) Example project schedule mapped to work breakdown structure.

Intergrated Master Schedule-MOSAICS			
ID	WBS Code	Outline Level	Task Name
1	**1**	**1**	**Requirements**
2	**1.1**	**2**	**Operational Requirements**
3	1.1.1	3	Evaluation & Recommendations (i.e. Site Charicterization Studies/Reports)
4	1.1.2	3	Requirements Development
5	1.1.3	3	Deliverable: Submit Operational Requirements
6	1.1.4	3	Integrated Management Team Re(IMT) Reviews Operational Requirements
7	1.1.5	3	Operational Sponsor Reviews and Validates Operational Requirements
8	1.1.6	3	Operational Sponsor Signed/Approved Requirements
9	1.1.7	3	Deliverable: Concept of Operations (CONOPS)
10	1.1.8	3	Integrated Management Team (IMT) Reviews CONOPS
11	1.1.9	3	Operational Sponsor Reviews CONOPS
12	1.1.10	3	Operational Sponsor Signed/Approved CONOPS
13	**1.2**	**2**	**Functional Requirements**
14	1.2.1	3	Evaluation & Recommendations
15	1.2.2	3	Functional Requirements Development
16	1.2.3	3	Deliverable: Submit Functional Requirements
17	1.2.4	3	Technical Management (TM) Reviews Functional Requirements
18	1.2.5	3	Integrated Management Team (IMT) Signed/Approved
19	**1.3**	**2**	**Technical Requirements**
20	1.3.1	3	Evaluation & Recommendations
21	1.3.2	3	Requirements Development
22	1.3.3	3	Deliverable: Submit Technical Requirements
23	1.3.4	3	System Engineering (SYSENG) Integrated Product Team (IPT) Reviews Functional Requirements
24	1.3.5	3	Technical Management (TM) Signed/Approved
25	**2**	**1**	**Security Engineering**
26	**2.1**	**2**	**Risk Management Framework (RMF) Lifecycle: Categorize**
27	2.1.1	3	eMASS
28	2.1.2	3	Stakeholder Requirements Definition

industry participants in the dialogue early in the design process, bringing together different viewpoints, building consensus, and leveraging advanced technology and component knowledge to inform system design approaches. The systems engineers perform specific activities to realize these benefits.

First, integrated product teams (IPTs) bring senior industry participants into the system conceptual design process at the earliest opportunity, "injecting their ideas during the formative stages of the development" (Alexander Kossiakoff, 2011). An IPT is composed of personnel from various competencies within an organization, bringing together expertise from appropriate functional specialties. Systems engineers need to leverage opportunities to engage senior industry participants in the conceptual development dialogue. The principal areas of input during this stage of development are to identify operational deficiencies and explore technological opportunities. There

Table 21.1(b) Example project schedule mapped to work breakdown structure.

Intergrated Master Schedule-MOSAICS			
ID	WBS Code	Outline Level	Task Name
31	2.2.1	3	Selection of Controls
32	2.2.2	3	Threat Analysis
33	2.2.3	3	Requirements Analysis
34	**2.3**	**2**	**Risk Management Framework (RMF) Lifecycle: Develop/Build**
35	2.3.1	3	RMF Input to Architecture Design
36	2.3.2	3	Implement Selected Security Controls
37	2.3.3	3	Map Controls to the Requirements to the Implementation
38	2.3.4	3	Verify Implementation Addresses the Conrols
39	**2.4**	**2**	**Risk Management Framework (RMF) Lifecycle: Assess (Test)**
40	2.4.1	3	Deliverable: Security Assessment Plan (SAP)
41	2.4.2	3	Perform Verification
42	2.4.3	3	Assess the Security Controls
43	2.4.4	3	Deliverable: Security Assessment Report (SAR)
44	**2.5**	**2**	**Risk Management Framework (RMF) Lifecycle: Deploy/Operate**
45	2.5.1	3	Deliverable: Plan of Action and Milestones (POA&M)
46	2.5.2	3	Submit the SAP to the Authorizing Official (AO)
47	2.5.3	3	Deliverable: Authority to Operate (ATO)
48	**2.6**	**2**	**Risk Management Framework (RMF) Lifecycle: Monitor**
49	2.6.1	3	Plan Maintenance
50	2.6.2	3	Configuration Management (CM) and Control Process for Information System (IS)
51	2.6.3	3	Plan Disposal
52	**3**	**1**	**Technical Development**
53	**3.1**	**2**	**Architecture**
54	3.1.1	3	Define the Architecture
55	3.1.2	3	Technical Capability Allocation
56	3.1.3	3	Generate Derived Requirements
57	3.1.4	3	Define System Interfaces
58	**3.2**	**2**	**Capability Source Selection**
59	3.2.1	3	COTS/GOTS Survey
60	3.2.2	3	Deliverable: List of Available COTS and Identification of Gaps (GOTS)
61	3.2.3	3	Developing the Evaluation Criteria and initial down select
62	3.2.4	3	Identification of Gaps (GOTS)
63	3.2.5	3	Deliverable: List of relevant criteria to address MOSAICS spiral specific capability and/or facility
64	3.2.6	3	Request for Information (RFI)

needs to be a leader among peers to achieve a consensus on the IPT. If there is no need for consensus among senior industry participants, then systems engineers could conduct market surveys to determine if a new system is technically and economically feasible (Ibid.). However, to realize the benefit of senior industry leaders, there needs to be high group interaction involving all IPT members for problem-solving. Key stakeholders must be involved in the project's key decision points. Secondly, "they bring together the different disciplines and specialty engineering viewpoints throughout the development"

Table 21.1(c) Example project schedule mapped to work breakdown structure.

Intergrated Master Schedule-MOSAICS			
ID	WBS Code	Outline Level	Task Name
65	3.2.7	3	Deliverable: Recommended COTS & GOTS based on selection criteria
66	3.2.8	3	Integrated Management Team (IMT) Signed/Approved
67	**3.3**	**2**	**Prototype Design**
68	3.3.1	3	Specific design elements for MOSAICS spiral and facility
69	3.3.2	3	Deliverable: Hardware development for MOSAICS spiral
70	3.3.3	3	Deliverable: Software development for MOSAICS spiral
71	3.3.4	3	Deliverable: Communications and data interface for MOSAICS spiral
72	3.3.5	3	Deliverable: Analysis of feasible design
73	3.3.6	3	[External to MOSAICS: Baseline Tool (Map-to-Model)]
74	3.3.7	3	System Engineering (SYSENG) Integrated Product Team (IPT) Reviews Design
75	3.3.8	3	Integrated Management Team (IMT) Signed/Approved
76	**3.4**	**2**	**Integration**
77	3.4.1	3	Deliverable: Integration plan for MOSAICS spiral and facility constraints
78	3.4.2	3	Deliverable: Internal MOSAICS spiral integration (with MOSAICS components)
79	3.4.3	3	Deliverable: External MOSAICS spiral integration (with facility)
80	3.4.4	3	Deliverable: User integration with MOSAICS spiral
81	3.4.5	3	Deliverable: Analysis of design integration
82	3.4.6	3	System Engineering (SYSENG) Integrated Product Team (IPT) Reviews Design
83	3.4.7	3	Integrated Management Team (IMT) Signed/Approved
84	**4**	**1**	**Test/Demonstration**
85	**4.1**	**2**	**Lab Test #1**
86	4.1.1	3	Deliverable: Lab Test #1 Test Plan
87	4.1.2	3	Deliverable: Lab Test #1 Report
88	**4.2**	**2**	**Lab Test #2**
89	4.2.1	3	Deliverable: Lab Test #2 Test Plan
90	4.2.2	3	Deliverable: Lab Test #2 Report
91	**4.3**	**2**	**Spiral Demonstration (Spiral 0)**
92	4.3.1	3	Spiral 0: Automation/Prove ability to automate sections of the ACI TTP

(Alexander Kossiakoff, 2011). There needs to be a clear need for an IPT to realize the benefits of various specialty viewpoints, and a WBS needs to be developed to identify the disciplines required for each work package. The IPT lead must ensure that the IPT operates as a cohesive unit with varying levels of expertise and is responsible for designing, developing, testing, and deploying systems. The expert members of an IPT must work collaboratively to make informed decisions regarding trade-offs and designs. The WBS supports the selection criteria for team members by demonstrating the necessary qualifications and skills. The IPT lead must be committed to the team's success and actively engaged with team members.

Thirdly, "[IPTs] capitalize on the motivational advantages of team collaboration and consensus-building" (Ibid.). To achieve this, the IPT should collaborate and adopt a consensus-based approach. All members of the IPT need to be included in the decision-making process. The mission and goals of the IPT must be communicated clearly to all stakeholders through an IPT charter. The IPT charter defines the description of needs, expected outcomes, authorities, membership, and skills. A work environment of respect needs to be practiced by everyone on the team, and there needs to be a commitment to the success of the IPT. Building a consensus environment is the responsibility of all team members who comprise the IPT.

Finally, "[IPTs] bring advanced technology and COTS knowledge to bear on system design approaches" (Ibid.). To realize the benefit of bringing advanced technology and COTS knowledge to bear, the work of the IPT requires resident expertise in the technical areas of consideration. Operational domain knowledge is essential for understanding how advanced technology and commercial off-the-shelf (COTS) solutions can be effectively applied to the problem set. The IPT needs to receive the appropriate resources to realize the advantages of system design approaches. Appropriately resourced means by people, tools, and funding.

The IPTs identified cyber capability gaps to better understand current and emerging cyber-related needs. The IPT then identified efforts to develop solutions that address the most critical gaps in support of system security and resilience. Because the priority ranking of gaps may lead to investment decisions involving multiple components, the process for determining priorities must be credible, transparent, and objective. This transparent and credible process helps instill confidence among component and external stakeholders that the technical leadership, represented by the program manager and systems engineer, is identifying and addressing critical cybersecurity needs.

21.5.1 IPT purpose

The purpose of the IPT is to bring together cross-organizational stakeholders to collaborate on a common goal. To ensure success, consider using the following steps to support the IPT:

- Define clear objectives and outcomes for the IPT
- Develop a standard process and approach for project team members
- Establish a governance structure that allows growth and improvement while maintaining foundational guidance and metrics to achieve targeted outcomes
- Execute IPT activities, which require gaining component consensus while maintaining IPT process integrity and

- Ongoing management and evaluation ensure the process remains effective over the long term

21.5.2 IPT process

The IPT process is designed to:

- Identify duplicative cyber domain activities and recommend ways to reduce duplication
- Provide an oversight platform to coordinate cross-component collaboration and track the investment profile of each site project to ensure progress, schedule maintenance, and
- This results in the development and transition of effective solutions to address priority cyber domain capability gaps across critical infrastructure systems and assets

21.5.3 IPT member qualifications

IPT members should be well-versed in the functional disciplines they represent. Field element managers designate individuals within their organizations who will participate in the IPT and certify the individual's competency after verifying their competency. A formal technical warrant holder (TWH) for the industrial control system (ICS) certification process can help ensure that the correct match of functional discipline abilities is achieved for at least one category of projects.

IPT members should possess specific knowledge, skills, and abilities, including understanding project-task goals and schedules, implementing procedures, safety and hazard controls, methods, requirements, and process metrics, as well as relevant skills. Project managers should define project- or task-specific training requirements. The goal of project/task-specific training is to ensure that each IPT member has at least the following:

- Working-level knowledge of the project's mission and how this mission relates to the sponsoring program
- A working-level knowledge of the procedures IPTs commonly utilize to fulfill their assigned roles and responsibilities
- A working-level knowledge of the Department's project management directives

- Familiarity level knowledge of non-project management directives that could impact the project
- A working-level ability to communicate findings and recommendations

21.5.4 Functional area competency

Specific functional area competency is needed about applicable commonly required standards for the International Energy Efficiency (IE), the International Organization for Standardization (ISO), International Code Council (ICC) such as the International Building Code (IBC), and the American National Standards Institute (ANSI) in the United States. Additional examples of specific functional area competency requirements associated with an IPT for critical infrastructure in the United States DOD are contained in the following standards:

- Electrical systems (DOE-STD-1170)
- Instrumentation and control (DOE-STD-1162)
- Fire protection (DOE-STD-1137)
- Industrial hygiene (DOE-STD-1138)
- Quality assurance (DOE-STD-1150)
- Environmental compliance (DOE-STD-1156)
- Facility representative (DOE-STD-1151)
- Waste management (DOE-STD-1159)
- Occupational safety (DOE-STD-1160)
- Mechanical systems (DOE-STD-1161)
- Deactivation and decommissioning (DOE-STD-1166)
- Safeguards and security (DOE-STD-1171)
- Safety software quality assurance (DOE-STD-1172)
- Criticality safety (DOE-STD-1173)
- Chemical processing (DOE-STD-1176)
- Technical program manager (DOE-STD-1178)
- Construction management (DOE-STD-1180)

- Facility maintenance management (DOE-STD-1181)
- Civil structural engineering (DOE-STD-1182)

21.5.5 Stakeholder involvement

The integrated product teams (IPTs) seek input from customers (i.e., owners and the user community), contractors, key suppliers, regulators, supporting functions, and non-governmental interest groups that may be affected by or have an impact on the project's eventual outcome. Although this input is necessary for the planning process, it should be recognized that stakeholders (other than the system owner) are seldom responsible for cost or schedule and may have little incentive to compromise or look beyond their self-interests. The challenge is to find a method of involving these groups without losing sight of the key concept that IPTs seldom find a means of satisfying all the stakeholders' desires and can only be effective when the members focus on what is best for the project from a combined frame of reference.

21.5.6 Technical leadership

Systems engineers in technical leadership roles, such as project managers, can significantly influence performance in various ways, including uniting team members around a shared vision and orientation. There are many things that the project manager can do to set the team on the right path toward success, such as:

- Clearly defining goals
- Improving and increasing communication to establish and enforce standards for information sharing among team members, creating an atmosphere within the project team that encourages frank and open exchanges
- Removing the dysfunctional behavior
- Aligning skills with clearly defined knowledge, skills, and attributes for positions they fill

"The first step in project team building is the selection of personnel to staff the project team" (Pinto, 2015). The knowledge, skills, and abilities (KSAs) for each work activity requirement must be defined so that the appropriate resources can be matched to the project. The organizational structure needs to conduct a realistic assessment of its current organization and work towards achieving cross-functional cooperation. Some of the sources of conflict between the functional and project managers include "competition for resources, violation of group or organizational norms, disagreements over goals or the means to achieve those goals, and so forth" (Pinto, 2015). The differentiation between the functional and project managers plays a role in the effectiveness of an organization. "The more profound the differentiation within an organization, the greater the likelihood that individuals and groups will divide into 'us' versus 'them' encampments, which will continue to promote and provoke conflict" (Pinto, 2015).

"Evaluation and control are part of every project manager's job. Control by 'wandering around' and/or 'involvement' can overcome most problems in small projects. However, large projects need some form of formal control. Control holds people accountable, prevents small problems from mushrooming into large problems, and keeps focus" (Gray and Larson, 2005). A technical lead can take the following steps to establish an effective project control process:

1. Set the baseline plan
2. Measure progress and performance
3. Compare the plan against the actual and
4. Take action

"The best information system does not result in good control. Control requires the project manager to use information to steer the project through rough waters. Control and Gantt charts are useful vehicles for monitoring time performance" (Gray and Larson, 2005). It is a key practice in systems engineering and program management to develop an integrated information system that establishes cost baselines, allowing for variance analysis and status reports. The project team needs to monitor and control the project's systems. An integrated information system is necessary to provide the team with all the data required to make informed and effective decisions. "Without matching time-phased budgets to the actual cost of work completed ... it is impossible to reliably measure cost performance" (Gray and Larson, 2005).

Setting the baseline presents an opportunity to introduce an integrated information system and an earned value system, which can help overcome cost variance issues. "The earned value system starts with the time-phased costs that provide the project baseline, which is called the planned budgeted value of the work scheduled (PV). Given this time-phased baseline, comparisons can be made with actual and planned schedules and costs using a concept called earned value. The earned value system uses data developed from the work breakdown structure, the project network, and schedule" (Gray and Larson, 2005).

The perspectives of external stakeholders should also be taken into consideration. System operators value fairness, courtesy, communication, and integrity. The company's readership base is interested in receiving a quality product that is delivered on time. Vendors and suppliers value trust and consistency and expect the relationship to develop positively for revenue generation. They also value assistance and understanding in challenging circumstances. Suppliers and contractors value fairness, courtesy, communication, and integrity.

21.5.7 End-state vision

The end-state vision is to effectively manage the critical infrastructure cyber mitigation project initiative by measuring progress and performance and providing the management team with performance evaluations.

21.5.8 Gap analysis

A gap analysis can help identify whether an implementation team lacks the necessary tools to manage the project initiative. If there is no effective, formal project control process in place, and no integrated information system exists that places costs in baselines and allows for necessary variance analysis and status reports to the management team. Setting a project baseline plan is the first step in establishing the project control process, which is essential to measuring the project's performance. "The baseline plan provides us with the elements for measuring performance" (Gray, 2005).

Project managers need to resist changes to the project baseline because deviations from the baseline are measured to determine the correct actions. Therefore, any unnecessary changes should be resisted to ensure an effective ability to compare plans with actual results. "The baseline (PV) serves as an anchor point for measuring performance" (Gray, 2005). There may be certain conditions under which a project manager might consider modifying a project baseline. For example, if it means that a project will fail without making the necessary changes, then the project manager should allow these changes. "[Project managers] should allow scope changes only if it is clear that the project will fail without the change, and the project will be improved significantly" (Ibid.). A project manager should not allow changes to a baseline when the adjustment is being proposed to cover for poor performance or changes that are retroactive to the actual work performed. Project managers should also be aware of scope creep, which is when small changes to a baseline are insidiously piled onto a project, potentially changing its overall security posture (among other problems).

In summary, applying project management concepts such as a project control process and an integrated information system can help a team achieve success. Some goals for the development team's consideration might be to establish a project control process to monitor time performance with a Gantt Chart, develop an integrated cost/schedule system, provide status reports to management, develop a contingency reserve, and establish a project baseline – these tools and the project control process help in problem framing.

21.6 Problem Framing

One purpose of problem-framing a solution described in a requirements document is to identify and describe all the stakeholders. An objective of developing and introducing a comprehensive cybersecurity solution is to deliver an affordable program, given very challenging budget constraints. There are many stakeholders interested in the development of the product. A stakeholder is a "[h]uman or other legal entity (company, etc.) playing a system role and thereby having a valid interest in the development of the system or product" (Ljerka Beus-Dukic, 2009). The onion model of stakeholders with a system role in solution development demonstrates the constraints and boundaries that may influence performance behaviors. Each stakeholder is essential in the product development. While stakeholders in the wider rings of the system boundary appear to have a "softer" impact, they nevertheless influence the design. The stakeholders identified in the stakeholder onion model can demonstrate the system boundaries based on the roles they play. For example, the operators are users of the solution. They are both normal operators and functional beneficiaries of the system product. "The system contains more than just the product" (Ibid.). The system also encompasses all stakeholders who play a role in the system. Stakeholders include owners of other products, individuals, standards, guidelines, and policies that can affect the system's performance. Documenting the stakeholder roles can bring together information about the system interfaces needed to make trade-off decisions by identifying the system boundaries.

A goal chart (Figure 21.1) identifies the goals each stakeholder wants to achieve. Identifying stakeholder goals is a crucial part of the process of discovering requirements and avoiding potential conflicts. "Detailed goals turn into requirements once they are ... fully verifiable, ... [and] prioritized by the

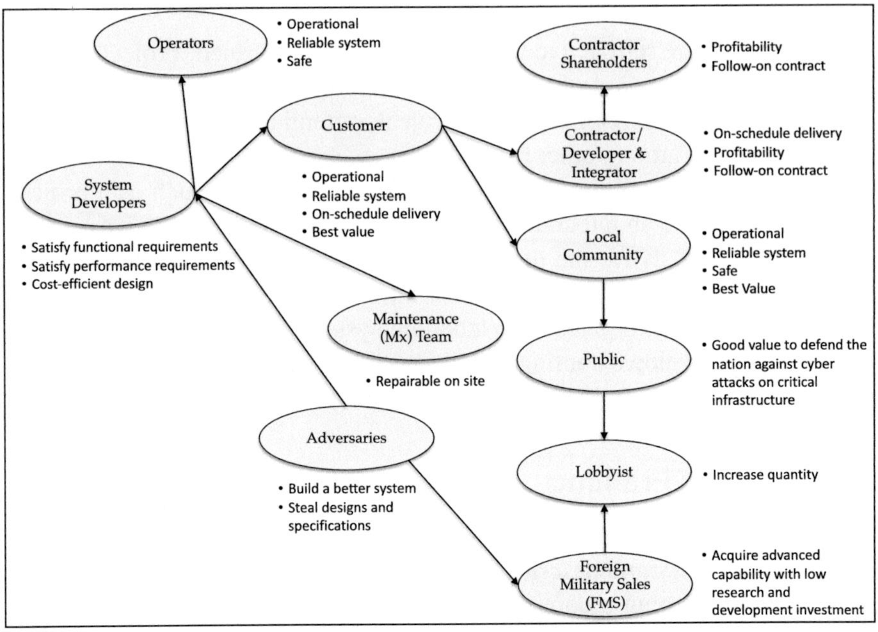

Figure 21.1 Goal chart.

project" (Ibid.). For example, system developers want to satisfy functional and performance requirements while developing a cost-efficient design. The operators want a safe, operational, and reliable system. The customer seeks a reliable system with on-time delivery that offers the best value for their money. The contractor, developer, and integrator want on-time schedule delivery, profitability, and a follow-on contract. The local community wants to acquire an advanced cyber defensive system with zero research and development (R&D) funds invested. The public wants a fair value in defending the nation against cyberattacks on critical infrastructure. The lobbyist wants to increase the quantity of systems purchased. Foreign military sales (FMS) also wants to acquire an advanced cyber defensive system with zero research and development (R&D) funds invested. The maintenance (Mx) team wants the system to be repairable on-site. The adversaries aim to develop a superior system and obtain its specifications. This process of problem framing is just one description of what happens during the design of an engineered system.

Problem frames can be used to discover a system's qualities and constraints. Problem decomposition is the process of partitioning a problem into distinct problem classes. Problem decomposition requires knowledge of problem frames and methods that can support the separation of concerns. These methods of problem framing can be used to describe the system. Examples of problem frames are required behavior, command behavior, information display, transformation, and simple workpieces. The required behavior model is the simple control of problems. The command behavior model controls behavior through commands at the request of an object's user. Information display presents information content representing a specific domain. Two types of transformation-frame models, Nondeterministic Finite Automata (NFA) and Deterministic Finite Automata (DFA), can be used to demonstrate the transformation of individuals into different individuals. Simple workpieces involve creating and changing something (processing) to meet a need. These two types of transformation-frame models can be helpful in evaluating new cyber defense projects when used in conjunction with financial screening models.

21.7 New Cyber Defensive Projects

Some criteria that should be used in evaluating all new projects before adding them to a current portfolio include the following six criteria for successful selection:

1. Realism. Examples include measurements against the organization's strategic mission and evaluations based on available technical resources
2. Analysis of current skill sets and required organizational learning curves
3. Capability. Examples are new technical expertise needed.
4. Flexibility. An example is prioritization, which involves screening against the current portfolio and other potential new projects.
5. Cost. An example is the required further training, available budget, and personnel for training.
6. Comparability. An example is aligning projects with portfolios, lessons learned, and corporate knowledge management to facilitate the transferability of knowledge.

21.7.1 Financial screening models

Excessive or exclusive reliance on screening models can lead to problems. For example, the present net value (NPV) model provides an advantage as a link to financial performance. However, the disadvantage is that this business model does not lend well to making accurate long-term predictions. NPV is a more realistic model than the internal rate of return (IRR) model.

One reason is that the NPV uses weighted values while the IRR does not. Generally, projects with higher internal rates of return (IRR) are considered better choices than those with lower scores. However, over-reliance on this approach also has drawbacks. For example, the IRR financial model does not have the actual rate of return for a project. An over-reliance on this model, if an organization "reinvested revenues only on lower-return projects, the real return on the project will be less than the calculated IRR" (Pinto, 2015). There are several other disadvantages of over-reliance on one financial model over another. Using NPV only can rank projects differently from IRR. IRR can only give conflicting results. Business managers need to understand the financial or other screening models being used, as well as appreciate both their advantages and disadvantages. Keeping this in mind will help the manager to make more optimal decisions and, thereby, avoid traps.

The effect of poor screening results in projects that are late, over budget, and inconsistently performing. Issues can be traced to the actual process of project selection. Issues can be attributed not to the main problems encountered nor the project management process itself, but to the way cyber projects are added to a company's portfolio" (Pinto, 2015). The steps should include preprocessing activities, process activities, and post-processing activities. One of the critical first steps is to select the screening methodology to be used in the evaluation. The process includes individual report analysis, screening, and selection, as well as portfolio adjustment (e.g., divestiture). A post-process consists of the development, evaluation, and completion (Ibid.). A formalized process was established to evaluate potential new projects against the company's strategic vision, identifying best practices that would benefit the company.

21.7.2 Case study: Porsche new product introduction (NPI) and cyber

There are many books and other resources available about project management, but there are few words on the subject regarding systems engineering cybersecurity efforts for critical infrastructure. Large-scale car manufacturing utilizes heavy machinery and incorporates both IT in manufacturing planning and process controls in the components and assembly industrial processes. A project or program manager must understand the critical success factors for critical infrastructure and ideally comprehend the entire value chain, from upstream raw materials to downstream after-sales services. Examples of critical success metrics that would be considered when tasked with designing a new system include the "iron triangle" (e.g., cost, quality, and schedule), information system (e.g., maintainability, reliability, validity, information quality, and fit for use), and organizational (e.g., strategic goals and increased profits) and stakeholder benefits (e.g., satisfied users and the economic impact to a surrounding community (Ibid.). These critical success metrics must be balanced and considered in conjunction with trade-off analysis and the law of diminishing returns to meet the threshold of organizational standards.

Recall from an earlier chapter that Germany's strategy is to be recognized as a technological pioneer in cybersecurity, with products known for their security. Germany focused on avoiding a "monoculture" dependency on single, large vendors, and the government sought policy goals of more open-source alternatives (Schallbruch and Skierka, 2018). German companies have followed this national strategy, as demonstrated by Porsche and Volkswagen (VW). In 2024, both companies allegedly changed the production of vehicle models for cybersecurity reasons. Porsche discontinued three models without a direct successor vehicle. The reason was based on United Nations Economic Commission for Europe (UNECE) Regulation 155, "Cybersecurity and Cybersecurity Management System." While the vehicle itself is not considered critical infrastructure, the automotive manufacturing industry is involved in the production of cars. Raw materials are processed using process control systems to stamp the body and engine parts. The engine control unit, component, and final product testing are all accomplished using process sensor measures to determine if the product is safe, reliable, and meets the required specifications. Therefore, the product change based on risks associated with cyber threats to vehicles is of particular interest.

Porsche cars have a virtually untarnished reputation and are considered among the finest performance vehicles in the world (Edmonds, 2008). This reputation is due to the manufacturer's automotive engineering leadership pursuing changing technology while remaining steadfastly customer-focused. The mid-1960s marked the introduction of newly engineered sports cars that further solidified Porsche's reputation as one of the world's premier automotive manufacturers with the launch of the now iconic model, the 911. By the late 1990s, Porsche had lost touch with its consumer base, and the stagnant introduction of technology to vehicles had left the company in a follower rather than a leadership position. "Because of a lack of dynamic product and tough market climate, Porsche was suffering financially" (Edmonds, 2008). Porsche reclaimed the company's position with customers with the introduction in 1997 of the Boxster, one of the now-discontinued models. The models Porsche is retiring are the Boxster, Cayman, and Macan.

A turning point in the company's history happened in 2003 when Porsche introduced the SUV model Cayenne – another of the now-discontinued models. Many Porsche enthusiasts believed the company had lost its purity as a sports car manufacturer. However, the Cayenne became a success that exceeded sales expectations, providing the company with the financial resources to remain positioned as a leader in the high-end automotive market segment. The Cayenne was carefully positioned in the high-end market segment and leveraged the Porsche brand to expand with a new product line outside of the company's original niche market strategy, leaving behind a history and enduring image built through decades of commitment to sports car purity. The cyber domain is adding a crimp in the platform strategy. However, Porsche, like Germany's national strategy, aligned itself as a cyber-safe company with products that meet strict cyber standards, as part of its customer relationship management (CRM) strategy. CRM has multiple meanings for different groups. CRM is used to collect information about customer expectations, providing insights to help manage customer experiences, control quality, and measure customer satisfaction in alignment with an organization's key success metrics. IT is a powerful tool in the industrial planning process, and the integration of IT applications into manufacturing is a key component of supply chain management.

The first prioritization of a critical success metric is the Porsche brand equity, specifically its quality. Each work activity subcomponent of the overall system delivery must meet the Porsche brand equity, which is known for its quality. The prioritization is to create cars that maintain the Porsche brand and ensure the future market aligns with the company's organizational strategic goals, thereby maintaining and increasing company profits. All work activities, including subcomponents of new cars, are measured against the criteria of the Porsche brand equity. The Porsche brand is known for its innovation. Therefore, building a bridge between the past and the future in the cyber world also requires the integration of highly innovative technologies and ideas. The customer is highly sensitive to maintaining the Porsche brand and is willing to invest in the necessary resources to get everything "just right" for a promising future. Client acceptance is dependent on attention to detail regarding brand equity. The Porsche brand equity entails the introduction of new ideas, creativity, and innovation that will charm and inspire all visitors for many decades to come.

The second priority of a critical success metric is the "Iron Triangle" (e.g., cost, quality, and schedule). The objective is to meet the schedule when new European cybersecurity laws come into effect in July 2024. The metric is that new passenger cars must be integrated with completely new electronic architectures by that date. The future of Porsche is closely tied to the car enthusiast community. Effective communication with the Porsche community must begin early to ensure buy-in for new models and to prepare the local distributor workforce for changes to the vehicle's electronic architecture. The local distributor community must be ready to act as an external stakeholder of Porsche, ensuring that owners have a pleasant driving experience from the time of the new car purchase, as well as in the event of the manufacturer discontinuing a model. In addition to the schedule, the quality of the Porsche is an expected success factor. Therefore, particular attention must also be given to potential cost overruns, as each subcomponent's work activities and products have quality measures that embody the Porsche brand. Each of the construction components will have to be carefully monitored, as will the purchase of materials. The program manager will need to oversee the customer program to ensure that these products are tested for quality and compliance with relevant regulations. A metric is that it meets the new regulation on time, on schedule, and at the agreed-upon budget.

A third critical success metric is the successful delivery of a complex, innovative information system, encompassing factors such as maintainability, reliability, validity, information quality, and fit for use. The project manager (PM) for the car model needs to be concerned with the engineering of the complex system of an "engineered product that contains diverse components and utilizes advanced technology" (Alexander Kossiakoff, 2011). Testing and evaluation are crucial components of integrating new cyber regulations. The intensity of test tracks with twists and turns requires the design, development, and integration of a complex information system with physical effects. Car builds require full lifecycle management planning for maintenance and reliability. Human factors must also be considered to ensure that the size and weight requirements for consumer drivers are met. The Porsche brand is renowned for its innovative ideas and commitment to safety. The program manager will need to ensure that all critical success factors are balanced against the company's brand equity and shareholder values.

Porsche's attention to detail is unique in that the company has long embodied the principles of systems engineering in its design of new concepts. Many manufacturers are introducing new models. However, it is Porsche that looks to utilize advanced technology in ways that involve risk and are central to the primary functions of ride and handling (Alexander Kossiakoff, 2011). For example, an owner's experience driving a Porsche is significantly more advanced than that of a typical car a person might drive. The integration of engineered complex signal and data systems is also unique to the Porsche brand. Porsche pays close attention to detail regarding the car's status and plans significant upgrades to protect against potential vulnerabilities.

As mentioned earlier in Chapter 3, "Cyber Policy, Regulations, and Guidelines," Germany's approach to the problem of cyber is to focus on a public–private partnership. The government is expected to protect its citizens in the cyber domain, just as private companies are expected to engineer and develop products or services that protect consumers broadly. Germany's official policy at the time of writing centered on the idea that the IT industry would reduce its dependency on untrustworthy providers and focus on the use of trustworthy solutions (Schallbruch and Skierka, 2018). The German perspective emphasizes the adage, "Economic security is national security." The economic impact of the cyber domain on national sovereignty is to ensure the country is producing economically impactful IT. We will examine how this strategy has been implemented since this book was written in a moment. However, the key point that cannot be missed leads one to ask if their respective country has a similar national strategy, and if so, why, and if not, why not. National Cybersecurity strategy, implementation, and tactics are essential conversations. There is a notion that cybersecurity professionals jump right in and keep everyone safe by following clear laws and statutes, while the laws and regulations are still in development.

Germany's strategic approach focused on being recognized globally as a country that develops and delivers safe cyber products and services. This idea of implementing technical, legal, and political measures ensures that a range of programs is available to strengthen German technologies. In the wake of the classified security information leaks, Germany went so far as to terminate contracts with some external communication providers and replace them with German providers as a measure to enhance the protection of private citizens' information (Ibid.).

Germany's approach, centered around cybersecurity, is expected. The results of Germany's cybersecurity strategy are demonstrated by two of Germany's automotive manufacturers, Volkswagen and Porsche, discontinuing vehicle models without indicating specific platform replacements. The cancellation of the models was announced to be directly tied to forthcoming European cybersecurity requirements. Both companies demonstrated that cybersecurity must be considered an integral part of the engineering process. The sub-theme echoes Germany's stance that commercial entities must consider cyber in engineering security risk analysis. Porsche essentially changed its development process due to cybersecurity constraints imposed by national regulations (Fluchs, 2024). Notably, the product managers determined that the potential for skilled hackers to conduct attacks against the product control system outweighed the sales of the product model. System process sensors must be safe and reliable. Moving a company's priorities to maintain its role as an industrial and technological leader is a significant undertaking that requires stakeholder alignment across the entire value chain. Stakeholder alignment also involves examining the organizational culture to include innovative thinking that may be outside traditional subcultures and company norms. At the same time, alignment with the German government and EU regulations is more important than maintaining brand equity. Automotive manufacturing is a highly competitive and lucrative industry, benefiting both the manufacturer and the country's economy.

The Porsche case study illustrates how companies and nations can gain a competitive advantage in a global economy. Manufacturers with government support can benchmark their products and services against those of international competitors and utilize cybersecurity to gain a competitive advantage. Cybersecurity has become a competitive advantage for maintaining and gaining market share in the worldwide economy. Customers expect trusted, safe cyber products.

21.8 Summary

Cyber is not a spectator sport. It is much more like the rough and physical sport of rugby, which involves frequent scrums, tackles, force, and, most of all, contact with others. It requires skill, strategy, and a critical level of decision-making. Just like in rugby, the captain leads the team. When engineering a system that operates with cyber interfaces, the systems engineer is responsible for making decisions in the field. The systems engineer is responsible for the technical team's performance and makes critical decisions that impact the team's overall success. A systems engineer should be a good team player and have a thorough understanding of the system and the sector in which it will operate. Other team members can also take on leadership roles. The program manager is the "fly-half" or the number 10 on the team. In rugby, the fly-half is the heartbeat of the team and another crucial leader. While the systems engineer has the most formal responsibility within the technical team and communicates directly with the program manager (or sometimes both roles), numerous trade-offs and give-and-take are involved in the decision-making process. Appropriate tools help keep the team working together smoothly. As Thomas Frisby wrote, "If we don't get bruised sometimes, we are sitting on the bench" (Frisby, 2015).

References

Alexander Kossiakoff, W. N. S., Samuel J. Seymour, and Steven M. Biemer 2011. *Systems Engineering Principles and Practice*, John Wiley & Sons, Inc. Publication.

Brotherton, S. A., Fried, R. T. & Norman, E. S. 2008. Applying the work breakdown structure to the project management lifecycle. *PMI® Global Congress 2008—North America.* Denver, CO: Project Management Institute (PMI).

Fluchs, S. 2024. No One Has Ever Changed a Development Process Just Because of Cybersecurity? *In:* linkedin.com/in/sarah-fluchs (ed.). LinkedIn.

Frisby, T. N. 2015. *Ain't All Good Ain't All Bad: That's The Way It Was*, CreateSpace Independent Publishing Platform.

Gray, C. F. & Larson, E. W. 2005. *Project Management - The Managerial Process*, McGraw-Hill.

Ljerka Beus-Dukic, I. F. A. 2009. *Discovering Requirements: How to Specify Products and Services.*

Pinto, J. K. 2015. *Project Management: Achieving Competitive Advantage.*

Scalco, A. W. & WINCE, K. 2012. A Proposed Methodology to Achieve Balanced Resilience and Risk. IE 566 Applied Systems Engineering: Iowa State University.

Schallbruch, M. & Skierka, I. 2018. *Cybersecurity in Germany,* Berlin, Germany, Springer.

22

Managing Risk

"[T]elling people to change their behavior when facing accidents
will not help because they will rarely believe
they are facing accidents"

– Willem A. Wagenaar (1941–2011),
Dutch cognitive psychologist

"If it works, it's out of date."

– David Bowie (1947–2016),
English musician

Abstract

Latent failures affect the psychological processes that shape system operators' actual behavior. Removing latent failures requires a thorough understanding of how the system shapes behavior (Willem A. Wagenaar, 1988). Phishing and spear phishing have emerged as the most prolific cyberattacks today. So, if telling people to change their behavior when facing danger (e.g., accidents or cyber risks) will not help because they rarely believe they are facing accidents, then given the complexity and the dependency on computer-aided activities of many human tasks and processes today, it is unlikely that humans will behave as desired to prevent cybersecurity breaches that may cause undesirable consequences (e.g., accidents or cyberattacks). Managing risk is complex and can be time and resource-consuming. Most people want a risk to be defined so they can manage it effectively. Ideally, most people want a quick and inexpensive fix provided to them. An adage about risk is: "Show me the smoking gun." If a risk has not occurred, then how much does someone really need to spend on mitigating the unclear risk? The systems engineer cannot wait for the final proof of a cyberattack. Recall from Chapter 1, "Understanding the Game," that cyber threats generated by phishing attacks, coupled with an advanced persistent threat (APT) payload, can potentially expose overlooked attack vectors. These adversarial tactics make detection

DOI: 10.1201/9788743809128-22

considerably more challenging and can remain undetected for extended periods. Managing risk in a cyber world involves understanding various behaviors, including system behavior, system operator behavior, and threat actor behavior, as well as risk management, the associated risks, and methods to mitigate cyber vulnerabilities.

22.1 Introduction

Risk is based on use cases against known vulnerabilities. Known is the keyword here. It is challenging to manage unknown vulnerabilities and resources for risk mitigation. In this chapter, we will share an example of how to budget for risk management and review some of the known cyber methods of system attack. The risk assessment process involves identifying critical risk events or processes, analyzing these events or processes to determine their likelihood of occurrence and potential consequences, and prioritizing the risks accordingly. In addition to the normal areas of design, testing, and integration, the risks associated with system safety and security considerations must be identified and analyzed by use case. An operational requirement for critical infrastructure to manage risk is to evaluate the severity and type of detected threats. A question to ask is who provides for the identification of new vulnerabilities and how they perform this function. The lack of failure rate data for vendor commercial off-the-shelf (COTS) products hinders traditional reliability analysis methods. However, an integrated solution may increase security and lower risk by partially mitigating the potential for supply chain tampering resulting from a compromise in a system component due to vendor diversity (Scalco and Wince, 2012).

22.2 Gone Phishing

Latent failures affect the psychological processes of the actual behavior of system operators. Removal of latent failures requires a thorough understanding of how the system shapes behavior. The idea that the system itself can affect the behavior of system operators is based on the research of Wagenaar et al. (Wagenaar, 1988). In an earlier article, Wagenaar and Groeneweg identified causal relationships between accidents and human factors. Some relevant facts about the article "Cognitive Engineering in Complex Dynamic Worlds" are now described. Firstly, the article was published in 1988, prior to the extensive integration of smart sensors, instruments, and other devices networked together with computer applications that we see today. "If one person can oversee the complete scenario, there is a possibility for correction; only the more complex scenarios remain unchecked. When the formation is distributed across more people, it becomes hard to realize the consequences of errors" (Willem A. Wagenaar, 1988). Secondly, Wagenaar made the observations prior to today's interconnected, highly computer-networked environments. The observations identify causal relationships between accidents and human factors, as well as human-introduced cyber vulnerabilities (HICV). Regardless of the amount and level of training, the cybersecurity training level was not then a reliable predictor of the system's defense strength against cyber exploitation attack vectors, such as phishing and spear phishing. Phishing and spear phishing have emerged as the most prevalent cyberattacks today, despite being virtually unknown in 1988.

One aspect of Wagenaar's research that still stands out today is that if telling people to change their behavior when facing danger (e.g., accidents) will not help because they rarely believe they are facing accidents, then given the complexity and the dependency on computer-aided activities it is unlikely that humans will behave as desired to prevent cybersecurity breaches that may cause undesirable consequences (e.g., accidents). "Accidents appear to be the result of highly complex coincidences which could rarely be foreseen by the people involved" (Willem A. Wagenaar, 1988). Further research is necessary in this area to understand the risks and dangers associated with operator-induced vulnerabilities in critical infrastructure control systems. A key question to ask is whether the organization can identify security risks in the first place, and who is responsible for designing and implementing the countermeasures to mitigate these risks.

Similarly, senior research scientist at Pacific Northwest National Laboratory (PNNL) Terry Merz observed a potential causal relationship with human factors during data collection efforts from the joint base architecture for secure industrial control systems (J-BASICS) initiative previously mentioned in Chapter 7, "Threats to Information-intensive Systems." Terry Merz observed in her research that users operating at traditional IT enterprise levels of industrial control systems (ICS) did not behave any differently than those ICS operators who had not received the same amount of cybersecurity training when confronted with phishing attacks (Terry Merz, 2019). Phishing exploits are a type of advanced persistent threat (APT) used against critical infrastructure. "Today, phishing, a human-focused exploit, constitutes 91% of successful attack vectors against federal assets. This means that human-introduced cyber vulnerabilities (HICVs) are the weakest link in cybersecurity. The success of these attacks also suggests that HICVs are not well understood nor mitigated" (Ibid.).

Another phenomenon that stands out in Wagenaar's work is that accidents tended to be attributed to the complexity of systems and scenarios. Given that the research pre-dated the extension of smart sensors, instruments, and other devices networked together, the possibility for correction in more complex scenarios is likely to remain unchecked. If this is correct, then the design of systems warrants greater computer-aided correction of the course of action (COA) or the use of automated systems. Interestingly, the most significant resistance to integrating automated systems is the perception that removing humans from the loop creates a greater potential for accidents or failures to achieve the desired effect. If this is correct, then a human-in-the-loop system design may not be the ideal approach. The ideal system design would be human out-of-the-loop, with fallback redundancy allowing for a human to intervene manually when an error is observed (Ibid.).

22.3 Risk

Risk is generally defined as a technical, cost, schedule, or performance risk. Each risk associated with a solution is identified, categorized, analyzed, mitigated, tracked, and rated based on its probability and its impact. Managing risk starts with identification. A key question to ask is whether playbooks with recommended courses of action are utilized, allowing for risk-aware, prioritized responses when a significant threat and device impact are detected (Vermilye, 2021; MOSAICS, 2025).

22.3.1 Risk identification

All identified risks must be documented in a risk management database, including a statement of the risk, a description of the conditions or situations that cause concern, and the context in which the risk occurs. A key question to ask is whether the capability is in place to monitor assets and network behavior to detect anomalous components and communication activity (Vermilye, 2021; MOSAICS, 2025). The technical and performance risks associated with the composed solution capability are identified in critical risk areas, including design and engineering, technology, logistics, and integration. The identified risk must be associated with at least one project element in the work breakdown structure (WBS). The risk may affect multiple elements and is therefore cross-referenced to the work breakdown structure (WBS) depending on the circumstances. "Since effective risk management requires a stable and recognized baseline from which to assess, mitigate, and manage program risk, it is critical that the program use an [integrated management plan (IMP)/integrated master schedule (IMS)] (IMP/IMS)" (Office of the Secretary of Defense (OSD) OUSD (AT&L) Systems and Software Engineering/Enterprise Development, 2006). The integrated master plan

(IMP) serves as the baseline planning tool for implementing the composed commercial suite of solutions.

22.3.2 Managing risk

The two primary components of risk management are risk assessment and risk mitigation. A risk assessment is the determination of the probability of failure to meet the desired outcomes of a project. Risk assessment includes risk planning and prioritization. Two examples of risk assessments are strengths, weaknesses, opportunities, and threats (SWOT) analysis and the Delphi technique. A strengths, weaknesses, opportunities, and threats (SWOT) analysis categorizes the project's top strengths, weaknesses, opportunities, and threats to assess the potential impact of risks. The Delphi technique, a forecasting method that utilizes structured rounds of questionnaire responses, is employed to gather information from subject matter experts (SMEs) and achieve a consensus on project risk. The second component is risk mitigation. Risk mitigation is taking steps to reduce the effects or impact of risks. Two examples of risk mitigation are redundancy and process control. Redundancy is the elimination of a single point of failure for each critical component of a system or subsystem. Process control involves designing controls built into processes to mitigate project or system risk, such as required approvals and gateways.

An example of how a risk management process for a new system development that proposes to use one or more components utilizing unproven technology would be to do the following:

1. Perform a SWOT analysis of the component technology
2. Use the Delphi technique to rank the technology readiness level (TRL) of the components
3. Design redundancy of the components utilizing unproven technology by the use of proven technology, and
4. Establish process control by reviewing component test results through gateway reviews and integrating the system's components through testing

Kossiakoff presents a method of combining two metrics into a single, combined metric for risk, using relative risk lives to assess the likelihood of high, medium, or low risk (Alexander Kossiakoff, 2011). This method of quantifying risk acknowledges that a single numerical scale is absent for the risk criticality. Therefore, the likely effects of the program are the same relative levels as the likelihood. The advantage is a simple single risk factor. However, the disadvantage is the potential of "the illusion of quantitative knowledge, which has no real basis" (Ibid.). Another approach is to combine two indices into one. However, the disadvantage is that a single score represents an abstraction of the overall performance. Finally, the highest level of criticality indicates that the project may lead to near-system failure, which would be programmatically unacceptable due to the implied design risk that would render the option unfeasible.

Further study is needed to build on other research and previous academic studies in the area of cyber resiliency for cyber-requirement-based systems found in critical infrastructure. A key distinction between research and existing research is the lack of research in the systems engineering community on transitioning solutions for critical infrastructure to the operational field, particularly in understanding the workforce available to address the problem statement and identifying those who possess expertise in cyber and critical infrastructure. Information is needed to support the development of prototypes by national research laboratories in the federal and commercial sectors.

22.3.3 Risk assessment

Systems engineers conduct risk assessments for each specific composed solution, function, and incremental phase. Risk assessments shall be updated, and the results will be presented at all functional reviews. The final update shall be prepared prior to presentation to the designated approval authority for approval decision. The user of a specific composed solution shall report identified risks to the integrated product team (IPT) lead so that lessons learned can be tested, validated, and incorporated into the composed solution library. Do not wait for perfection. Start by assessing the current resources, knowledge, and circumstances.

22.3.4 Budgeting to manage risk

Managing risk requires a budget. An example scenario in risk analysis is as follows:

Imagine you currently spend 3% as much on countermeasures to prevent attacks as the reported expected cost of the attacks themselves (that you are presumably preventing, that is, you spend USD 3 for every USD 100 of total risk). Your manager wants this percentage of spending to be only 2% next year (i.e., spend 2% as much on security countermeasures as the cost of the attacks if they were not prevented). You find reliable data showing a 240% increase in costs associated with attacks (if successful, i.e., the risk will increase by 140% of the current value) next year, due to improved threat intelligence among agencies worldwide. Should your budget grow or shrink? By how much? If you currently have 20 employees, how many should you hire or furlough?

One approach might be to address the budget shrinkage by 1%, as requested by management, to meet the 2% target. This would allow all 20 employee billets to be maintained, assuming a total budget of USD 3.00 per USD 100 of risk buy-down, including employee-loaded labor costs. A key variable is the ratio of labor (encompassing knowledge, skills, and attributes) to the tools employed. Due to the increased threat agents pose around the world, the cost of these security countermeasures is USD 1.25 per USD 100 or 1.25%.

Here is the rationale: The current spend is 3%, or USD 3, allocated for protection, covering a total risk of USD 100. Due to improved threat agent skills in exploiting system vulnerabilities worldwide, the risk will increase the current value by 140%. Therefore, every USD 3 spent protects USD 240 in total risk. The budget can be reduced to USD 1.25 per USD 100, or 1.25%, to achieve the same protection. This calculation is based on the following: USD 3/240 = x/100; 240x = 300; x = USD 1.25. Therefore, USD 3.00 – USD 1.25 = USD 1.75 is the amount the budget can be decreased by from USD 3 to achieve the same risk buy-down. If the risk increases to USD 2.40 for every USD 1 this year, then 3% of USD 100 is USD 3, while 2% of USD 240 is USD 4.80. Therefore, a 60% increase in budget (from USD 3 to USD 4.80) is justified.

To further defend the position, someone might promote defense in depth and defense in breadth, allocating a budget for continuous cybersecurity training to keep pace with business and IT risks, maintain cybersecurity skills, and cover the costs of cybersecurity certifications, such as CISSP. Personnel would be encouraged to pursue these certifications for promotable roles and responsibilities as part of overall career and skills development. Understand what you want to protect and find cost-effective ways to protect the data or obfuscate it using solutions that leverage innovative technologies, such as software-defined networking (SDN) and advanced behavioral analytics. An organization can achieve better risk reduction through defense-in-depth, which involves multiple layers, aiming to make the adversary pay more to attack. Two different security mechanisms, employed in a layered approach, may be more effective and cost-effective than a single layer of defense. The government uses Type 1 security, which is very expensive for encryption. However, two different layered commercial products may be more secure. I would ask the team to conduct a risk assessment process to identify assets, assess potential attack vectors, and prioritize actions to mitigate cybersecurity risks. The best practice is to follow the National Institute of Standards and Technology (NIST) Framework for improving critical infrastructure cybersecurity (Technology, 2018; Stouffer et al., 2023). The effort is to identify unmet budgetary needs essential to managing cybersecurity risk, including understanding the threat, identifying capabilities, assessing risk, remediating risk, and mitigating risk.

Furthermore, a threat-based approach can facilitate informed decisions regarding cyber risk. This approach enables the organization to create a cyber threat mitigation heat map, prioritizing where resources should be allocated to address the most critical threats. Security is not a one-and-done process. Many variables influence the decision-making process beyond a straightforward mathematical financial approach.

22.4 Cybersecurity Awareness

Numerous resources, translated into various languages, are available for cybersecurity awareness. However, we cannot have a book on systems engineering for critical infrastructure to overcome cyber vulnerabilities without a few words about general cybersecurity. Managing risk for critical infrastructure to overcome cyber vulnerabilities requires organizational security awareness (and cybersecurity, increasingly, cyber-physical security) training. Two examples of how security training can be made more effective are continuous staff training and promoting security awareness to make cybersecurity a habitual practice for all employees within an organization. The concept of training repeatedly reinforces learning and retention when using effective materials and changing these training methods more frequently helps avoid monotonous repetition (Simske, 2020).

22.4.1 Security policy

Small or medium-sized companies can prevent data loss and unauthorized access to data by implementing a comprehensive security policy. An operational requirement is to identify both internal and external data flows and connections relative to the critical infrastructure target environment. Data loss and unauthorized data access prevention can be implemented for small or medium-sized companies by having technical security controls in place, along with regular enforcement of security policies for all accounts (i.e., frequent password changes and high password strength), and by continuously providing education through security awareness training. Two examples are keeping data safe by regularly backing up data and protecting the network by securing remote access (e.g., with VPNs), safeguarding wireless access points, and using strong passwords. Data loss protection should be employed at all levels of the ziggurat (the shape in a defense-in-depth chart, stacking all security zones from perimeter to network to endpoint to application to data to mission-critical assets). A question to ask is whether sensitive network connections between traffic sources and points of encryption are monitored on the organization's network systems. Defense-in-depth is a multi-layered security strategy that incorporates redundancies to enhance security measures, protecting data across multiple layers rather than relying on a single layer. For example, the National Security Agency (NSA) commercial solutions for classified (CSfC) program is part of the commercial cybersecurity strategy, which utilizes layered solutions to provide adequate protection for classified data. Even if attackers bypass firewalls and intrusion detection systems, sensitive data should still be protected using encryption at all levels of the defense-in-depth (DiD) stack. The implementation of DiD may increase the time it takes a hacker to steal data, or an insider threat may be discovered or slowed down, but the risk of data loss remains. Therefore, data loss protection should be designed at all layers of defense-in-depth, with each technology and policy implemented to serve the purpose of protecting data from an event occurring as well as reducing the severity of a breach event.

22.4.2 Behavioral analytics

Analytics on network traffic can help identify both internal and external attacks. Monitoring user and entity behavior using analytics for all system administrators' activities is a best practice that simplifies the discovery process for detecting vulnerabilities and assessing safeguards, continuously learning about new threat vectors. Evaluating network traffic enables organizations to generate security alerts and identify potential security threats and intrusions. The analytics would allow stakeholders to make informed security decisions and provide network situational awareness, thereby enhancing security. Search analytics enables the ability to "review reports and identify any known or reported exceptions from network and antivirus security tools" (Diogenes and Ozkaya, 2019). A second way is by using machine analytics, "enabling [organizations] to autonomously scan large amounts of data and give brief and simplified results to people to further analyze" (Ibid.). A key question to ask is how user accounts and credentials are managed within the organization, and what type of authentication approach is used, such as modern authentication, cloud identity authentication, or federated authentication.

Additionally, it is essential to determine whether decision-makers understand the approach being taken.

22.4.3 Network traffic analysis

Wireshark is an example of a network traffic analysis tool that can be used to analyze packet captures or capture live network traffic packets. Wireshark provides information about the current network state and past network intrusions. What makes Wireshark such a threatening piece of technology is that it can easily steal authentication details from traffic sent out of a network in a few steps (Ibid.). Wireshark is easy to use. It decodes, records, and displays packet capture (PCAP) data in a graphical user interface (GUI)-based utility and sniffs low-bandwidth connections. A PCAP is an application programming interface (API) that enables different software applications to communicate with each other, allowing them to capture network data packets and share information about network events. A user can view each protocol layer for each packet. Finally, it is free. Wireshark is also a threatening piece of technology because it is readily available at no cost to nefarious people. The user can utilize a reference architecture to gain a better understanding of how different computer systems communicate over the network. Reference models are further discussed in Chapter 23, "Abstraction Models."

22.4.4 Social engineering

As mentioned earlier in this chapter, many companies grapple with issues related to social engineering, such as phishing, pharming, spoofing, stolen accounts, and blackmail. Social engineering is an attack surface that can be used against a system or product, which can quickly spread the attack surface beyond the initial compromise. Data and information may be exploited through excessive social internet access to obtain sensitive information that can be used in passwords or as answers to security questions to reset accounts or escalate password privileges. "Users are known to use weak passwords due to laziness or lack of awareness about the threats" (Ibid.). It is possible that user information and details can reveal sensitive information about a user, providing an attacker with access to other organizational accounts using similar information or by *clicking on* or opening attachments sent via social media. Malware can then be unleashed on the computer and spread to other systems.

"The single greatest vulnerability is people (untrained, unmotivated, or malicious insiders)" (Simske, 2020). There is a tendency for network administrators to use weak password settings and fail to install patches in a timely manner, which creates security vulnerabilities in the system. These security-related tendencies may be addressed by training security staff, which differs from the threat surface presented by underpaid security staff members. The potential in the latter is that the perceived underpayment might lead to the security staffer being "unmotivated" or becoming an active insider threat (i.e., purposely sharing information with unauthorized individuals or intentionally causing other damages). Either of these presents exploitable weaknesses that need to be addressed as part of the overall risk management plan to safeguard assets. Most breaches involve weak, stolen, or infrequently changed passwords. Therefore, many involve "insider" actions, so considering the "underpaid" and "undertrained" personnel as part of the business cybersecurity risk is critical.

An example of the importance of managing cyber risks can be found in the Oldsmar water treatment facility incident, which illustrates the adverse consequences of cybersecurity vulnerabilities in physical systems and control systems. On February 5, 2021, someone tried to poison the water supply of the city of Oldsmar, Florida, through a cyberattack (CISA et al., 2021). By chance, a facility supervisor saw the hacker's mouse move across the screen in an attempt to make unauthorized changes to settings, averting an increase in the amount of sodium hydroxide ("lye") used in the water treatment process that would have made the chemical a caustic hazard to humans. The supervisor happened to be in the right place at the right time, recognized that something unusual was afoot, and took action. A facility supervisor, knowing the physical system and possessing cybersecurity awareness, averted disaster. A subsequent joint cybersecurity advisory about the Oldsmar water treatment compromise, co-authored by the Department of Homeland Security (DHS) Cybersecurity and Infrastructure Security Agency (CISA), the Department of Justice (DOJ) Federal Bureau of Investigation (FBI), the Environmental Protection Agency (EPA) and multi-state Information Sharing and Analysis Centers (ISACs), reported the cyber actors likely exploited fundamental cybersecurity weaknesses in the water treatment facility. Water treatment facilities have systems with lifecycles of 30 years or more, which is significantly longer than the lifecycle of IT components and systems. Therefore, it is not surprising that computer networks were running on Windows 7 operating systems (OS) with an end-of-life status, which were exploited to gain access. An outdated operating system (OS), such as Windows 7, is susceptible to exploitation because security updates for the end-of-life OS are no longer supported, leaving it vulnerable to emerging vulnerabilities. One question an organization might ask is whether outdated operating systems are used in its operations (Aleksandra Scalco, 2021).

Physical security measures, such as installing independent cyber-physical safety systems with limited cybersecurity capabilities, provide additional safety protection to limit damage. Limiting who can release caustic chemicals and restricting the amounts that can be released are safety controls – another important cyber hygiene mitigation, as well as the security of remote access control to the system. Remote party access is commonly employed as a means for legitimate remote control of computer systems. The remote capability can also allow unauthorized entities to gain access to control systems.

Other methods of cyberattack include actions such as SQL injection. The injection is an exploit that attacks application code, web components, networking hardware, and other endpoints. For example, SQL injection is an attempt to run unauthorized code using an unauthorized path for database manipulation.

Software developers create dynamic database queries that allow user-supplied input, which guarantees free pathways for injection attack tool success (e.g., Safe3SI, SQL Ninja, Pangolin, and The Mole). Dynamic SQL is interesting as otherwise time-consuming tasks can be split into runtime variables, which is a security threat if not cleansed appropriately prior to use. The intention of a dynamic database can be easily hacked, returning additional data from the database for an attacker to steal. The most effective way to prevent SQL injection is to ensure that data is verified and sanitized before it is inserted into the database. SQL can be prevented by using known coding techniques, code reviews, and testing to "sanitize" user inputs (Ibid.). "[A]lways look at the [Open Web Application Security Project] OWASP Top 10 for the latest update in the list of most critical web applications" (Diogenes and Ozkaya, 2019). Installing security plugins, regularly updating websites, and utilizing an SQL injection scanner can also help prevent SQL attacks.

Some consider an attack to be "successful" once it has reached the exfiltration stage (Ibid.). However, we consider success to be at the reconnaissance stage, during which system vulnerabilities are discovered. Once a delivery is made, the system is compromised. The threat actor may conduct a test run of exfiltration to buy time for another, even more harmful attack, moving beyond data and software attacks to compromise the hardware and lower-level processes of a system within an organization (Ibid.).

Distributed denial of service (DDoS) attacks can have devastating market effects on organizations when the availability of online resources is interrupted, resulting in loss of productivity and potentially eroding customer trust. DDoS attacks can have a significant financial impact, resulting in substantial revenue losses for businesses. DDoS can lead to the disruption of services (i.e., complete mission interruption), loss of sensitive data that may result in fines (i.e., regulatory penalties), and otherwise pose a financial threat to business services. Smaller organizations can really be negatively impacted financially. An example is the 2019 DDoS attack on Amazon Web Services (AWS), which rendered it impossible for users to connect when AWS categorized their queries as malicious. DDoS, although it may not lead to a data breach, can still pose a substantial financial threat due to lost revenue from reduced business traffic and the financial burdens of IT replacements.

Extensive logging and backup of logging are essential records for security controls. Logging provides visibility into what actually occurs in a network on a per-user or per-application basis (Ibid.). There are two key concepts to consider when it comes to logging. The first is for forensics, and the second is for monitoring and actively using logs for trend analysis. Maintaining extensive logging (and backup of logging) records for security controls is essential to understanding security incidents during a forensic analysis. Extensive logs are also helpful in establishing baselines that demonstrate the value of the system and reveal operational trends of users and administrators during audits and forensic analysis. An operational requirement to create and manage audit logs protects the system from cyber events. A key consideration in the critical infrastructure environment is whether data is collected from various cyber defense tools (e.g., intrusion detection system (IDS) alerts, firewalls, and network traffic logs) to analyze events that occur within the target critical infrastructure environment, with the goal of mitigating threats. Another key concept of extensive logging (and backup of logging) is actively using these logs to monitor security-related activities. Maintaining logs is essential for post-incident forensic investigation, as is monitoring and analyzing (in near real-time) to detect attacks and deploy appropriate countermeasures.

22.5 Software Threat Modeling

Identifying system entry and exit points in software threat modeling is essential. Software threat modeling is a critical process that helps improve software by identifying potential threats and vulnerabilities, allowing them to be addressed and fixed. System asset access and exit points are potential attack surfaces. For example, an open port protocol or authentication mechanism needs to be protected from misuse. Characterizing the correlation and data flows represents where and how data is stored or in transit. Identifying system entry and exit points as part of the system baseline under configuration management (CM) helps to understand the system better and address potential threats. The system's entry and exit points identify potential ways attackers can exploit the system's software application. Entry and exit points serve as interfaces with other internal system components. Identifying potential vulnerabilities and threats can inform security design decisions, enabling the implementation of more effective security measures. For example, actively review logs and security alerts to detect anomalous activity for all system components.

22.5.1 Intrusion detection systems (IDS)

Intrusion detection systems (IDS) utilizing artificial intelligence (AI) techniques, such as flow packet analysis, are based on temporal relationships during data preprocessing in machine learning (ML), which enables unsupervised learning and reduces the likelihood of false alarms. AI techniques help address the challenges of IDS, including high false alarm rates and low detection accuracy. The innate "temporal flaw" of the system is that while merit refers to the ability to perform intelligent actions, the organization using the technique must carefully state the requirements for appropriate responses and acceptable characteristics. The argument that AI will help solve intrusion detection systems (IDS) assumes that the past approach to IDS will remain valid and that what has been true in the past will continue to be true.

Technical leadership must ensure that everyone understands how various attack methods can be used to deliver malicious code. Viruses, worms, and trojans are all types of malware. A computer virus attaches itself to a computer program or file, enabling the virus to spread from one computer to another via an executable file. A "virus is malware that, when executed, tries to replicate itself into other executable code (CRISPR Code 9); when it succeeds, the code is infected. When the infected code executes, so does the viral code" (Simske, 2020). A virus cannot actually infect the asset without the malicious program executing, which traditionally requires human action to spread. Given today's automated environment, an automated course of action (COA) could potentially replace a human in launching the executable file. A virus attaches to an executable file and (typically) requires human access to spread.

A worm has the potential to move laterally from computer to computer without human assistance. It is similar to a virus but has the capability to travel using information transport features on a system without human action. A worm can replicate itself on a system. A "worm is a program that runs independently and propagates a complete working version of itself onto other hosts on a network" (Ibid.). For example, a worm could send a copy of itself to everyone listed in an Email address book. Then, everyone listed in the receiver's address book further manifests itself. Worms do not require human action to spread and can replicate themselves on a system.

A Trojan horse is a malicious software code that may superficially appear helpful but is actually malware designed to do damage by deleting files and destroying information on a system. A "Trojan horse is a 'useful' program that has a hidden and possibly malicious function that evades surveillance … [It] often uses legitimate actions to launch the hidden attack code" (Ibid.). Trojans are also used to create computer backdoors that give unauthorized system access. Trojans require human action but do not self-replicate.

Exploit kits are automated toolkits or frameworks designed to identify and exploit vulnerabilities, delivering a malicious payload to a machine. As a friend and author, Rick Howard has wisely said, "*You can't bring boots to a cyber fight.*" Rick Howard's book, "Cybersecurity First Principles: A Reboot of Strategy and Tactics," is a must-read for anyone in the cybersecurity profession who wants to gain a deeper understanding of strategies and tactics (Howard, 2023). Attackers use automation to attack a system, and automation is required to mitigate these automated attack courses of action (COA). Attackers use exploit kits that are sold or made available on invitation-only forums to avoid discovery by law enforcement. Over time, there appears to be some decline in the prevalence of exploit kits. Two reasons why there may be a declining prevalence of exploit kits are the increased mitigation strategies such as antivirus products with updated, current signature sets used to detect and prevent exploit kit threats and better network defender implementation of attack surfaces (i.e., firewall rules, policy measures such as denying HTTP requests over non-standard ports, blocking DNS domains outside of core business functions, application whitelisting, patching); and improved user education and response training focusing on web browsing policies such as limiting browsing only to business relevant sites to lower chance of exposure to sites hosting exploit kits. All these security defense activities make exploit kits less profitable and less interesting to the criminal market. If we can all work together, then the rising tide will raise all boats. Rick Howard also said, "*Let's not make the internet safer. Let's make it safe.*" The same should be said for critical infrastructure.

22.6 Governance and Policy

What are some key governance and policy principles that should be considered to manage risk and mitigate cyber vulnerabilities? Blocking incoming packets with an internal host address is used to counter IP address spoofing, a technique that involves using a false source IP address to impersonate another computer system. Packets sent using forged source addresses are a form of packet spoofing, which is used to obscure the source of an attack or to intercept network traffic, causing replies to be sent to another system. Blocking outgoing packets without an internal host source to defeat zombies or zombie botnets. During a zombie attack, a command is sent to each zombie to flood the victim with packets or trigger a coordinated attack. Blocking outgoing packets without an internal host source address helps to defeat the zombies.

Honeypots or honeynets can be used as part of the strategy. The purpose of a honeypot/honeynet is to distract (by attraction) an attacker from a legitimate target by deceiving them into believing they are attacking a computer system that is not a legitimate target. A proxy server is used to triage all traffic through the server and analyze the packet contents, determining whether they are from a legitimate source to forward. The proxy server can be used to deceive attackers and to learn about the attacker's methods (e.g., tactics, techniques, and procedures (TTPs)). Typically, a proxy server intercepts and analyzes packet contents using a set of firewall rules to determine whether they are forwarded.

Personnel should never unsubscribe from spam emails. Spoofers will modify email headers to make malicious traffic look legitimate. As a countermeasure, it is better to block spam instead of unsubscribing. Unsubscribing is a response to an auto-reply message, which informs spammers that the email account is live and will not result in being removed from spam mail. Instead, it will demonstrate an active email or potentially also connect to a malicious website that downloads malware onto the asset. A bigger threat than email might be texting. As organizations move to hybrid work environments, the use of mobile devices for texting becomes problematic. Text phishing uses text services to retain access to sensitive information such as usernames, passwords, and credit card details disguised as a trustworthy entity in electronic communications. Texting data is under the vendor's control, and web-based texting can be accessed from anywhere and at any time. Despite what vendors may claim, text messages are not encrypted or protected. Text messages are transmitted from a telephone to a cellular tower and then typically again from a cell tower to the mobile provider (via unencrypted transmission). This data is then processed and stored in a database and sent in reverse order (via unencrypted transmission) to the recipient. Recall that an operational requirement for critical infrastructure is to identify internal and external data flows and connections of the environment. Sensitive network connections between traffic sources and points of encryption must be monitored on the organization's network systems to protect data in transit and data at rest from unauthorized access or manipulation.

22.7 Summary

As previously stated in Chapter 7, "Threats to Information-intensive Systems," critical infrastructure and information-intensive systems are highly susceptible to cyber-induced failures. Cyber-induced failure can result from system behavior, system operator behavior, or threat actor behavior. Therefore, the systems must be designed to overcome all the challenges that cyber presents, including dependencies, constraints, and risks to the target system, as well as any other influences that affect the system's enterprise and elements throughout the lifecycle. Managing risk is a method to overcome cyber vulnerabilities in the systems engineering toolkit. Autonomous systems, robotics, advanced communication, machine learning (ML), artificial intelligence (AI), and biometrics, as described in Chapter 19, "Biometrics," and other technologies, can be leveraged to enhance risk management.

References

(CISA), C. I. S. A., Justice, D. O., Agency, E. P. & Isac, M.-S. 2021. Joint Cybersecurity Advisory — Compromise of U.S. Water Treatment Facility. Cybersecurity & Infrastructure Security Agency (CISA).

Alexander Kossiakoff, W. N. S., Samuel J. Seymour, and Steven M. Biemer 2011. *Systems Engineering Principles and Practice*, John Wiley & Sons, Inc. Publication.

Diogenes, Y. & Ozkaya, E. 2019. *Cybersecurity — Attack and Defense Strategies,* Birmingham, UK, Packt Publishing Ltd.

Howard, R. 2023. *Cybersecurity First Principles: A Reboot of Strategy and Tactics,* Hoboken, New Jersey, John Wiley and Sons.

Mosaics 2025. More Situational Awareness for Industrial Control Systems (MOSAICS) Block 1 Specification. *In:* CIO, N. H. (ed.). NAVFAC HQ CIO: NAVFAC HQ CIO.

Aleksandra Scalco, S. S. 2021. Championing Cultural Change and Control Systems Cyber Security. *AIAA-INCOSE Wasatch Aerospace and Systems Engineering Mini-Conference.* AIAA Utah Section.

Scalco, A. W. & Wince, K. 2012. A Proposed Methodology to Achieve Balanced Resilience and Risk. IE 566 Applied Systems Engineering: Iowa State University.

Simske, S. J. 2020. Cybersecurity Lectures. *SYSE 569 Course.* Fort Collins, Colorado, USA: Colorado State University.

Stouffer, K., Pease, M., Tang, C., Zimmerman, T., Pillitteri, V., Lightman, S., Hahn, A., Saravia, S., Sherule, A. & Thompson, M. 2023. NIST Special Publication NIST SP 800- 82r3. *Guide to Operational Technology (OT) Security.* NIST: NIST.

Technology, N. I. O. S. A. 2018. Framework for Improving Critical Infrastructure Cybersecurity. 1.1.

Terry Merz, C. F., Aleksandra Scalco 2019. A Context-Centred Research Approach to Phishing and Operational Technology in Industrial Control Systems. *The Journal of Information Warfare.*

Willem A. Wagenaar, J. G. 1988. Cognitive Engineering in Complex Dynamic Worlds. *International Journal of Man—Machine Studies.*

Vermilye, C. More Situational Awareness for Industrial Control Systems (MOSAICS) Requirements. MOSAICS Industry Day #3/TechConnect, October 19, 2021 2021 Washington, D.C.: MOSAICS JCTD, 13.

23

Abstraction Models

"Don't panic"

– Douglas Adams,
The Hitchhiker's Guide to the Galaxy
(Adams, 1979)

Abstract

In the novel, *The Hitchhiker's Guide to the Galaxy*, the phrase "Don't panic" is on the cover to prevent interstellar travelers from panicking because a system looks too "insanely complicated" to operate (Adams, 1979). The phrase is a reassuring instruction used when things go wrong, and panic might be someone's more immediate response to an unexpected, negative situation. Such a situation as the massive power utility blackout in New York City in July 1977, which lasted for two days, or the major blackout that occurred across Portugal and Spain in April 2025. It is a reminder to us all that a small amount of preparedness can help us manage through whatever occurs. This phrase is useful when applied to systems in a cyber world. Much of the critical infrastructure built 20 or 30 years ago was designed without cyber in mind. Critical infrastructure is designed for functionality. A water system was designed to deliver water. It is helpful to know what architecture is used and consider the knowledge, skills, and abilities (KSAs) expected of the workforce in relation to the architecture. For example, suppose an organization utilizes the Purdue Enterprise Reference Architecture (PERA) Model for a water system. In that case, it is beneficial to understand the types of knowledge the workforce requires regarding system physical vulnerabilities, system cyber vulnerabilities, specific skills such as software proficiency or hands-on experience with production machines, and the ability to analyze situations and problems presented. The abstraction model is a simplified conceptual representation of a complex system, highlighting essential aspects of a physical system's design, data flows, and underlying principles. A hierarchy model of knowledge, skills, and abilities (KSAs) for the transition of cyber mitigations in

DOI: 10.1201/9788743809128-23

industrial control systems (ICS) is presented in this chapter, illustrating how a relatively traditional type of survey can empower the workforce. Examples of abstraction models are described for some of the typical reference architectures, such as the Open Systems Interconnection (OSI), the transmission control protocol (TCP)/internet protocol model (TCP/IP Model), the PERA Model, and the International Society of Automation ISA112 supervisory control and data acquisition (SCADA) systems management lifecycle model.

23.1 Introduction

Critical infrastructure encompasses systems that are essential for societal functions such as water, transportation, and healthcare. Systems Engineers use an abstraction model as a simplified representation of a complex system found in critical infrastructure, such as power, water, transportation, and manufacturing. Therefore, it is notable when expected system functions are disrupted, as was the case about midday on April 28, 2025, in Portugal and Spain. Regardless of the cause, operators had to restore operations using a complex "black start" process to recover the electric grid and restore power. Grid operators continually work to maintain the balance between power generation and load, ensuring reliable and stable operations (Research, 2025). It takes power to perform black start operations, either through a generator that can supply enough stored energy back to the grid or through a smart microgrid, as described in Chapter 20, "Foundations of Systems Engineering," which can help a failed grid recover. Black start voltage recovery must be highly coordinated and controlled to minimize voltage disturbances. Behavioral modeling and simulation using abstraction models of the system enable a technical team to evaluate black start configurations and test the black start plan.

The increasing reliance on digital technologies makes critical infrastructure systems insanely complicated. As noted in Chapter 8, "Use Cases and Operational View (OV) Illustrations," critical infrastructure is fragile, with cross-sector dependencies and inter-system dependencies. An example of a use case in the healthcare and hospital vertical demonstrates the need for a paradigm shift in engineering approaches to designing secure system solutions: "Don't panic." Systems Engineers use an abstraction model as a simplified representation of a complex system. It is easier to analyze, simulate, and predict system behavior in a model. Some examples of types of models are the Open Systems Interconnection (OSI), the internet protocol suite framework, also known as the "transmission control protocol (TCP)/ internet protocol model" (TCP/IP Model), the Purdue Enterprise Reference

Architecture (PERA) model, and the International Society of Automation ISA112 supervisory control and data acquisition (SCADA) model. These models help to reduce the complexity of systems that collect real-time data from sensors and transmit data between field devices and central control systems. Abstraction models are a key to systems engineering. The abstraction model makes what seems insanely complicated less complicated to understand when used.

23.2 Hierarchy of Knowledge, Skills, and Abilities (KSA) Model

Abstraction models reveal underlying principles about essential aspects of a system. An organizational challenge for critical infrastructure in a cyber world is determining what training or upskilling is required for various roles. Professionals want to stay relevant in their field, which may be a career field they do not fully understand. Technological advancements in critical infrastructure introduce cyber risks. As a result, organizations often look to personnel from computer science, software development, or cybersecurity fields to help mitigate these risks. However, these professionals, while with the best intentions, have limited experience and perspective on what is required to protect critical infrastructure. The clear separation of the "us" and "them" perspectives between computer science (generalized) and engineering and technician (generalized) fields hinders the organizational ability to assign the right person to the right job for an industrial control system (ICS). The expectation is that the cyber professional can understand and succeed in any environment, regardless of their engineering knowledge of a system. A general lack of knowledge about a physical system makes it more challenging for talented personnel to implement cyber mitigations in industrial control systems (ICS). At the same time, engineers and technicians often expect the cyber community to protect systems from any vulnerabilities.

A different approach is needed. A more effective strategy is to develop a hierarchy of knowledge, skills, and abilities (KSAs) within an organization's architecture model to visualize the specific KSAs required for a position. The approach to using abstraction models to understand needed KSAs enables cyberculture to develop organically within the organizational environment, taking our prior research a step further to make recommendations for organizational strategies, initiatives, and workforce training programs that can be quantifiably measured based on requirements derived from the advanced cyber industrial control system tactics, techniques, and procedures (ACI TTP) for Department of Defense (DOD) industrial control systems

(ICS) described in Chapter 7, "Threats to Information-intensive Systems" ((USCYBERCOM), 2018).

In prior research, we were the first to take a relatively traditional type of survey and turn it into analytical insights for many people. Using the mathematical model and methodology, organizational leaders can pinpoint areas of uncertainty in an environment where they are likely to exist, particularly those related to cyber vulnerability. We transitioned from a simple survey to being able to identify uncertainty in it and then postulated that uncertainty is "vulnerability," which is applied to cybersecurity in the control system environment. The vulnerability induced by the misalignment of two different parties or groups in an ecosystem may be greater than the innate system design vulnerability. It is possible to measure the uncertainty of agreement through statistical analysis and an analytical model, identifying pain points where different stakeholders disagree. That same measure can be traced back to stakeholders' sources, starting with a vulnerability assessment to help drive better alignment and, ultimately, agreement. This analytic approach is a repeatable process that an organization can use to identify uncertainty among professionals that leads to system vulnerability (Scalco, 2022).

In this hierarchy of knowledge, skills, and abilities (KSA) model for the transition of cyber mitigations for industrial control systems (ICS), we take the research a step further by illustrating the placement of some survey questions used in our study within an abstraction hierarchy model. This hierarchy model provides a starting point for what the workforce needs to consider in various work role areas. For example, foundational knowledge of engineering concepts and operational procedures is at the bottom of the hierarchy. This knowledge encompasses both the physical system and cybersecurity. Questions about the operational and control layer include those related to awareness and hands-on experience, such as software proficiency in collecting near-real-time data from production systems. At this level, in addition to potential economic damage and risk of failure to critical infrastructure, there is also the potential risk to personnel and facility safety operations.

Questions about knowledge of physical systems include:

- "I have access to the industrial control system (ICS)/supervisory control and data acquisition (SCADA) capabilities."
- "I have access to the physical network topology."
- "I know the infrastructure supporting these critical assets."
- "Our organization has hazardous operations."

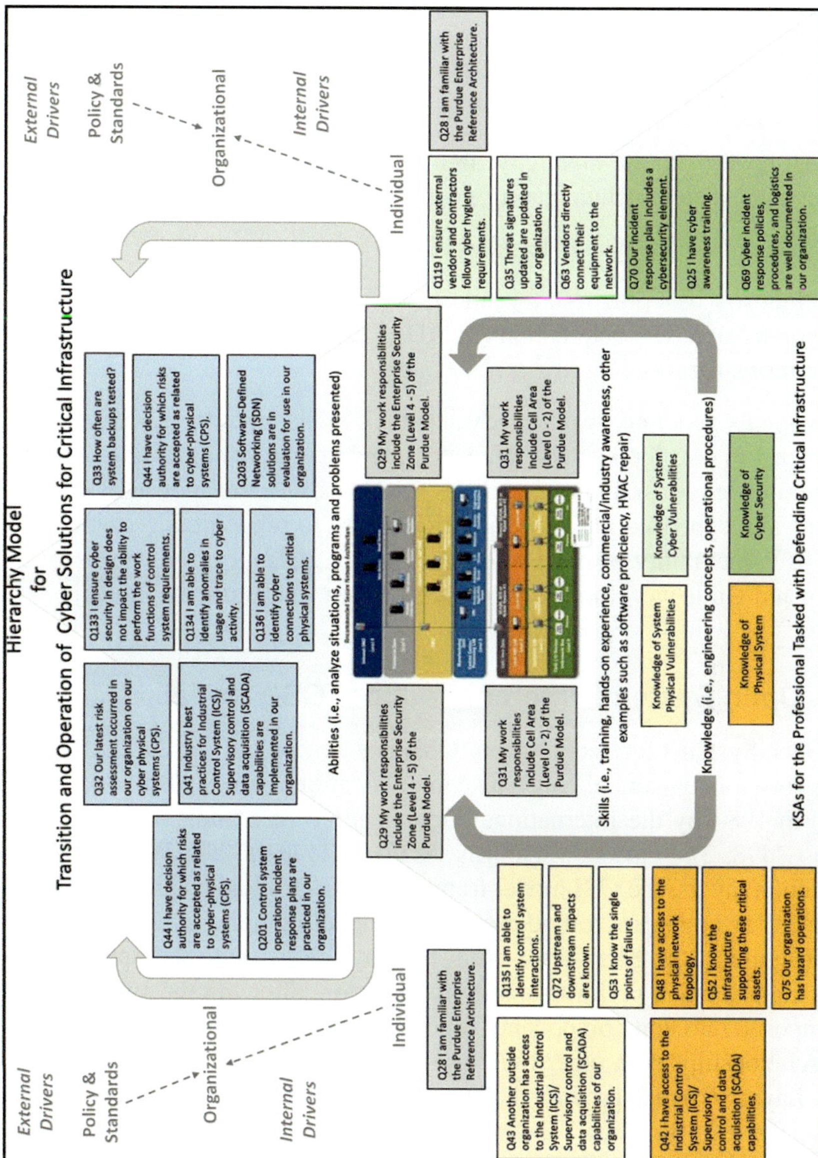

Figure 23.1 Hierarchy of knowledge, skills, and abilities model for the transition of cyber mitigations for industrial control systems.

Questions about knowledge of cybersecurity include:

- "Our incident response plan includes a cybersecurity element."
- "I have cyber awareness training."
- "Cyber incident response policies, procedures, and logistics are well documented in our organization."

Moving upwards in the hierarchy, abilities needed to analyze situations and problems presented include questions such as:

- "How often are system backups tested?"
- "Industry best practices for industrial control system (ICS)/supervisory control and data acquisition (SCADA) capabilities are implemented in our organization."

Professionals generally want to advance their skills, and organizations generally want to develop leaders, innovators, and operators possessing the necessary knowledge, skills, and abilities (KSAs) to secure industrial control systems in a cyber world. These questions, used to pinpoint uncertainty that leads to vulnerability in prior research (Scalco, 2022), can be employed to inform and guide workforce development within an abstraction model.

23.3 Open Systems Interconnection (OSI) Model

The Open Systems Interconnection (OSI) is a conceptual standard reference framework for conceptualizing data communications between networks, published in 1984 by the International Organization for Standardization (ISO) as ISO 7498. The current version is ISO/IEC 7498-1:1994. With its seven abstraction layers, the OSI model framework offers a logical and systematic approach for engineers to understand and model complex system architectures (Figure 23.2). Each layer, from the physical layer to the applications at the highest level, performs a specific function, allowing the system to be decomposed into unit components. Understanding component interfaces and their relationships with other functionalities is key to effective system modeling. Layers 6 and 7 provide the business value. Layers 4 and 5 are where the concept of "Big Data" abstraction is found. Layer 3 is Cloud computing. Layers 1 and 2 are the underlying computing resources that connect to devices and sensors generating data, utilizing decentralized edge computing or Fog distributed computing. The OSI Model consists of seven layers,

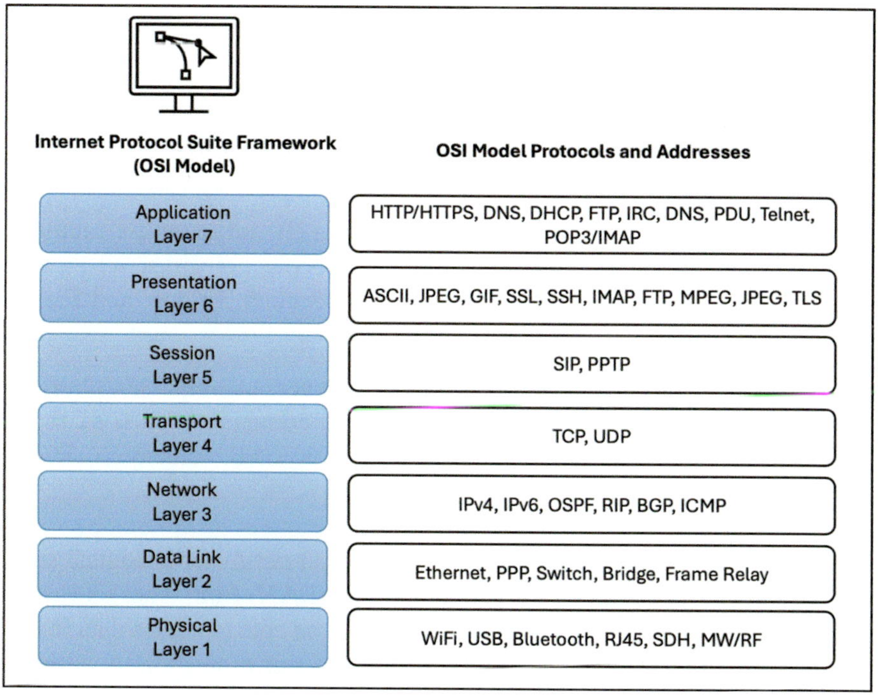

Figure 23.2 International Organization for Standardization Open Systems Interconnection Model.

arranged from top to bottom: the application layer, the presentation layer, the session layer, the transport layer, the network layer, the data link layer, and the physical layer.

- Layer 7 – Application layer. The application layer is the closest to the end-user. It encompasses the people, devices, sensors, and processes that transform decision-making through "thing" applications and data. Protocols include HTTP, DNS, DHCP, FTP, IRC, DNS, PDU, Telnet, POP3/IMAP, WiFi, USB, Bluetooth, RJ45, SDH, and MW/RF.
- Layer 6 – Presentation layer. The presentation layer includes custom applications built using "thing" data. Protocols include ASCII, JPEG, GIF, SSL, SSH, IMAP, FTP, MPEG, JPEG, and TLS.
- Layer 5 – Session layer. The session layer encompasses data analytics, including reporting, data mining, and machine learning (ML). Protocols include SIP and PPTP.

- Layer 4 – Transport layer. The transport layer includes end-to-end connections for data ingestion, such as Big Data, and storage of "thing" data. Protocols include TCP and UDP.
- Layer 3 – Network layer. The network layer encompasses global infrastructure, including cloud services (public, private, hybrid, and managed). Protocols include IPv4, IPv6, OSPF, RIP, BGP, and ICMP.
- Layer 2 – Data link layer. The data link layer encompasses connectivity and edge computing communications, protocols, networks, Wi-Fi, and telecom. Protocols include Ethernet, PPP, Switch, Bridge, and Frame Relay.
- Layer 1 – Physical layer. The physical layer encompasses physical structure devices, sensors, controllers, and connectors, such as coax, fiber, or wireless. Protocols include Wi-Fi, USB, Bluetooth, RJ45, SDH, and MW/RF.

The demilitarized zone (DMZ) is either a physical perimeter or logical subnet that separates an internal local area network (LAN) from other untrusted networks. An external network node should only be able to access data in the demilitarized zone (DMZ). The DMZ adds a layer of protection to network enclaves or assets from external, untrusted networks on the internet or other public or commercial wide-area networks, preventing unauthorized access to internal networks and sensitive data. A network enclave boundary protection is located between the local area network (LAN) and the system it protects, keeping it secure from outside the protected system boundaries by a firewall. The enclave provides a level of security for confidentiality, integrity, and data security within the enclave from potential attacks.

23.4 Internet Protocol Suite Framework (TCP/IP Model)

The internet protocol suite framework is also known as the "transmission control protocol (TCP)/internet protocol model" (TCP/IP Model). It is the model that was used in the development of the Internet. The TCP/IP Model (Figure 23.3) is an open standard for conceptualizing data communication link protocols between networks. It is widely used as the de facto standard for the internet and transport layers. It is an implementation of the Open Systems Interconnection (OSI) model. The TCP/IP model maps to the internet's network structure and protocols. The model is used to define how devices

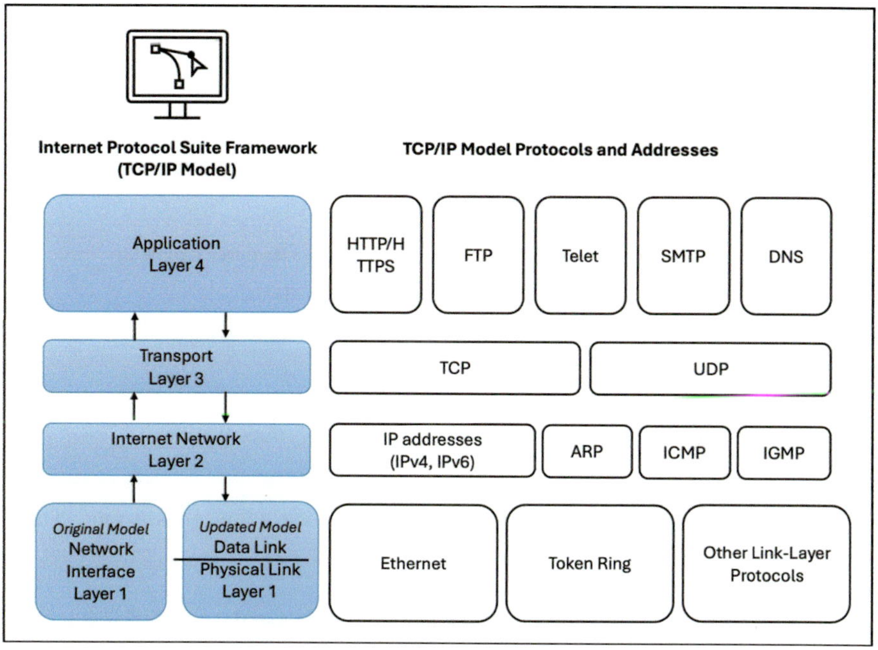

Figure 23.3 Transmission control protocol/internet protocol model.

transmit, receive, and exchange data over networks and large distances. Essentially, the model illustrates how devices connected to the internet communicate with each other. The TCP/IP model was developed prior to the OSI model and was created by the International Organization for Standardization (ISO). Both are abstractions of similar models for communication infrastructure. There are four layers in the TCP/IP model, from top to bottom, as shown in figure: Protocol suite framework (TCP/IP Model): the application layer, the transport layer, the internet network layer, and the network interface.

The application layer generates data and initiates communication requests. This layer is responsible for user interfaces, application services, and processes. The transport layer establishes the data connection, splits data into smaller packets, and acknowledges receipt of these packets. The internet network layer provides the IP address, sends the packets, and routes data to the correct network between devices. The network layer adds the MAC address, sends the data between applications over the network, and handles the physical infrastructure. The TCP/IP does not show the lowest network interface layers.

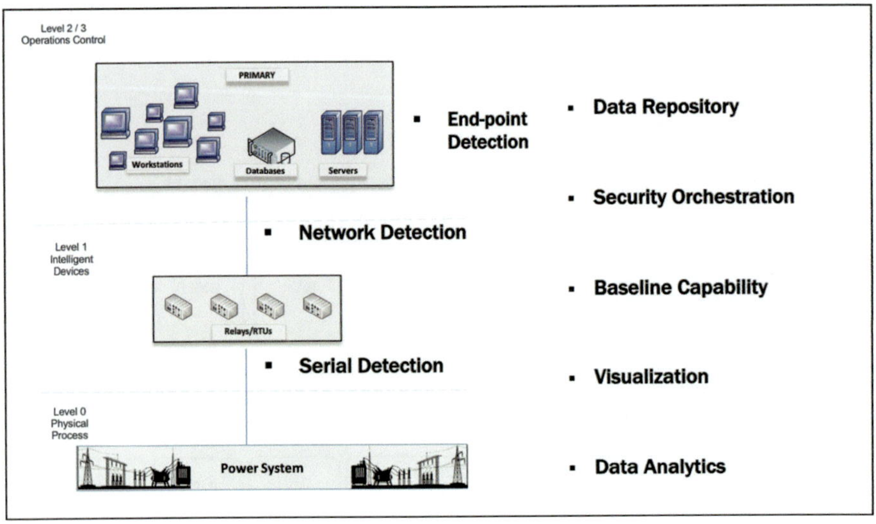

Figure 23.4 More Situational Awareness for Industrial Control Systems using the Purdue Enterprise Reference Architecture model.

23.5 Purdue Enterprise Reference Architecture (PERA) Model

While the OSI model consists of seven layers, the Purdue Enterprise Reference Architecture (PERA) model, which is often used for industrial control system security, consists of five layers. The PERA model, applied to MOSAICS in Figure 23.4, is frequently also referred to as the "Purdue" model. The layers in the PERA model are arranged from top to bottom: the enterprise layer, the demilitarized zone (DMZ) layer, the operation and control layer, the control layer, the process or intelligent device layer, and finally, the physical process layer, where sensors and actuators are located. Processes and workflows at the lowest machine levels operate in milliseconds. Communication between machine sensors and actuators, as well as the supervisory control and data acquisition (SCADA) systems that operate in milliseconds, is vulnerable to cyber-driven downtime.

- Layers 4 and 5 – Enterprise layer. The enterprise layer encompasses primary business functions such as enterprise resource planning (ERP) systems. Typically, in the Purdue model, mail and Web servers are located at the enterprise level.
- Layer 3.5 – Demilitarized zone (DMZ) layer. Security systems, such as firewalls and proxies, are typically located in the demilitarized zone (DMZ) layer. The DMZ is used to prevent lateral movement of threats

between the business enterprise layer and the operational technology found at the lower layers of the architecture. Bidirectional data flow between the operational technology and business enterprise layers provides business operations data that organizations can use to increase business efficiencies, as monitored in ERP systems, such as production schedules in manufacturing, material usage, and inventory levels. Conversely, the bidirectional data flow can introduce cyber risk to the system, making it particularly vulnerable to potential malware attacks as described in Chapter 8, Use Cases and Operational View (OV) Illustrations. A layered approach is much more effective than a single, flat network design.

- Layer 3 – Operation and control layer. This layer manages production workflows, such as those found in manufacturing operations, and facilitates the collection of near-real-time data from production systems. At this level, in addition to potential economic damage and risk of failure to critical infrastructure, there is also the potential risk to personnel and facility safety operations.
- Layer 2 – Control systems layer. The control systems layer encompasses the supervisory control and data acquisition (SCADA) software, which controls physical processes and aggregates data to send to the data historians located in the operation and control layer. This layer encompasses human–machine interfaces (HMIs) that connect distributed control systems (DCSs) and programmable logic controllers (PLCs), as well as DCSs deployed locally in manufacturing that perform supervisory control and data acquisition (SCADA) functions.
- Layer 1 – Process/intelligent devices. This layer encompasses instruments such as programmable logic controllers (PLCs) and remote terminal units (RTUs) that send commands to devices, including sensors and actuators, located at the lower physical process level.
- Level 0 – Physical process layer. This layer encompasses the sensors and actuators that directly control the physical processes. In today's digital world, modern devices communicate directly with monitoring software located in the cloud via remote access, which presents a cyber risk to the system.

23.6 Summary

The increasing complexity of critical infrastructure necessitates understanding a system as a whole. Abstraction models help describe the system and its relationships with the various combinations of people, machines, technologies, and modern, sophisticated functions integrated into the system. The PERA model was developed in the 1990s as an abstraction of information and operational technology for industrial control systems (ICS) and remains widely used today. It serves as an analysis and design tool but does not directly address cybersecurity. The International Society of Automation (ISA) ISA112 supervisory control and data acquisition (SCADA) systems management lifecycle model provides a framework that offers greater detail for industries operating and maintaining SCADA systems. The model architecture promotes consensus-based technical standards focused on reliability and integrity, tailored to industrial control systems (ICS) found in various industries, including power, water and wastewater, oil and gas, and others. The International Society of Automation (ISA) 62443-2-1 Standard for Industrial Automation and Control Systems (IACS) provides more detailed guidance on how to cybersecure near-real-time critical infrastructure environments. These frameworks enable organizations to become more cyber-resilient and implement effective cybersecurity strategies.

Cybersecurity strategies, such as Zero Trust, described in Chapter 1, "Understanding the Game," require an understanding of the system abstraction model for the workforce to apply strategy pillars effectively. Recall from Chapter 1 that the first DOD Zero Trust reference architecture represents a paradigm shift in the philosophy of how infrastructure, networks, and data are secured. The key tenet of Zero Trust is that everything and anything attempting to establish access to a system, whether from within or outside the system perimeter, must be verified (John Kindervag, 2012). It is essential to understand the current system architecture to apply new game-changing strategies, such as Zero Trust, effectively. Abstraction models help the workforce do this without overly complicating what is already complex in practice by putting special attention on key functions and boundaries that separate processes and humans involved in the execution of those functions and processes. Terms related to critical infrastructure, such as SCADA, can have varying definitions across different industries, regions, and specific organizational contexts, depending on the user. A SCADA network in the manufacturing industry typically refers to the supervisory controllers found in the PERA model at Levels 2 or 3. In contrast, in a water facility, SCADA systems often refer to the controller network located below Level 1 in the PERA model. These represent two differing understandings of the terms used that can be addressed by a data dictionary specific to the organization and sector.

When building cyber defenses for industrial control systems, numerous variations of models and standards exist. A robust network architecture requires protection at all levels of industrial processes, from the most sensitive data to system control points. Cyber needs to be treated equally to ensure safety in critical infrastructure. Cyber requires a paradigm shift. Abstraction models help the workforce understand the system for detecting attacks, as well as how to budget, plan, and integrate changes to the architecture. Abstractions help simplify what would otherwise be an exceptionally complicated system.

References

Adams, D. 1979. *The Hitchhiker's Guide to the Galaxy,* New York, Harmony Books.

John Kindervag, S.b., Kelley Mak 2012. Build Security Into Your Network's DNA: The Zero Trust Network Architecture. Forrester: Forrester.

Research, I. F. E. 2025. What Caused Spain and Portugal's Massive Power Outage?

Scalco, A. 2022. *Measuring Disagreement in Segments of the Cybersecurity Profession as a Means of Identifying Vulnerabilities,* Colorado State University.

24

Creating a Model-based Organization

"Hockey is a metaphor for life. You have to be willing to get knocked down and get back up."

– Mike Eruzione (1954),
1980 Winter Olympics
US National Ice Hockey Team Captain

Abstract

Returning to the metaphor about cybersecurity introduced in Chapter 1, "Understanding the Game," which can be drawn from the game of ice hockey, organizational strategies must be understood as much as individuals need to possess technical skills. The US National Ice Hockey Team's Coach, Herb Brooks, was able to leverage individual player skills while motivating each player to play for the team. Teamwork was key for a group of mainly college-level players to overcome the odds of winning against a near professional-level team. A team needs a mutual understanding of the game and a shared commitment to the vision. Similarly, a model-based architecture can support systems engineering activities and contribute to the successful development of systems. The model-based architecture helps manage complexity, facilitates customer dialog, executes technical tasks, captures authoritative design, and manages team and subcontractor relationships. Systems engineering activity linkages to the evolving architecture complemented by a good architecture model facilitate understanding and communication by and among its users to both physical and virtual (simulation) prototypes, rigorous traceability through the flow of architecture development from customer requirements to the elements of a delivered solution, support for the implementation of a service-oriented architecture (SOA) (Borky and Bradley, 2019). In this chapter, we examine the principles of model-based systems engineering (MBSE) and how, when combined with model-based systems analysis and prediction (MBSAP), they deliver benefits for both developers and users of a system.

DOI: 10.1201/9788743809128-24

24.1 Introduction

Model-based systems engineering (MBSE) formalizes aspects of systems engineering. Model-based systems architecture practice (MBSAP) further formalizes and allows for reuse of design patterns such as the operational view (OV), which is the conceptual data model (i.e., data discovery), the logical view (LV), which is the logical data model (i.e., data management), and physical view (PV), which is the physical data model (i.e., data management) (Borky and Bradley, 2019). These viewpoints and activities are associated with the traditional systems engineering Vee diagram, mapped to the left side of the Vee, which places the design of cybersecurity practices early in the design process. The nature of the transitions is decomposed flow-down from the top-level partitioning content, primary system functions, overall information categories, domain services, supporting information of functional components and interface specifications, class and class collaboration functions, systems information entities, specific functional services, and supporting information to the hardware and software and component specifications.

Model-based systems engineering (MBSE) formalizes aspects of systems engineering. The nature of the transitions is decomposed flow-down from the top-level partitioning content, primary system functions, overall information categories, domain services, and supporting information to the logical view (LV) content of functional components and interface specifications, class and class collaboration functions, systems information entities, specific functional services, and supporting information to the physical view (PV) hardware and software and component specifications. Recall from Chapter 6, "Cyber Vulnerabilities Start at the Hardware Level," that each specific hardware element has to be protected down to the lowest level in the architecture. A key question to ask is whether the network topology is known and whether an inventory baseline exists at this level. Initial standards provide and forecast database tables, physical service definitions, and support information (Herber, 2020). Systems engineering activities used in the process flow can be supported by architecture, including linkage of the evolving architecture to both physical and virtual (simulation) prototypes, rigorous traceability through the flow of architecture development from customer requirements to the elements of a delivered solution, support for implementation of SOA, complemented by a good architecture model that facilitates understanding and communication by and among its users (Borky and Bradley, 2019).

An architecture model can be used to implement a configuration management (CM) process. This CM process is a formal method for defining, controlling, and documenting the system's design baseline. CM is part of the architecture lifecycle support, supported by a set of tools and data to collect, store, manage, update, analyze, and present data about configuration items and their relationships, thereby documenting the design baseline of a system. An architectural model can be used to implement a CM process by establishing a linkage between the evolving architecture and physical configuration items, and by ensuring rigorous traceability throughout the architecture development process to the elements of a delivered solution. CM helps development teams control the process of changes to configuration items in the system's baseline. The configuration items in the baseline indicate approved items as attributes of a system at an agreed point in time, which serves as the basis for defining baseline changes.

24.2 Model-based Systems Engineering (MBSE) Tenets

The five tenets of MBSE and how these features yield benefits to the developers and users of a system include reference architecture (RA), enterprise architecture (EA), system architecture (SA), software architecture, and hardware architecture. Each can convey a lot of information that translates well into terms that each group using the model can understand. A reference architecture (RA) can help reduce development costs and maintenance costs (i.e., better understanding of the system), ease development, and increase productivity. "[A] RA is instantiated by the addition of relevant detail to achieve an architecture to satisfy a specific set of requirements within the domain" (Herber, 2020). An enterprise architecture (EA) provides a holistic view that helps to manage data-driven complexity, increases agility, and speeds development and delivery. System architects design the embedded system product from the top level, taking into consideration the hardware and software components. The system architecture (SA) is used to review the design with both hardware and firmware architects. System test developers utilize the system architecture to understand how components are intended to interact with one another and to develop effective test plans. The SA helps address perennial difficulties with complex systems (i.e., lack of interoperability with and among systems) (Herber, 2020). A software architecture represents a common abstraction that developers can use to make design decisions governing the system, such as software reuse and modularity. A hardware architecture benefits the developers and designers of the firmware by ensuring that the firmware can implement the required features (i.e., during the specification and design cycle). The roles architecture can play in a program and how these roles can contribute to successful system development include complexity management, customer dialog, technical execution, authoritative design capture and configuration management, and team and subcontractor management.

Complexity management is designed to address a complex entity and its components, maintain traceability from requirements to physical components, and ensure that all system behaviors are captured and mapped to solution elements. Customer dialog brings the need and the solution to life in a clear and intuitive way, supporting rapid prototyping, what-if scenarios, man-in-the-loop simulations, technology demonstrations, requirements refinement, prioritization, and informed design choices. Technical execution represents the technical conscience of the program and program management, with its best insight into the status and potential problems. Authoritative design capture and configuration management are used for trade studies, performance and risk assessments, and product evaluations. Team and subcontractor management using modeling tools is beneficial in creating a shared, configuration-controlled design repository, normalizing key processes (e.g., models, metrics, versions, and reports), while allowing for local options (e.g., choice of compiler).

24.3 Use Cases

Use cases provide a systematic and intuitive way to identify, organize, and begin to describe system behavior, especially in terms of the observable functions that satisfy customer needs. "[T]hey identify system behaviors and provide a convenient graphical layout that summarizes this fundamental aspect of an architecture in one or a few easy-to-read diagrams" (Borky and Bradley, 2019). Some ways interaction fragments can be used in a sequence diagram include representing occurrences, state invariants, interaction uses, and combined fragments.

Considering the functions of a system architect or architecture team leader and the members of such a team, an example organization chart for a power utility SAS including program management, engineering management, and teams responsible for products, integration, test, and system configuration management, and the relationships of the architect and architecture team to these other organizational entities is shown in Figure 24.1.

While the Department of Defense (DOD) successfully demonstrated more situational awareness for industrial control systems (MOSAICS) at a power utility, and the Navy was designated as the transition lead for using the MOSAICS framework in operations, a common service strategy has to be designed and developed. The development of an overarching service strategy

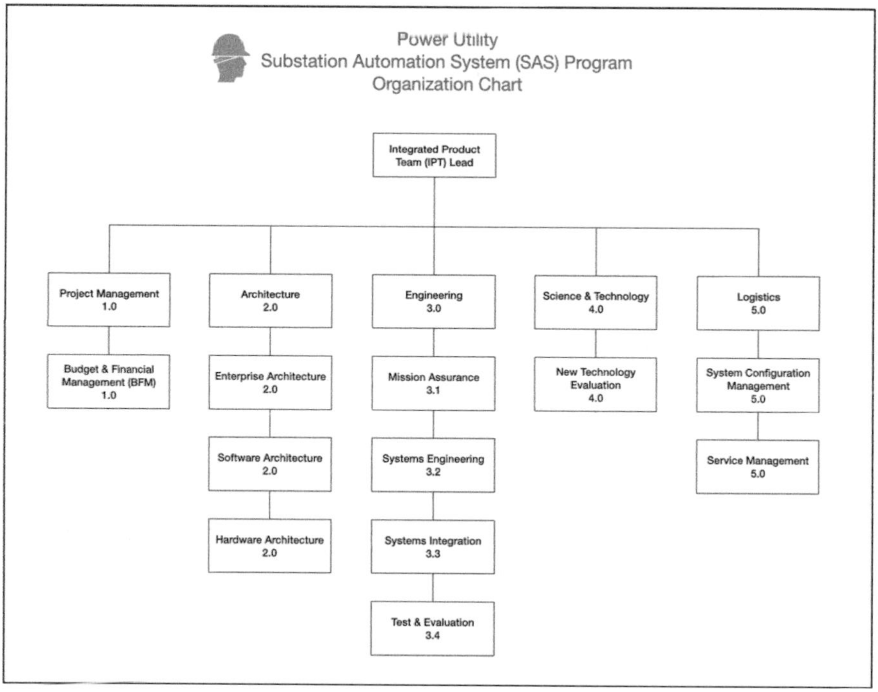

Figure 24.1 Power utility substation system (SAS) program organizational chart.

engages additional stakeholders in the overall systems engineering process. A data strategy has to be created that, in turn, informs the design of the sensor strategy. Both data and sensor strategies depend on the capability of the MOSAICS framework to baseline the system inventory and passively, or ideally, safely and actively monitor the system to the lowest level of the stack, the programmable logic controller (PLC). The mission thread and cyber vulnerabilities must have visibility from the cloud to the programmable logic controller (PLC) for full system Command and Control (C2) and to overcome cyber vulnerabilities.

24.4 Object Orientation (OO)

Some reasons for preferring object orientation (OO) as the approach to defining and developing complex, information-intensive systems include:

- The inherent power of object-based decomposition to break down complex entities and relationships so that the architect can deal with comprehensible and coherent subsets of components and resources, along with their functions and content, without losing the essential perspective of the whole

- The use of abstraction to discover and exploit underlying patterns and common characteristics across seemingly diverse entities
- The benefits of encapsulation include localizing design and implementation issues, simplifying system integration, minimizing the need for large-scale testing, minimizing the system-level impacts of specific design changes, and promoting the reusability of designs and products
- The focus on interfaces among system elements and between a system and its environment is inherent in [object-oriented design] (OOD)
- The fact that these and other aspects of OOD collectively provide a basis for achieving an open system (Borky and Bradley, 2019)

Some key principles of object-oriented (OO) design that are important in systems architecture include abstraction, encapsulation, modularity, generalization and inheritance, aggregation and composition, interfaces, and polymorphism. A Package is a collection of logically related UML elements. "The primary way of organizing most UML (and SysML) models involves Packages, which are essentially containers in which related model elements can be placed. Most UML tools use Packages in a Browser to create the internal structure of a model" (Borky and Bradley, 2019).

24.4.1 Modeling languages

Engineers use graphical modeling languages to visualize system designs, structures, behaviors, and interactions. Modeling helps systems engineers create models of complex systems to support model-based organizations, which foster a common understanding of these systems, their hardware and software components, and the behaviors of these systems, including human interactions. An *actor* in a unified modeling language (UML) or a model in a UML extension, systems modeling language (SysML), is "anything that is outside the boundary of a system but with which the system interacts or that in some way influences system behavior is modeled as an Actor. Actors are most commonly associated with use cases, which are part of the behavioral aspect of an architecture [sic] they have an important role in structure as well because they are used to explicitly declare relationships between a system and external entities" (Borky and Bradley, 2019).

The two primary uses of a deployment diagram are to model the hardware components, or "boxes," of a system, which are referred to as Nodes in UML, and to illustrate the physical connections among them. The diagram also allocates software components to the nodes where they will be loaded and executed. Some typical systems engineering tasks that SysML explicitly supports are the specification, analysis, design, verification, and validation of systems that include hardware, software, data, personnel, procedures, and facilities (Borky and Bradley, 2019).

There are several significant differences between SysML and UML 2. UML is software-centric. SysML is more engineering systems-oriented. The replacement of classes and objects by blocks and their Instances is the most significant difference between UML and SysML. SysML is particularly useful when an architecture model needs to deal with complex design and implementation details of both hardware and software. UML 2 is a rich and complex modeling language, and SysML further enhances an architect's ability to address requirements, performance, design, integration, and other software engineering processes and tasks. SysML defines several types of action pins, represented by small squares that are visually reminiscent of Ports in a structural diagram. A UML class diagram can specify mathematical relationships when defining class methods. Still, the parametric diagram used in SysML provides an intuitive portrayal of where values originate and where they are directed, as well as the way various constraints interact to determine system behavior (Borky and Bradley, 2019).

"In SysML, the primary structural entity is a block, which is a modular unit of a system and generally has structural and behavioral features. The replacement of classes and objects by blocks and their instances is perhaps the largest difference between UML and SysML" (Borky and Bradley, 2019). "A UML Class has three basic partitions or compartments: identifier, attributes, and operations. A SysML block can have many more compartments and can be listed in any order. Typical examples include:

- *Properties, namespace* – shows blocks that are defined in the namespace of the block
- *Structure* – shows connectors and other structural elements of the block
- *Allocation* – shows model elements that are allocated to the block
- *Ports* – shows ports that have been created on the block
- *Other* – additional compartments, often defined by individual modeling tools" (Borky and Bradley, 2019)

SysML is a widely supported language standard that enables precise and unambiguous mathematical modeling (Herber, 2020). SysML offers significant benefits in modeling physical design due to its enhanced semantics and ability to represent structures and their relationships, such as procurement specifications, build specifications, and product data. Initial blocks that are part of a functional design are fleshed out with specifications, product data, and other details. Flows of information and physical quantities are quantified. Ports are described with interface documentation. Any significant information on the implementation of the architecture is included either directly on the diagrams or in the specifications for the various model elements. Details can be included in the model in multiple ways, including nodes directly shown on a diagram and links to supporting documentation such as product specifications (Borky and Bradley, 2019).

24.5 Example Model-based Systems Engineering (MBSE) and Model-based Systems Engineering Practice (MBSEP)

The purpose of this example is to model the initial defensive capability of a composed cyber solution for critical infrastructure, specifically for use in a power utility substation automation system (SAS), utilizing modeling-based systems engineering (MBSE) and model-based systems architecture practices (MBSAP). Cybersecurity presents engineering design and architecture challenges for complex, information-enabled enterprises that rely on critical infrastructure to protect sensitive information and processes. While the initial demonstration of MOSAICS aimed to examine the military utility of new technology for potential transition to operational use in a power utility, the framework for the capability can be applied to other critical infrastructure sectors (e.g., water, wastewater, HVAC, and facilities) in the broader DOD enterprise context and transitioned to commercial industry for further development and advancement of the capabilities.

It is worth reiterating that there is an operational need for cyber defense capabilities to protect mission-critical infrastructure from cyberattacks (Scalco and Simske, 2021). Critical infrastructure systems, including power, water, wastewater, facilities, and safety controls, have a significant impact on physical environments. These systems traditionally rely on physical security, such as physical access control. Mission-critical systems perform functions essential to the continuation of operations, and the failure or disruption will result in mission failure. The operational need is to mitigate adversarial means of non-kinetic disruption of mission-critical infrastructure operations. Key performance characteristics to satisfy the user needs are: to provide a cyber vulnerability baseline of the substation automation system (SAS); enhance asymmetric threat indicators and warnings; anomaly detection; and near real-time information sharing capability within an automation framework that enables time-sensitive (near real-time) semi-automated response actions to disrupt attack kill chain and recover to restore normal operations; and non-time-sensitive sharing of threat indicators to degrade adversarial reuse of attack methods against system behaviors.

The MOSAICS capability concept aims to automate existing procedures for detecting, mitigating, and recovering from cyberattacks, leveraging commercial technologies such as analytics, visualization, decision support, and information sharing within the broader context of a joint "enterprise" between the DOD and commercial industry partners. "System studies identified three initial MOSAICS capabilities: (1) an operational ability to enable defense of control systems; (2) an industrial control systems (ICS) baselining tool for programmable logic controller (PLC) sensors; and (3) tailored visualizations, analytics, and automated cybersecurity orchestration," (Scalco, 2020). As MOSAICS was initially a demonstration rather than a program of record (POR), typical DOD program artifacts, such as network architectures, are presented in very general, high-level OV-1 drawings (Figure 24.2).

An initial model of MOSAICS, detailed using SysML, will significantly benefit the DOD critical infrastructure community, particularly once a system utilizing the framework is transitioned to operational use. Given a scenario of an environment susceptible to unpredictable, high-consequence events of recursive complexity, the recommended solution architecture is a "design for resilience" that "provide[s] robustness and timely recovery to a minimally functional state" (INCOSE, 2015). Context-sensitive critical infrastructure dynamic classes are interpreted by the operational technology (OT) or IT operator's view, within the critical infrastructure sector, and dynamically classified at the time of operation, rather than as a static set of classes. The OT operator manages physical processes and machinery, while the IT operator manages digital data flows (Scalco, 2020). The distinction between static and dynamic classes of critical infrastructure systems is substantial. "Each critical infrastructure sector is dynamic, and within each sector, every cyber-physical system is dynamic. The potential risk introduced by the context-sensitive critical infrastructure dynamic classes must be addressed as early as possible and revisited throughout the systems engineering lifecycle" (Scalco, 2020). MBSE will contribute to a better understanding of MOSAICS' diversity, connectivity, interactivity, and risk as the system is applied to an increasingly complex spectrum, capturing design from the enterprise to the component level.

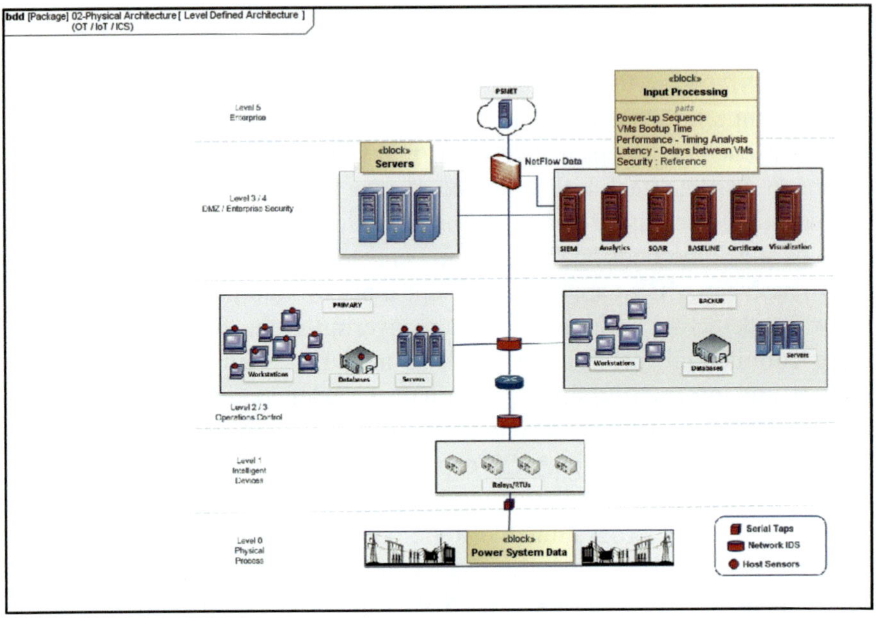

Figure 24.2 Initial MOSAICS conceptual design.

There is value in understanding what distinguishes a system architect from other kinds of engineers. The system architect uses a formal representation of an entity, such as its structure, behavior, or rules, to describe it. Systems engineers are generally viewed as the central coordinators of system development. The architect remains involved throughout the design and development of a system, utilizing an architectural model to address overarching issues such as the diverse concerns of a system's stakeholders, stated and derived system requirements, planned system evolution, policies, and other constraints, as well as the operational suitability of the delivered system. The system architect brings together stakeholder concerns at different levels of abstraction. The system architect promotes modular, technology-independent designs to manage complex, information-intensive systems, thereby reducing risk, facilitating stakeholder dialog, and enabling the incremental development of capabilities (Herber, 2020).

The fundamental premise of model-based systems engineering (MBSE) is that the discipline formalizes systems engineering using models and accomplishes SE tasks and processes based on the foundation of a formal architecture model, thereby transitioning from a "document-centric" to an "architecture-centric" systems engineering approach. MBSE aims to capture the system design from the enterprise to the component level. MBSE can address the perennial difficulties with complex systems (Herber, 2020).

The system architecture definition used in John M. Borky and Thomas H. Bradley's book, "Effective Model-based Systems Engineering," is closely tied to the systems engineering processes that occur throughout a system's life cycle, providing a foundational, structured approach and methodology using model-based systems engineering (MBSE). The Department of Defense (DOD) and the National Aeronautics and Space Administration (NASA) utilize model-based systems engineering (MBSE) to more effectively trace operational needs to technical requirements in structured, unambiguous descriptions, enabling the model to achieve reuse and interoperability (Herber, 2020).

The architecture is the core of model-based systems engineering (MBSE), providing traceability throughout the entire architectural model. The systems engineering flow is anchored in the architecture model. (Herber, 2020) Architecture is the core of MBSE. "MBSE formalizes [systems engineering] using models to ensure rigor, repeatability, and producibility in [systems engineering] processes, promote quality, completeness, and correctness in system designs, reduce risk in requirements analysis, design, integration, and test, and other activities, enhance communication and synchronization of activities across organizations and disciplines" (Borky and Bradley, 2019). The system architecture provides a structured, unambiguous system description to address complexity. Unambiguous language enables the approach to complexity management, including decomposition, composition, modeling, reuse, and metrics. "Architecting involves the search for optimum solutions in problem spaces with many dimensions, and the architect must balance competing interests while keeping at the forefront the needs of the ultimate customer. Clear and consistent terminology helps in meeting that challenge" (Borky and Bradley, 2019).

Enterprise architecture applies to the top level of an ensemble of nodes, systems, elements, or other resources that collaborate to fulfill the functions of an overall organization or business process. System architecture applies to an ensemble of elements (ultimately, hardware and software components) that collaborate to meet defined requirements allocated to a node or system, implying that a clear system boundary and user interfaces are defined. Service-oriented architecture (SOA) describes the interactions among elements and the external environment, where capabilities are exposed as services that are invoked by users or among elements, typically in a request-response model using messaging. Executable architecture represents a design in models that can be run as simulations to quantify and analyze behaviors. Characteristics of an effective open architecture contribute to the development and use of a system by the ability to:

- Select and assemble "best of breed" products from various sources without expensive modifications and integration efforts
- Selectively replace, modify, or upgrade at the level of relatively fine-grained components, such as a processor or software application, with minimal (ideally, no) impacts on the remainder of the system other than using the new or better functionality
- Scale system capabilities by the addition of modular units of hardware and software, achieving performance commensurate with the installed assets; similarly, create new system components that can be reused for equivalent functions in other systems, and reuse such components from other systems
- Interoperate with networks and other systems with minimal (ideally, no) added effort to provide adapters or other special-purpose components
- Make extensive use of commercial-off-the-shelf (COTS) products to minimize development effort and enjoy mass-market pricing (Borky and Bradley, 2019)

24.6 Configuration Management

An architecture model can be used to implement a configuration management (CM) process, which is a formal method for defining, controlling, and documenting the design baseline of a system. CM is part of the architecture lifecycle support, which is supported by a set of tools and data to collect, store, manage, update, analyze, and present data about configuration items and their relationships, thereby documenting the design baseline of a system. An architectural model can be used to implement a CM process by linking the evolving architecture to physical configuration items and ensuring rigorous traceability throughout the architecture development process to the elements of a delivered solution.

CM helps development teams control the process of changes to configuration items in the system's baseline. The configuration items in the baseline indicate approved items as system attributes at an agreed point in time, which serves as the basis for defining baseline changes. Five tenets of MBSE and how these features yield benefits to the developers and users of a system:

- *Reference architecture (RA)* – An RA can help reduce development and maintenance costs, providing a better understanding of the system, easing development, and increasing productivity. "[A] RA is instantiated by the addition of relevant detail to achieve an architecture to satisfy a specific set of requirements within the domain" (Herber, 2020)

- *Enterprise architecture* – Provides a holistic view that helps to manage data-driven complexity, increases agility, and speeds development and delivery
- *System architecture (SA)* – System architects design the embedded system product from the top level, taking into consideration the hardware and software components. The SA is used to review the design with both hardware and firmware architects. System test developers use the SA to understand how comments are intended to interact with each other and develop effective test plans. The SA helps address perennial difficulties with complex systems, such as the lack of interoperability among systems (Herber, 2020)
- *Software architecture* – A software architecture represents a common abstraction that developers can use to make design decisions governing the system, such as software reuse and modularity
- *Hardware architecture* – A hardware architecture benefits firmware developers and designers by ensuring that the firmware can implement the required features (e.g., specification and design cycle)

The roles architecture can play in a program and how it can contribute to successful system development include:

- Complexity management to deal with a complex entity and its elements, maintain traceability from requirements to physical components and ensure all system behaviors are captured and mapped to solution elements
- Customer dialog to bring the need and the solution to life clearly and intuitively, support rapid prototyping, what-if, man-in-the-loop simulations, technology demos, requirements refinement and prioritization, and design choices
- Technical execution is the program's technical conscience, providing program management with the best insight into the status and potential problems
- Authoritative design capture/configuration management for trade studies, performance and risk assessments, and product evaluations

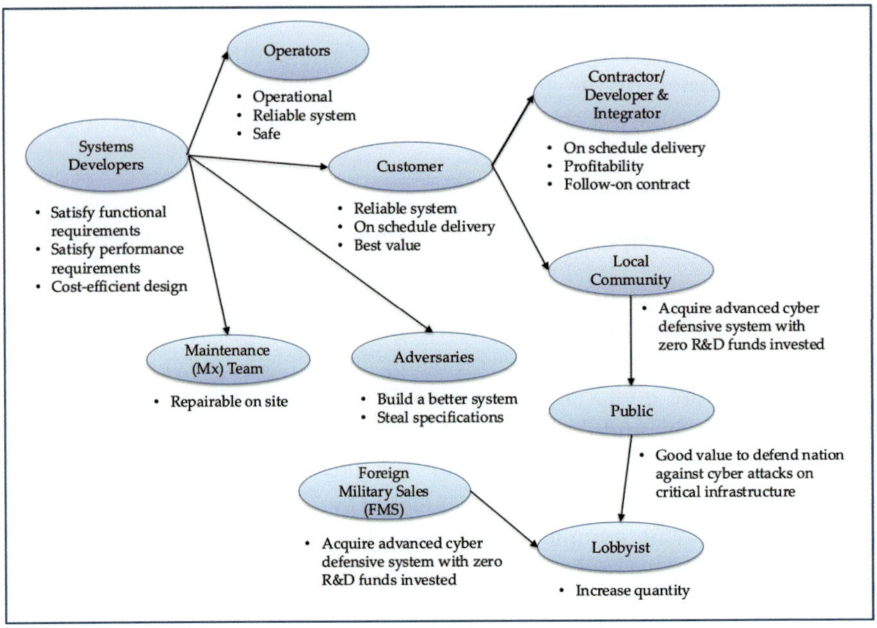

Figure 24.3 Stakeholders for a power utility substation automation system.

- Team and subcontractor management to create a shared, configuration-controlled design repository, normalizing key processes (e.g., models, metrics, versions, and reports) while allowing local options (e.g., choice of compiler)

Considering the functions of a system architect or architecture team leader and the members of such a team, an example of stakeholders for a power utility substation automation system (SAS) including program management, engineering management, and teams responsible for products, integration, test, and system configuration management, and the relationships of the architect and architecture team to these other organizational entities is shown in Figure 24.3. The engineering organization is responsible for transitioning the new solution into operations and designing and developing the service strategy to support both transition and ongoing operations.

24.7 Object-oriented Coding

Hardening computing environments complements secure coding. Object-oriented coding can be utilized as an element in a security software strategy, as objects are self-contained or encapsulated instances of a class, which are created instead of writing a program from scratch. Object-oriented coding enhances security by incorporating it into a solution from the outset, thereby improving cybersecurity and establishing a stronger security posture. Object-oriented coding is modular, allowing cybersecurity engineering teams to work simultaneously on functionality without conflicting efforts. Object-oriented code also provides for ease of reuse, which is an advantage in maintaining attributes and functions instead of rewriting extensive new code. The reuse allows for inheritance, as well as the use of parent classes and polymorphism. Better code quality testing upfront reduces costs and results in less rework, making it easier to meet deadlines and aid in cybersecurity. Some reasons for preferring object orientation (OO) as the approach to defining and developing complex, information-intensive systems include:

- "The inherent power of object-based decomposition to break down complex entities and relationships so that the architect can deal with comprehensible and coherent subsets of components and resources, along with their functions and content, without losing the essential perspective of the whole
- The use of abstraction to discover and exploit underlying patterns and common characteristics across seemingly diverse entities
- The benefits of encapsulation in localizing design and implementation issues, simplifying system integration, minimizing the need for large-scale testing, minimizing the system-level impacts of specific design changes, and promoting reusability of designs and products
- The focus on interfaces among system elements and between a system and its environment that is inherent in OOD
- The fact that these and other aspects of OOD collectively provide a basis for achieving an open system" (Borky and Bradley, 2019)

Some key principles of object-oriented (OO) design that are important in systems architecture include abstraction, encapsulation, modularity, generalization and inheritance, aggregation and composition, interfaces, and polymorphism. A package is a collection of logically related UML elements. "The primary way of organizing most UML (and SysML) models involves packages, which are essentially containers in which related model elements can be placed. Most UML tools use packages in a browser to create the internal structure of a model" (Borky and Bradley, 2019).

An actor in a UML or SysML model is "anything that is outside the boundary of a system but with which the system interacts or that in some way influences system behavior is modeled as an actor. Actors are most commonly associated with use cases, which are part of the behavioral aspect of an architecture [sic] they have an important role in structure as well because they are used to explicitly declare relationships between a system and external entities" (Borky and Bradley, 2019).

Use cases, as illustrated in Chapter 8, "Use Cases and Operational View Illustrations," provide a systematic and intuitive way to identify, organize, and begin to describe system behavior, especially in terms of the observable functions that satisfy customer needs. "[T]hey identify system behaviors and provide a convenient graphical layout that summarizes this fundamental aspect of an architecture in one or a few easy-to-read diagrams" (Borky and Bradley, 2019). Interaction fragments can be used in a sequence diagram to represent occurrences, state invariants, interaction uses, or combined fragments.

The two primary uses of a deployment diagram are to model the hardware components, or "boxes," of a system, which are referred to as Nodes in UML, and to illustrate the physical connections among them. The diagram also allocates software components to the nodes where they will be loaded and executed. Some typical SE tasks explicitly supported by SysML are the specification, analysis, design, verification, and validation of systems, including hardware, software, data, personnel, procedures, and facilities (Borky and Bradley, 2019).

24.8 Requirements

The starting point for the MOSAICS architecture is a requirements baseline derived from the participant DOD Combatant Commands. A concept of operations (CONOPS) is developed that clearly defines the user needs and operational context. The CONOPS is a description that precedes or supplements a specification. "A CONOPS expresses the needs and assumptions of the operational end user; describes processes, missions, and constraints; and provides context for trade studies and design decisions. It includes all the functions, behaviors, and characteristics a system must exhibit to be satisfactory in its intended usage" (Borky and Bradley, 2019). The performance specifications provide design guidance and may define the minimum acceptable thresholds and levels of desired performance objectives. Performance specifications effectively establish the design trade space "within which performance, cost, schedule, risk, reliability, and other variables can be analyzed to optimize the solution" (Borky and Bradley, 2019).

The principles of systems engineering and the model-based systems engineering (MBSE) methodology are used to categorize and verify requirements through rigorous analysis and management in a database or model. Using a tool to capture and manage a baseline or an alternative SysML requirements diagram is an approach for verification. A database or a model is a prerequisite for traceability from early conceptualization through the design, engineering, testing, and delivery of the system solution.

Four broad categories of architecture drivers may be considered: high-level requirements, business constraints, technical constraints, and quality attributes. Performance requirements are verified by measuring the appropriate parameters and comparing them to required values, and [non-functional requirements] NFRs may be verified through inspection of the system and its documentation, analysis of compiled data on things like failure and repair rates, feedback from operational users, design audits against applicable standards, and many other sources (Borky and Bradley, 2019).

24.9 Operational Viewpoint Artifacts

The operational viewpoint (OV) is assembled from content drawn from the requirements, use cases, structural and behavioral packages, conceptual data model (CDM), and context and general information packages (Borky and Bradley, 2019). The task of putting together the data discovery should begin from the operational viewpoint (OV). The *Architecture Summary and Overview* document can be used in the early stages of a program to coordinate the OV "among stakeholders, especially between the architecture teams and program management, [and] can significantly facilitate the discussion over scope, products, resources, methodology, priorities, and other essential aspects of the architecture. If it can be brought to a point where all (or at least most) stakeholders are prepared to sign off that it represents an adequate architecture approach" (Borky and Bradley, 2019). Systems engineers can use the model to garner stakeholder input and discussion. The model is valuable during an architecture kickoff workshop, where goals are negotiated among stakeholders, supporting an effective development effort. The fundamental aspects are: (Herber, 2020)

- Identification of information entities, which may be concrete or abstract
- Definitions of the characteristics of these entities, including identity, location, time parameters, content and format, ownership and pedigree, data precision, and similar properties

- Identification of relationships among entities based on interactions, dependencies, inheritance, structural relationships, and so forth
- Where data entities are used in domains, where they are stored, both temporarily and permanently, when and by what processes they are created and destroyed, and how they are exchanged (e.g., by publish/subscribe)

24.9.1 Structural perspective

A foundation class is an abstract, broad information category that defines the characteristics of the abstract information elements. Foundation classes are stereotyped (i.e., <<Abstract, InfoElement>>. System classes are no longer abstractions but represent the real data entry and can be instantiated elsewhere in the model. There may be multiple inheritance system classes (Herber, 2020).

24.9.2 Services perspective

The service taxonomy (Table 24.1) is mapped to domains such as sensor management, control system components, IT components, security information and event management (SIEM), and automation. Anticipated benefits of the MOSAICS framework are mapped to specific system services, use cases, domains, and domain services, as shown in Table 24.1.

- Enhance understanding of risk to critical infrastructure and supported operational capabilities
- Detect control system threats faster – from months to minutes
- Enhance situational awareness, inform real-time decisions, and facilitate cyber defender response
- Disrupt the adversary's kill chain in mission-relevant time
- Limit adversary re-use of attacks through enhanced sharing of indicators and mitigations (Aleksandra Scalco, 2019)

24.9.3 Contextual perspective

The contextual perspective in the model illustrates the initial defensive capability of a cyber-physical system (CPS)/control system (CS) for mission-critical infrastructure in a power utility substation automation system (SAS),

Table 24.1 Service taxonomy.

System service	Use cases	Domains	Domain services
Sensing	Perform sensing	Sensor management	Control sensors Maintain sensor schedule Process sensor reports Collects data from the system
Detect	Employ detection	Control system components	Collect IT component information Collect control system component information Matches data to known anomalies Sends alerts to the operator
Protect	Implement safeguards	Control system components, IT components, security information and event management (SIEM)	Initiate collection of event information (e.g., startup, recovery, and restoration from emergencies, operations) Monitor operations from a periodic sampling rate Control and configure core cybersecurity devices protecting the facility Implement protection on identified facility components Actively query control system components for status Actively query control system networks for status Actively query control system IT components for status
Respond	Present courses of action (COA)	Automation	Sends courses of action (COA) to security orchestration automation and response (SOAR) SOAR sends actions to the control system components SOAR sends actions to IT components

utilizing modeling-based systems engineering (MBSE) and model-based systems architecture practices (MBSAP). Intelligent devices (e.g., programmable logic controllers (PLCs)) and physical devices (e.g., flow and pressure meters) are susceptible to cyberattacks. The system's objective is to defend the utility at the lower levels of the architecture.

24.9.4 Behavioral perspective

The sequence diagram illustrates the behavioral perspective of the system. The objective is to monitor critical infrastructure facilities and alert operators if a cyber event or cybersecurity issue requires attention, based on a baseline analysis of components, such as field devices. The collaboration behavior in the baseline of the field devices group of blocks (i.e., control system, device interrogation, network ingest, SIEM, and analysis) provides support awareness, decision-making, and workflows across multiple human roles, predefined expertise, and details for further action: Has something gone wrong? Who needs to know? Who is affected? How should responses be prioritized?

The sequence diagram for a programmable logic controller (PLC) shows the component's behavioral perspective. The behavioral diagram illustrates an example of non-stateful behavior in detecting an attack on a programmable logic controller (PLC). The sensor collects data from log sources, correlates events, alerts on security incidents, generates security and compliance reports, and archives logs for forensic analysis and review.

Specification Interfaces for services might include the following:

- Service A – New technology development, test, and validation (T&V)
  - Hardware and software lists
  - Specifications
  - Security classification guide
  - Test plans
  - Test reports
- Service B – Transition management
  - Acceptance criteria
  - Design and interface specification
  - Transition plans

  - Operation plans and procedures
  - Licenses
  - Cost model
  - Training plan
  - Change schedule
  - Known errors
  - New capability processes ("Playbooks")
- Service C – Maintenance and operations
  - Service design packages
  - Utility and warranty details
  - Recovery procedures
- Service Consumer – Operations
  - Facility engineering command operations
  - Operate system

24.10 Domain Diagram

24.10.1 Use case diagram with actors and specifications

Key performance characteristics to satisfy the user needs are to provide a cyber vulnerability baseline of the system; enhance asymmetric threat indicators and warnings, anomaly detection; and near real-time information sharing capability within an automation framework that enables time-sensitive (near real-time) semi-automated response actions to disrupt attack kill chain and recover to restore normal operations; and non-time-sensitive sharing of threat indicators to degrade adversarial reuse of attack methods against system behaviors.

24.10.2 Use case 1 – Detect with specifications

General: Developing and implementing appropriate safeguards to ensure delivery of mission-critical services. MOSAICS will collect the facility's state through data either directly from CS and IT components or indirectly from tools such as security information and event management (SIEM) that interface with the components, logs, and data.

Precondition 1: New facility events occur (e.g., startup, recovery, restoration from emergencies, and operations) that initiate the collection of information

Precondition 2: During normal facility operations, MOSAICS will monitor the operations from a periodic sampling rate (to be determined)

Postcondition 1: MOSAICS successfully (actively) queries and receives replies from all active CS/IT components and understands that protection works against threat data/tests

Postcondition 2: MOSAICS successfully (passively) collects from all passive CS/IT components

Postcondition 3: MOSAICS successfully triggers alerts on known alerting logic statements and refines as needed to address threat activities

Actors involved: MOSAICS, CS components, IT components, security information and event management (SIEM), external sharing organizations, facilities operator, Site CIO, and threats.

User roles: TBD

Primary data objectives:

Primary – Normal mission:

- MOSAICS collects data from the system
- MOSAICS determines conditions that are out of normal limits (e.g., malicious activities, anomalies, or unauthorized configuration changes)
- MOSAICS receives incidents of interest
- MOSAICS matches data to known anomalies and alert patterns
- MOSAICS sends alerts to the operator
- MOSAICS performs a threat assessment of the facility based on alert conditions
- The cyber and application support lead (on duty), the facility engineering operation center (FEOC) manager, or the duty desk officer (after duty hours) may consult with external organizations to analyze the incident data received.

- MOSAICS may retrieve external organization data to compare local incident data
- MOSAICS determines the severity of the threat to the facility based on the data received
- MOSAICS determines the potential course of action (COA) for the operator

24.11 Cybersecurity Quality Attributes

Quality attributes for security, assessment factors, and some safeguard issues are addressed throughout the system to the lowest level. Baseline the system, including details concerning system interfaces and maintenance of proper security configurations. All software and hardware products used should be included as factors for scoring the system under baseline configuration control (high, medium, and low). One challenge is that in a substation automation system (SAS), components at the lower levels are not IP-based and are therefore difficult to baseline. Data classification, encryption, and backup are applicable controls to ensure the availability and integrity of data. Factors for scoring include an audit that verifies all data is backed up, as well as processes for accessing and restoring backup data (rated as high, medium, or low). The challenge is establishing an effective security classification guide.

A comprehensive incident response plan (IPR) is critical to recovering from an attack. An organization has a legal obligation to report an incident and implement a recovery plan in the event of an attack. Factors for scoring are an audit (high, medium, low). Challenges are the cost-benefit analysis of the recovery plan (i.e., the "smoking gun" plan). Patch management is a crucial practice for mitigating risks associated with zero-day attacks. Hackers seek opportunities to exploit vulnerabilities, including security defects in vendor software. Factors for scoring are an audit (high, medium, and low). A challenge is ensuring that personnel maintain effective patch management. Security automation is an emerging solution to address this challenge.

Perimeter defense protects the network from attacks originating from the Internet and defends against online threats, including those that may arise from a telework policy that allows employees to work remotely from home using virtual private networks (VPNs). Factors for scoring are an audit (high, medium, and low). Challenges are effectively categorizing sensitive assets. Employee training and awareness should include insider threat awareness, acquisition and use of approved software, effective password management, and the expectation of compliance with security policy. The training should consist of the personal use of social media sites to prevent attacks executed through these platforms. Factors for scoring include an audit, which involves completing training and certifications. Recall from Chapter 7, "Threats to Information-intensive Systems," that a significant challenge is that human-introduced vulnerabilities remain a high attack vector. Humans are vulnerable to phishing attacks regardless of the amount of training implemented.

24.12 Summary

The model-based architecture helps manage complexity, facilitates customer dialog, executes technical tasks, captures authoritative design, and manages team and subcontractor relationships. Systems engineering activity linkages to the evolving architecture, complemented by a well-defined architecture model, facilitate understanding and communication among its users. Model-based systems engineering (MBSE) formalizes aspects of systems engineering. The principles of systems engineering and the model-based systems engineering (MBSE) methodology are used to categorize and verify requirements through rigorous analysis and management in a database or model. Using a tool to capture and manage a baseline. Models are foundational to an organization, enabling it to make informed decisions about complex systems and processes. An organization's structure must be aligned with the model and representative of the system's complexity, including its internal and external interactions, as well as human behavior and interactions with the system. Building a model-based organization helps address some of the complexity that cyber introduces and aligns stakeholders. In the next chapter, we explore a more effective approach by utilizing a model for stakeholder alignment as a means of identifying vulnerability induced by misalignment that may be greater than the inherent vulnerability of the system design (Scalco, 2022).

References

Aleksandra Scalco, M. J., Steve Simske 2019. More Situational Awareness for Industrial Control Systems (MOSAICS) Joint Capability Technology Demonstration (JCTD): A Concept Development for the Defense of Mission Critical Infrastructure. *Homeland Defense & Security Information Analysis Center.*

Borky, J. M. & Bradley, T. H. 2019. Protecting Information with Cybersecurity. *Effective Model-Based Systems Engineering.* Springer International Publishing.

Herber, D. R. 2020. Systems Engineering Architecture Lectures, SYSE 567. *System Architecture and Model-Based Systems Engineering.* Colorado State University: Colorado State University.

Incose 2015. A Complexity Primer for Systems Engineers. INCOSE.

Scalco, A. 2020. Cyber-Physical System (CPS) and Control System (CS) Architecture for Cyber Defensive Capability — MOSAICS. *Using Model-Based Systems Engineering (MBSE) and Model-Based System Architecture Process (MBSAP) Methodology.* Colorado State University Systems Engineering Architecture, SYSE 567: Colorado State University.

Scalco, A. 2022. *Measuring Disagreement in Segments of the Cybersecurity Profession as a Means of Identifying Vulnerabilities,* Colorado State University.

Scalco, A. & Simske, S. 2021. Digital Transformation of Cyber-Physical Systems and Control Systems. *Journal of the Homeland Defense & Security Information Analysis Center (HDIAC),* Vol. I and II.

25

Stakeholder Alignment

"I'd agree with you, but then we'd both be wrong."

– An adage

This section was previously published (Scalco and Simske, 2022) (Scalco and Simske, 2023) (Scalco, 2022b).

Abstract

Disagreement exists among professionals with different roles due to variances in engineering practice, paradigms, processes, and culture for people in these roles. Understanding what can be done to improve effective collaboration among these professionals is a continuous science and engineering challenge – stakeholder disagreement results in measurable misalignment that leads to vulnerability. The vulnerability induced by stakeholder misalignment may be greater than any innate system design vulnerability. It is possible to measure the uncertainty of agreement through statistical analysis and use an analytical model to identify pain points where different sets of stakeholders disagree. The same measure can be walked back to stakeholder sources, beginning with a vulnerability assessment to help drive better alignment and, eventually, agreement. It is the disagreement that ultimately leads to vulnerability (Scalco, 2022a). This chapter describes an analytic model and methodology as a new means of assessing uncertainty and interpreting Likert scores to understand, manage, and overcome control system cybersecurity vulnerability.

25.1 Introduction

Cyber introduces new capabilities to control systems. However, professionals' uncertainty and lack of agreement about the cyber domain create system

DOI: 10.1201/9788743809128-25

vulnerability. Such disagreement leads to misalignment, which in turn creates vulnerability. Why is this important? Reports of significant cyber incidents targeting infrastructure have increased globally over the past several years. Advanced persistent threat (APT) attacks on critical infrastructure sectors, such as power grids or water treatment facilities, can result in severe consequences. Thus, there is a tremendous effort to design the next generation of capabilities to reduce incidents and inform entities about known security issues, vulnerabilities, and exploits in industrial control systems (ICS) and supervisory control and data acquisition (SCADA) systems (CISA, 2022). The Department of Homeland Security's Cybersecurity and Infrastructure Security Agency (CISA) releases cybersecurity advisories. It provides sector-specific cybersecurity framework guidance to help inform stakeholders about vulnerabilities and how to remediate them. As a result of increasing awareness of the cybersecurity needs for control systems, the expected global market for operational technology (OT) security, as measured by the compounded annual growth rate (CAGR), is 42% during the forecast period of 2025, and the market is expected to reach USD 3.53 billion by 2025. However, a lack of expertise and the complexity of systems are likely to hinder the growth (Market Research Future Report, September 2020).

Generally, we know that when people disagree on the certainty of a subject in a field, that uncertainty is a potential source of exploitable vulnerability. For example, the 2021 cyberattack on the Oldsmar, Florida, water treatment facility highlighted the cybersecurity vulnerability of control systems that often use legacy software and hardware (FBI, 2021). The continued use of outdated, unsupported operating systems that contain well-known vulnerabilities means that even if a system is perceived as not being vulnerable to internet-facing threats, it can still be an easy target. Therefore, we have already established that there is vulnerability. Can we go further? Are there different types of vulnerability and exposure that can be measured? Measuring disagreement in segments of the cybersecurity profession in the control system field can identify different vulnerability types. Some areas need to be advanced to have a robust operational defense of infrastructure that can mitigate actions and remove the ability of an asymmetric attack.

Furthermore, stakeholders question where to allocate resources to address the complexity of the challenges, as demonstrated by the Fiscal Year (FY) 2022 National Defense Authorization Act (NDAA). Section 1505 of the NDAA FY22 requires the Department of Defense (DoD) to report on the effectiveness of protecting Department of Defense (DOD) mission-critical control system (MCCS) assets and operational technology (Congress, 2021). Close collaboration between the DoD and agencies such as CISA, along with other relevant agencies and departments, including the Department of Energy (DoE), as well as commercial industry and vendors, is crucial for effectively understanding the technical and business needs of stakeholders, system behavior, and threat-informed vulnerabilities. DoD-led efforts, such as the more situational awareness for industrial control systems (MOSAICS), provide a framework for a holistic, vendor-agnostic, threat-informed, integrated approach to control system defense. However, the nuanced differences in reliance on operational technology (OT) rather than information technology (IT) are not widely understood among organizations responsible for making resource allocation decisions.

The problem is that if two entities disagree entirely on where the system is most vulnerable or uncertain, there is overall vulnerability. For example, in 2020, a global pandemic highlighted the importance of essential critical infrastructure workers as key infrastructure sectors underwent a significant transformation worldwide. In addition, the global COVID-19 pandemic forced many workers to quickly transition to online work environments without adequate cybersecurity training, processes, plans, or tools, resulting in a lack of adherence to fundamental cybersecurity rules during the remote work transition (Tasheva, 2021).

25.2 Why is this Model Valuable?

When the data shows no correlation between two professional groups, that is a vulnerability. When that vulnerability is identified, future research can lead to resources being allocated, given the known disparities between how professionals perceive a control system's cybersecurity functioning (Scalco, 2022a). The onset of the pandemic marked a significant shift in the levels of teleworking compared to those previously observed. More changes lie ahead as control systems expand their Internet connectivity. Remote access to operations increased significantly in 2020 (Ribeiro, 2021). For example, Nozomi Networks reported that some customers extended remote control access to operations from 9% to 60% in 3 months (Networks, 2021). Additionally, according to Nozomi, ransomware attacks on industrial organizations increased by 500% between 2018 and 2020 (Ribeiro, 2021).

Furthermore, the incidents increased by 116% between January and May 2021. Predictions are that global ransomware damage costs will exceed USD 265 billion by 2031 (Braue, 2021). Presidential Directive 41 (PPD-41), the United States Cyber Incident Coordination, outlines the federal government's response to cyber incidents for both government and private sector entities. The guiding principles include shared responsibility between individuals, the private sector, and government agencies; risk-based response actions; respect for safeguarding details about affected entities; unity of government effort; and enabling restoration and recovery. (Obama, 2016).

The integration of IT capabilities into traditional operational technology (OT) environments in recent years has yielded extraordinary direct monitoring and control capabilities of physical devices and processes at the smallest component level (Aleksandra Scalco, 2021). However, OT is still not widely understood for the many reasons described throughout this book. A challenge is that the control system context differs significantly from that of an IT system. As a result, in addition to operating within a diverse environment, OT operators respond to operational anomalies differently than IT users (Scalco and Simske, 2020). Engineering solves technical problems by applying principles of physics and mathematics to systems in the physical domain. Cybersecurity encompasses every aspect of integrating IT capabilities into critical infrastructure control system environments, including the spectrum of applications used, information accessed, networks utilized, operational processes employed, data encryption, access control, end-user training, and disaster recovery. Advanced capabilities, such as programmable event and alarm filters, enable air traffic controllers and maintenance personnel to interface IT with OT physical systems. This allows engineers and personnel to monitor and make desirable system changes from remote locations (Aleksandra Scalco, 2021). Systems engineering methodology and practice can offer equilibrium to complex, context-sensitive, and dynamic classes of control systems (e.g., mission critical control systems (MCCS)). Therefore, the proposed analytic model can provide a valuable tool to overcome the uncertainty that leads to control system cybersecurity vulnerabilities.

Presidential Directive 21 (PPD-21) identifies 16 critical infrastructure sectors (February 2013) to establish a national policy on critical infrastructure security and resilience, advancing a national unity of effort to strengthen and maintain secure, functioning, and resilient critical infrastructure (Obama, 2013). These sectors may have various entities with oversight authority, sub-sectors, and varying dependencies. Critical infrastructures are highly interconnected within the infrastructure sector, and in many cases, provide information or control over critical systems. Dependencies may be functional, such as electric power-generating plants needing fuel. They may be geographic, for example, a petroleum facility connected to a specific electric power substation. Dependencies often include human decision-makers and contractual, treaty, or regulatory constraints. Mental models are insufficient to capture the complexities, so carefully designed simulation models are helpful. Consistency among stakeholders in describing control system challenges and attributes will help the industry better understand challenges. For the government, authorities are distributed among agencies such as the Department of Energy (DOE) for power infrastructure; the Department of Homeland Security (DHS) for state, local, and tribal environments; within the Department of Defense (DOD), and US Cyber Command (USCYBERCOM) for cyber effects; as well as other agency entities. Similarly, there are varying oversight and governance authorities in the sector of commercially owned assets.

As access to the Industrial Internet of Things (IIoT) and discussions about cybersecurity and critical infrastructure protection expand, so does the opportunity to apply transformational capabilities to control systems. Key drivers include increased flexibility and ease of use for operators, scalability, modularity, enhanced memory access, evolving features such as wireless connectivity, and faster speed solutions. The role of industrial controls in controlling machinery is changing as increased software, information, and communication technology expands the capabilities of these machines. This change has a profound impact on the security posture of mission-critical infrastructure, introducing resiliency challenges. The approach of "security by obscurity" is a failed strategy. Critical infrastructures are highly interconnected at various levels, ranging from within individual infrastructure sectors to across multiple sectors, such as electric power and communications. Infrastructure sectors also rely on one another. Communication systems require the ability to operate, while electric power systems rely on communication services. Readily available information enables threat actors to quickly identify control system assets connected to the internet, creating a "perfect storm" of easy access to unsecured assets, open-source information about devices, and commonly used exploit frameworks.

Wolfgang Schwab and Mathieu Poujol provided insight through a 2018 survey on the state of industrial cybersecurity. Their findings indicated that most companies surveyed considered cybersecurity for OT/ICS a significant priority. However, IT and OT professionals possess varying goals, processes, tools, and language (Schwab and Poujol, 2018). While most thought the company would be a target of a cybersecurity attack in the OT industrial control systems (ICS) space, as likely, only 23% were compliant with minimal mandatory industry standards or government guidelines and regulations (Schwab and Poujol, 2018). While half of the participants stated that the company had not experienced a cyber incident in the previous 12 months, Schwab and Poujol observed that participants may not have recognized if they had, as most could not detect an event or track it. The data showed that 8% revealed they do not know, and 10% do not measure cybersecurity incidents with control systems. Another interesting finding is that IT and OT professionals possess varying goals, processes, tools, and language (Schwab and Poujol, 2018). The term "cybersecurity" is the least understood within the Department of Defense (DOD) (Span et al., 2018). No reinforcement of complementary safety and cybersecurity measures (i.e., activity logging for attack detection and accident anticipation); friction where safety and cybersecurity requirements are conflicting (Kavallieratos et al., 2020). Confusion arises as new terms emerge to describe emerging technologies and capabilities without clearly distinguishing the relationships between them (e.g., Industrial Internet of Things (IIoT), supervisory control and data acquisition (SCADA), and industrial control systems (ICS) (Kayan et al., 2021)).

25.3 Use of the Model to Measure Disagreement as a Means of Identifying Vulnerability

If two agents disagree entirely on where the system is most vulnerable or uncertain, there is an overall vulnerability. Where the correlation is poor, these professionals will be at odds. If people disagree on certainty in the field, that is a source of vulnerability. Disagreement among professionals regarding cybersecurity leads to misalignment, which in turn creates a vulnerability that may be greater than the inherent vulnerability of the system design. The model we describe is an approach to measuring disagreement in cybersecurity professionals' segments, aiming to identify those vulnerabilities (Scalco, 2022a). Disagreements exist among professions due to variations in engineering practice, paradigms, processes, and culture. Understanding the entire picture and identifying areas for improvement is a continuous challenge in science and engineering. This challenge holds particularly true for systems that control the physical world, such as power systems. Disagreement leads to misalignment, which in turn creates vulnerability. It is possible to measure agreement uncertainty through statistical analysis and an analytical model, which helps identify pain points where different stakeholders disagree.

The same measure can be walked back to stakeholder sources, beginning with a vulnerability assessment to help drive better alignment and, eventually, agreement. It is the disagreement that ultimately leads to vulnerability. We have an overall vulnerability if two entities disagree entirely on where the system is most vulnerable or uncertain. Measuring disagreement among professions can help resolve transition readiness gaps for new capability acquisition into operations, enabling a response to threat intelligence, alignment with strategic guidance, informed investment decisions, and budget and funding allocation. A mathematical model and analytical methodology are employed to measure disagreement within segments of the cybersecurity profession, aiming to identify vulnerabilities through statistical uncertainty analysis of Likert semantic differential scales. This model enables one to pinpoint areas of potential disagreement, which can lead to misalignments and vulnerabilities, and then walk that back through the model from exposure to alignment and agreement. The analytic approach correlates with the agreement, and a lack of agreement is a vulnerability. Recall from Chapter 22, "Managing Risk," that risk management aims to identify and implement effective controls against the most significant risks. The goal is to reduce the attack surface to the feasible minimum. What if there is disagreement among professionals regarding the treatment of cybersecurity in control systems? The results can include the possible loss of mission, physical injury, loss of life, economic impact, and undesirable social consequences, such as loss of public confidence. No system can be perfectly secure. The goal is to mitigate suitable risks in the correct order to achieve an acceptable residual risk level while incurring allowable costs and operational impacts. First, however, we must identify the risks and sources of vulnerabilities and agree on how to address them. Agreement on how to treat vulnerabilities requires that cyber threats be identified and analyzed to develop an effective strategy for achieving a resilient cybersecurity posture. Understanding where there may be disagreement among professional groups helps minimize the risk of vulnerability induced by misalignment.

A questionnaire was developed as a starting point to collect representative data for statistical inferences about the population. Questions from the questionnaire are shared throughout this book and are also included in Appendix A, "Stakeholder Alignment Survey Questions," for reference. Statistical significance indicates that a pattern exists. R-squared is the strength of the pattern. Recall that R^2 ranges from 0 to 1, where the value represents the amount of variation in the dependent variable, explainable by the variation in the independent variable. The vulnerability induced by misalignment may be greater than the innate system design vulnerability. A numerical value was assigned to each item. For example, if the Likert scale includes the response options: "Yes," "No," or "I do not know." Responses are coded to have "yes" = 1, "no" = 2, and "I do not know" = 3 so that a higher score reflects a higher level of uncertainty in each item. After entering the individual scores, the mean score for the whole group, calculated for each question, was determined. For example, participants were asked 20 questions about infrastructure, such as:

- I have access to the physical network topology.
- I have access to the logical network topology.

- Critical assets related to control systems in our organization have been identified.
- The critical assets of our organization are well-defined.
- I know the infrastructure supporting the critical assets.
- I know the single points of failure.
- I know the infrastructure supporting single points of failure.

Data in each category was given both a Likert Entropy (LE) score and a Likert Coefficient of Variation (COV) rank. If everyone agrees and knows, "this is where we should start," we are in pretty good shape. However, we are trying to show that we are not in pretty good condition and not in pretty good shape. Calculating the coefficient of variation (COV) as a measure of dispersion is helpful when comparing scores with different units of analysis or means. This measure allows us to understand the disbursement of scores from participant to participant. The percentage of COV generated is a dimensionless ratio of the standard deviation divided by the mean (eqn (2)). The ranking is based on questions and an overall comparison of ranks. Entropy and COV are ranked and summed to determine a total overall rank of uncertainty, which is illustrated in a scatter chart. The measures of dispersion of a response dataset relative to the mean are first calculated as the square root of the variance, by determining each data point's deviation from the mean, to generate the charts from the survey results. In other words, the measure of the dispersion of individual participant responses to the questions is calculated by placing all participant responses to questions in bins based on the nine sections of questions described earlier, which cover aspects of multi-concern assurance. Some questions received highly similar responses, indicating the stability of the answers, such as "critical assets related to control systems in our organization have been identified" and "the critical assets of our organization are well-defined."

For example, a set of questions might be a block of questions about network systems, infrastructure, incident response, or training and certifications. Responses to the collection of questions are analyzed by x- and y-variables such as occupation (i.e., engineers, computer scientists, and technicians), by employment sector x- and y-variables (i.e., federal, non-federal, and commercial sector), or by critical infrastructure sector x- and y-variables (i.e., transportation, communications, and IT). The mean Likert value for a question and the standard deviation of the Likert value are then calculated. For example, the standard deviation of the response count from all engineers to questions about network systems is calculated for each question. The mean and coefficient of variation (COV) are calculated for each question using all responses from each employment sector, such as engineers. The same calculation is performed for each sector (i.e., computer scientists, technicians, and safety personnel).

Finally, the value of the function y is plotted against x to see if one variable's response is fully explained by the other. A scatter plot is used as a tool for analysis to examine the relationship between two variables, such as engineers and computer scientists, and safety occupations, and to show how much the two occupational variables correlate, providing a visual sense of the data. The strength of the relationship is established by how close the data points are plotted on a straight line. Data points on a straight line from the origin to high x- and y-values show a positive correlation. Entropy measures the amount of uncertainty, or unknown, in the data, as visualized by the correlation explained through the measurement of data related to x- and y-variable pairings, which is represented by the amount of space occupied on the scatter plot. The experimental points are plotted and later reviewed to see if there is an agreement in the relative ranks using linear regression. The experimental points plotted are: {(rank_Entropy(x)+ rank_ COV(x)), (rank_Entropy(y)+rank_COV(y))}, which is the graphical representation of the relative values of X=(rank_Entropy(x)+rank_COV(x)) and Y=(rank_Entropy(y)+rank_COV(y)). The researchers later examined whether there was an agreement in the relative ranks using linear regression. As participant certainty or knowledge of the state of the whole cybersecurity of control systems increases, the entropy goes down. As uncertainty increases, the entropy goes up. The entropy is measured by pairings between variables, such as by profession (i.e., engineers and computer scientists, engineers, and safety professionals) or critical infrastructure sector (i.e., communications and energy, or communications and IT).

For example, participants were asked 23 questions about network systems, such as: "User accounts and credentials are managed in our organization by the following authentication approach," graphically represented as the point of origination (10, 11) in Figure 25.1 engineer (y-coordinate 11) agreement with computer scientists (x-coordinate 10) about network systems. X-coordinate 10 is the total rank sum of the rank entropy and rank COV of the computer scientists' response to the question, x=(rank_Entropy(x)+rank_COV(x)). Y-coordinate 11 is the rank sum of the rank Entropy and rank COV of engineers' responses to the question, y=(rank_Entropy(y)+ rank_COV(y)).

Similarly, the total rank-sum of rank entropy and rank COV is plotted for the other questions in the bin as {(rank_Entropy(x)+ rank_ COV(x)),

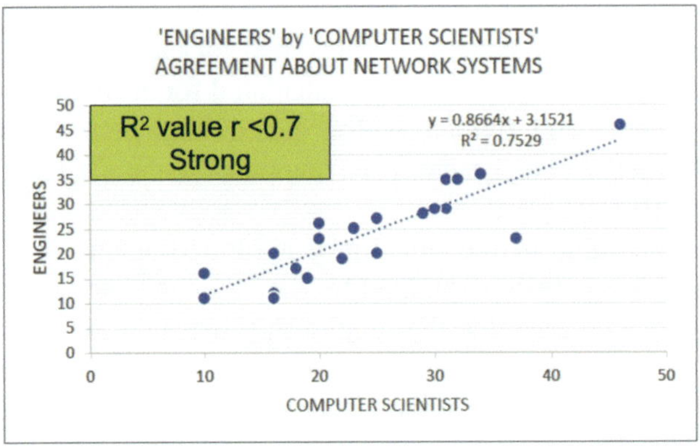

Figure 25.1 Engineers (y-coordinate) by computer scientists (x-coordinate) about network systems.

(rank_Entropy(y)+ rank_COV(y))}. The scatter chart shows the total overall rank of uncertainty about an aspect of multi-concern assurance. It is illustrated in a scatter chart along with the rank sum of entropy and COV response points to other questions, such as: "I have decision authority for which risks are accepted as related to cyber-physical systems (CPS)," shown as point (28, 29). X-coordinate 28 is the total rank sum of the rank entropy and rank COV of the computer scientists' response to the question. Y-coordinate 29 is the rank sum of the rank entropy and rank COV of engineers' responses to the question. The correlation of agreement between engineers and computer scientists is plotted as point (28, 29), as shown in Figure 25.1, correlation between engineers and computer scientists for network systems questions.

Another example is the response to the network systems question: "Our organization uses threat detection capabilities," demonstrated as point (37, 23) for the y-coordinate sum of rank entropy and rank COV for engineers (23) and the x-coordinate sum of rank entropy and rank COV for computer scientists (37). Higher entropy represents the perceived information entropy, or level of uncertainty, between the two variables, graphed in this case as the correlation between the variables of engineers and computer scientists. The graphical representation is effective for the complexity of cybersecurity for control systems.

The test was run by viewing variables such as engineering as the y-variable and computer scientists as the x-variable and flipped around to engineering as the x-variable and computer scientists as the y-variable to observe the relationship. Entropy and COV were each ranked in their respective bins for each question. This approach incorporates non-parametric elements into the analysis, which helps overcome potential non-Gaussian distributions. The ranking of scores is from 1 to N, and the maximum domain and range of the chart is 2N, assigned to the data. Finally, a total rank-sum of entropy and COV was calculated and plotted in a scatter graph for the set.

The correlation coefficients indicate the strength and direction of the predicted values of the pattern in relation to the observed value of the linear relationship between the movements of two variables and quantify the strength of the relationship. A value of 1 implies that the equation perfectly describes the relationship between two variables. The correlation coefficient is denoted by r. R-squared (r^2) is the proportion of explained variance or the strength of the pattern. The r^2 is the statistical measure of the proportion of variance for the dependent variable explained by the independent variable in the mathematical model. Data were assessed by occupation, employment sector, and critical infrastructure sector. The questionnaire's occupational field options included engineering, computer science, industrial management, mathematics, physical sciences, safety, and technicians. Employment sector field options included federal, non-federal, Federally Funded Research and Development Centers (FFRDC), University Affiliated Research Centers (UARC), commercial industry, academia, student, and military service.

For example, in Figure 25.1, we see that the dependent variable is engineers, and the independent variables are other occupations such as computer scientists or technicians. Figure 25.1 compares engineers (y-coordinate) and computer scientists (x-coordinate) in a similar manner about infrastructure and has an r^2 value of 0.7529. An r^2 of 0.75 for engineers and computer scientists is relatively high for the data set. It means that a 75% variation of one variable is entirely explained by the other. It is a big deal to account for three-fourths of what we are examining. This single variable explains 75% of the uncertainty in agreement, without controlling for any other factors.

When we examine the agreement of engineers with safety professionals about network systems, as seen in Figure 25.2, we know an r^2 of less than 4%. That is enormously significant. This single variable cannot account for the variation. We want to let the data talk and avoid any spurious correlation by connecting variables that may appear causal but are not due to an unseen third factor not apparent in the initial examination. Two variables may be statistically related but not directly causally related and may falsely appear to be associated. The variation in data from engineers and computer scientists, as well as data from engineers and safety professionals, highlights the uncertainty surrounding agreement on cybersecurity for control systems. The extension of connectivity in the digital transformation of engineering to critical infrastructure introduces new challenges in system engineering.

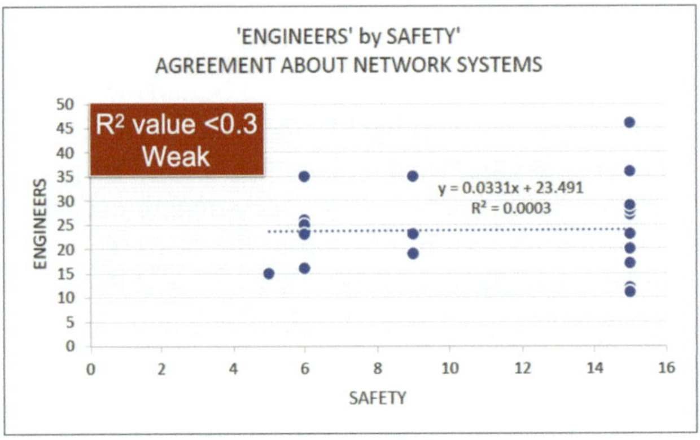

Figure 25.2 Engineers (y-coordinate) by safety (x-coordinate) agreement about network systems.

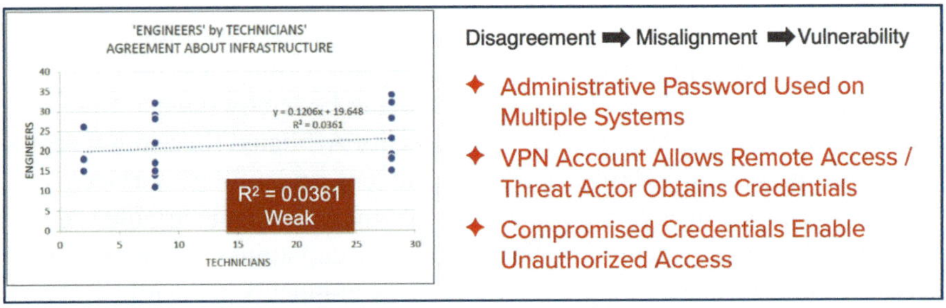

Figure 25.3 Disagreement leads to misalignment, which is a vulnerability.

Implementing policy or adding new capabilities to defend control systems will not work if there is disagreement among professionals.

An example of how the model may be used to understand better these types of vulnerabilities is shown in Figure 25.3. Walking forward through the model, when there is disagreement between two sets of professionals, such as engineers and technicians, about the infrastructure, the resulting misalignment leads to vulnerability. The system may be designed to mitigate cyber risks. However, a threat actor can obtain credentials if the same administrative password is used on multiple systems, and VPN accounts allow remote access. The compromised credentials can be used to enable unauthorized system access.

Walking back through the model, when incident information or known vulnerabilities are shared, the data can be used to inform stakeholders and achieve alignment and agreement. For example, as shown in Figure 25.4,

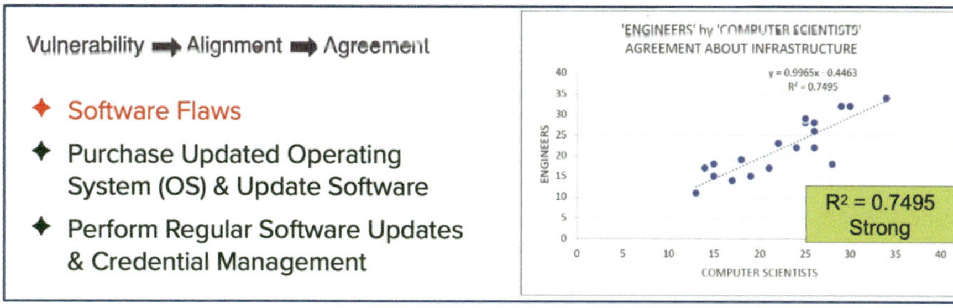

Figure 25.4 Use of the model to inform and align stakeholders.

identifying software flaws can lead entities to purchase an updated operating system (OS) and update software. Additionally, alignment can ensure that professionals perform regular software updates and enforce credential management. OT personnel come into their field through backgrounds and education in engineering disciplines such as electrical engineering, mechanical engineering, or industrial engineering, in addition to on-the-job training (OJT). IT personnel enter their field predominantly through computer science, vendor-specific or non-vendor-specific training and certifications, and on-the-job training (OJT). Knowing what the compromises are in a system is a sign of hope. Not knowing what it is and not agreeing on the state of a field is a vulnerability because each group will be complacent in one area and non-complacent in another.

Moreover, they will be competing. Skills found in OT occupations generally require more extraordinary experience with older technologies than those found in IT enterprises (Harp and Gregory-Brown, 2014). If sets of professions compete, that leads to vulnerability. We need to reach a common understanding, or we will not be able to build a system from the ground up that addresses both cybersecurity needs and the system's functional requirements. Let alone retrofit cybersecurity into a system. If there are A and B components of a system that need to be secured, and if one group thinks system A is in good shape and system B is in terrible condition. Moreover, group two believes that A is in terrible condition and B is in great shape. The two groups will not only compete but also undermine the work done in each area. A common understanding is needed.

Furthermore, retrofitting IT onto OT poses a considerable risk because the OT systems were designed without knowledge of future IT developments. Retrofitting security is problematic. A better approach is an early collaboration, agreement, and alignment between IT security and operations teams, or SecOps, to enhance security by integrating tools, processes, and technology from the ground up into the system. Currently, standards are still emerging and are voluntarily implemented. Understanding the entire picture and identifying areas for improvement is a continuous challenge in science and engineering.

While governments are trying to bring key components together, a genuine concern is that the problem of vulnerabilities will not be resolved until stakeholders agree on several key issues, such as resource allocation, training, and standardization. Despite all the valiant efforts, there will continue to be increased vulnerability if disagreement and misalignment on what constitutes a cybersecurity vulnerability are not addressed. It is concerning when professionals hold competing views. When the data shows no correlation between two professional groups, that is a vulnerability. Future research can allocate resources when that vulnerability is identified, given the known disparities in how professionals assess a control system's cybersecurity. The professionals will compete against one another and undermine the work to secure the system. A shared understanding is necessary, or we will not be able to develop a comprehensive design from the ground up to address cybersecurity. Let alone retrofit security into a system. The professionals will be working from different assumptions and will not come together. The process presented is a new approach to addressing uncertainty using measurable data, which would otherwise lead to vulnerability. We do not have an alignment if someone is trying to defend something that someone else is not attacking, using the Red Team/Blue Team analogy. We know people are spending money on these things without fixing them. If you have ever had a conversation within your organization about where and how much to allocate resources for cybersecurity, you know that securing stakeholder agreement is a complex challenge. Implementing policy or adding new capabilities to defend control systems will not work if there is disagreement among professionals.

25.4 Emerging Policy

Implementing policy or adding new capabilities to defend control systems will not work if there is disagreement among professionals. It has been more than a decade since the Stuxnet Worm appeared on the scene as the first publicly known digital weapon targeting equipment and operations that control the physical world in 2010. As discussed in Chapter 13, "Needs Analysis," Stuxnet specifically targeted industrial software and equipment for uranium enrichment. Since 2010, attacks on control systems have been on the rise, creating a need to champion cultural change in the cybersecurity of these systems. These are computing devices that control physical processes in the world today. A typical computer is estimated to have a lifespan of a couple of years, with regular updates and software fixes expected to be made frequently. Control systems devices are different. The life cycles can span up to 30 years or more, while the time and speed of a cyberattack are measured in seconds. In addition, the systems often rely on legacy operating systems with meager resistance against common attack techniques, even as conveniences such as remote access are introduced. Getting all agents to agree on where the system is most vulnerable and where to allocate resources and energy to prevent and identify vulnerabilities is a complex challenge.

New legislation underscores the importance of evaluating and understanding aspects of control system cybersecurity to strengthen and maintain secure, functioning, and resilient critical infrastructure. The Strengthening American Cybersecurity Act of 2022 presents a package of cybersecurity measures that require critical infrastructure operators and federal civilian agencies to report cyber incidents to the Cybersecurity and Infrastructure Security Agency (CISA). HR 1833, introduced in March 2021, grants the Cybersecurity and Infrastructure Security Agency (CISA) greater authority to protect critical systems against cyberattacks. Executive Order (E.O.) 14028 removes barriers to sharing threat information, thereby accelerating efforts to deter, prevent, and respond to incidents. The action also enables more effective defense of agencies' systems and data collected, processed, and maintained by or for the Federal Government. In addition, on January 19, 2022, President Biden signed a National Security Memorandum (NSM) to improve the cybersecurity of National Security, Department of Defense, and Intelligence Community Systems, as required in his Executive Order (E.O.) 14028, Improving the Nation's Cybersecurity. The National Security Memorandum (NSM) establishes voluntary cybersecurity goals for owners and operators of critical infrastructure.

Furthermore, the National Security Memorandum (NSM) requires that, at a minimum, National Security Systems employ the same network cybersecurity measures as those required of federal civilian networks under Executive Order 14028. The National Security Memorandum (NSM) builds on the Biden Administration's efforts to protect the nation from sophisticated, malicious cyber activities by nation-state actors and cybercriminals. The Memorandum establishes timelines and guidance for implementing these cybersecurity requirements, including multi-factor authentication, encryption, cloud technologies, and endpoint detection services.

25.5 Summary

When the data shows no correlation between two professional groups, that is a vulnerability. When that vulnerability is identified, future research can lead to resources being allocated, given the known disparities between how professionals perceive a control system's cybersecurity functioning (Scalco, 2022a). The extension of connectivity in the digital transformation of engineering to critical infrastructure introduces new challenges in system engineering. Implementing policy or adding new capabilities to defend control systems will not work if there is disagreement among professionals. While the government is trying to bring critical components together, a genuine concern is that throwing money at this problem does not resolve vulnerabilities until stakeholders agree on the resources. Despite all the valiant efforts, there will continue to be increased vulnerability if disagreement and misalignment on what constitutes a cybersecurity vulnerability are not addressed. It is concerning when professionals hold competing views. When the data shows no correlation between two professional groups, that is a vulnerability. When that vulnerability is identified, future research can lead to resources being allocated, given the known disparities between how professionals perceive a control system's cybersecurity functioning. The professionals will compete against one another and, worse yet, undermine the work done to secure the system. A shared understanding is needed, or we will not build a design from the ground up to address cybersecurity effectively. Let alone retrofit security into a system. The professionals will be working from different assumptions and will not come together.

A better approach is to utilize the model described as a means of identifying vulnerability induced by misalignment that may be greater than the inherent system design vulnerability. The analytic model and process presented are a new means of addressing uncertainty using measurable data that would otherwise lead to vulnerability. Measuring disagreement in segments of the cybersecurity profession in an organization is an impactful means of identifying vulnerabilities. Once an entity understands this vector better, more informed decisions can be made to form an effective strategy. For example, a technology strategy may use an automation initiative to address the vulnerability. An action plan may include deploying automated courses of action (COA) to enhance the response security posture. Example metrics may be the degree of automation, operation, and maintenance (O&M) cost, count number of COA (i.e., playbooks), operational availability (i.e., uptime vs. downtime), number of cyberattacks detected, number of cyberattacks mitigated, or staffing requirements (i.e., number of full-time employees (FTE), education and skill level of operators, and labor categories). The first step involves developing a strategy for identifying vulnerability. An initiative is to use the analytic model. An action plan is to use the analytical model and methodology to measure disagreement in segments of the cybersecurity profession as a means of identifying vulnerabilities, then remeasure if other strategies, such as a technology strategy, are working and how well, using an r^2 statistical significance score that indicates that there is a pattern. This approach will provide quantitative information to inform your strategy and systems engineering practices for critical infrastructure in a cyber world.

References

Aleksandra Scalco, S. S. 2021. Championing Cultural Change and Control Systems Cyber Security. *AIAA-INCOSE Wasatch Aerospace and Systems Engineering Mini-Conference.* AIAA Utah Section.

Braue, D. 2021. Global Ransomware Damage Costs Predicted to Exceed $265 Billion by 2031. *Cybercrime Magazine.*

CISA 2022. Implement Cybersecurity Measures Now to Protect Against Potential Critical Threats. *In:* Defense, D. O. (ed.). Washington, D.C.: Department of Defense.

Congress, T. 2021. National Defense Authorization Act (NDAA) for the Fiscal Year 2022. *In:* CONGRESS, T. (ed.). Washington, D.C.

FBI, C., EPA, MS-ISAC 2021. Compromise of U.S. Water Treatment Facility. *In:* ADVISORY, J. C. (ed.).

Kavallieratos, G., Katsikas, S. & Gkioulos, V. 2020. Cybersecurity and Safety Co-Engineering of Cyberphysical Systems—A Comprehensive Survey. *Future Internet*, 17.

Kayan, H., Nunes, M., Rana, O., Burnap, P. & Perera, C. 2021. Cybersecurity of Industrial Cyber-physical Systems: A Review. *arXiv:2101.03564*, 32.

Networks, N. 2021. What You Need to Know to Fight Ransomware and IoT Vulnerabilities, Including Recommendations for Enhancing Cyber Resilience. *OT/IoT Security Report.* nozominetworks.com: Nozomi Networks, Inc.

Obama, P. B. 2013. Presidential Policy Directive/PPD-21, Critical Infrastructure Security and Resilience. *In:* Secretary, O. O. T. P. (ed.). The White House: The White House.

Obama, P. B. 2016. Presidential Policy Directive/PPD-41, United States Cyber Incident Coordination. *In:* HOUSE, T. W. (ed.). Washington, DC: Executive Office of the President of the United States.

Ribeiro, A. 2021. Ransomware strikes rise sharply, fueled by profit potential. *Industrial Cyber.*

Scalco, A. 2022a. *Measuring Disagreement in Segments of the Cybersecurity Profession as a Means of Identifying Vulnerabilities,* Colorado State University.

Scalco, A. 2022b. *Measuring Disagreement in Segments of the Cybersecurity Profession as a Means of Identifying Vulnerabilities,* Colorado State University.

Scalco, A. & Simske, S. 2022. Measuring Stakeholder Alignment to Overcome Control System Vulnerability. Department of Homeland Security (DHS) CISA Industrial Control Systems Working Group.

Scalco, A. & Simske, S. 2023. A model for measuring multi-concern assurance of critical infrastructure control systems. *Systems Engineering*, 12.

Schwab, W. & Poujol, M. 2018. The State of Industrial Cybersecurity 2018. *In:* LAB, K. (ed.). Munich, Germany: CXP Group.

Span, M. T., Mailloux, L. O. & GrimailA, M. R. 2018. Cybersecurity Architectural Analysis for Complex Cyber-Physical Systems. *The Cyber Defense Review,* 3.

Tasheva, I. 2021. Cybersecurity post-COVID-19: Lessons learned and policy recommendations. *European View*.

26

Conclusion

"If you please – Draw me a sheep!"
– Antoine de Saint-Exupéry,
The Little Prince (Saint-Exupéry, 1943)

"So long, and thanks for all the fish"
– The Hitchhiker's Guide to the Galaxy
(Adams, 1979).

Abstract

In the novella *The Little Prince* (Saint-Exupéry, 1943), the prince asks the aviator and narrator of the story to draw him a sheep. The aviator tries, but with each attempt, the prince rejects his attempts, criticizing each sketch for not being quite right. In his frustration, the aviator draws a crate with three circles (holes in the crate) and tells the prince the sheep is inside the crate. This drawing of a sheep inside a crate satisfies and makes the *Petit Prince* happy. The drawing represents an abstraction that ideas vary from person to person (Saint-Exupéry, 1943). Similarly, in model-based systems engineering (MBSE), digital engineering (DE) abstracts aspects of a system to simplify an underlying conceptual model. These help us better understand a system and communicate information about elements of the system and relationships between elements at various semantic levels among various stakeholders, making the representation easier to use. The model simplifies the complexity of large systems, hiding the complexity of detail and showing what needs to be understood from a particular view. In Chapter 2, "History of the Term Cyber," the term cyber has predominantly become associated with IT. This cognitive shortcut to IT loses the impact of the term itself, which is more broadly the application of IT in organizational theory. Cyber is a cross-disciplinary field, extending readily to the myriad forms of cyber-physical security, as demonstrated in this book. The addition of technologies such as artificial intelligence (AI) and biometrics has further evolved cyber as

DOI: 10.1201/9788743809128-26

a cross-disciplinary field, integrating it into our lives and bringing significant technological enhancements. Organizations and people have come to rely extensively on digital technology. Systems engineering is a steward of these technologies and steers the team into alignment of understanding in the cyber medium.

26.1 Introduction

Model-based systems engineering (MBSE) involves the digital modeling of a system, abstracting its aspects to simplify the underlying conceptual model. These help us better understand a system and communicate information about elements of the system and relationships between elements at various semantic levels among various stakeholders, making the representation easier to use. The model simplifies the complexity of large systems, hiding the complexity of detail and showing what needs to be understood from a particular view.

In the novella *The Little Prince* (Saint-Exupéry, 1943), the pilot must fix the plane's engine before his water supply runs out and leaves him stranded in the Sahara Desert, and the Little Prince must return to take care of his planet. The pilot and the prince are in a struggle of abstraction and responsibility. Certainly, there is much more to appreciate in parables found in *The Little Prince* than an analogy to cyber. Both the aviator and the prince want to go home. However, there is a little loneliness in cyber that creates an urge for people and teams to connect and understand each other's perspectives. The internet and technology devices, like different planets, can stifle meaningful human interactions by creating physical connections that replace genuine human interactions with technology. In the abstraction of the sketch of a box, the prince sees the exact sheep he needs.

26.2 Have We Pushed the Envelope Too Far?

Recall from Chapter 2, "History of the Term Cyber," that cybernetics evolved as a field of study of control systems, communications, and the interface between humans and machines. The term cyborg originates from the terms cybernetic and organism. A cyborg is an integration of organic and artificial components into a single entity. Many people ask the question and wonder if we have pushed the cyborg envelope too far. Technology devices enable instant communication to audiences of millions (Scalco, 2010). In "Ethics for the Information Age," author Michael J. Quinn writes, "It's fair to ask: Do we drive technology, or does technology drive us" (Quinn, 2009)? While some may think the adoption of information technology tools has reached a level that pushes the cyborg envelope too far, the reality may be in the number of variables the human can actually process. Already in 1970, in his book "Future Shock," futurist Alvin Toffler described how some people exposed to too much rapid change develop symptoms of "future shock" syndrome (Toffler, 1970). Toffler suggests that there are adverse effects on people who receive too much information at an incredible speed. This observation was made within a couple of years of the internet's development.

If intellectually advanced groups, researchers, and managers are exposed to the effects of a nervous breakdown when exposed to rapid changes prior to the general adaptation of network connectivity, then today's cyber exposure would seem to cause people and organizations to fall ill. The reverse may be true. It may be that the connectivity enables the sharing of new ideas and information that were previously unknown. "With the internet, we are inundated with others, with information about ideas and customs well beyond our parochial experience" (Townsend and Bennett, 2001). Apparent anxiety and insecurity can be replaced by a positive vision of the future that supports developments in science, politics, culture, technology, and the arts. Making data more relevant from a vast amount of immediate information is an abstract phenomenon that has become increasingly visible in the information age.

26.2.1 Do we know less and think less?

The abstraction of making vast amounts of information relevant is challenging to grasp. The trouble is that what is efficient for one person at a moment is not necessarily what is best for the organization as a whole. For example, in 2007, it took several hours for the complicated emergency alert system to warn students that there was a shooting near the Virginia Tech campus. Messages sent throughout the day left students, instructors, and police confused and vulnerable. What is needed is "not so much newer and better sources of information; what we need is a better way to filter the information that we're getting" (Orzel, 2009). People needed valuable information quickly during the crisis unfolding near the Virginia Tech campus.

26.2.2 How much information is too much?

So, when does instant access to large amounts of information become "too much" information? As it turns out, psychologists at the University of Queensland conducted research to discover that. In the study "How Many Variables Can Humans Process," Graeme Halford and his colleagues concluded that humans can only process relations between four variables (Halford, 2005). The study tested 30 academics to their mental limits. Graeme Halford provided information about interactions in bar graphs with incomplete descriptions between variables and then asked participants to complete and correctly describe the graphs. "At the level of the four-way interactions, participants made comments such as 'Everything fell apart and I had to go back,'" Professor Halford said. "Only chance levels of performance were obtained for five-way interactions" (Halford, 2005). People can comprehend levels of abstraction up to about four-way interactions; after that, it becomes more challenging. Cyber technologies used in systems raise the levels of interaction to much higher than four-way, which suggests that we need to find better ways to augment a team's understanding of these interactions.

The cyborg envelope conceptually represents an additional dimension. Accessing information using information technology tools introduces another variable to the processing capacity. The Halford study concluded that "[w]hile all levels of complexity are logically possible, the evidence suggests that they are not cognitively manageable ... We found that four dimensions are the most humans can conceive of" (Halford, 2005). It is easier to access more information than previously imagined, which is good. The human brain may just not be hardwired to process it all at once. It will be interesting to witness the skills and abilities of the new leadership that emerges from a lifetime of exposure in the information age. Perhaps through the natural evolution of species, these new leaders will develop changes in the brain that enable them to conceptualize the fourth and fifth dimensions. Four dimensions are about the limits of our current human understanding. After that, it becomes an abstraction like the sheep in the box. Systems engineering principles and practices can help to clarify expectations of the box, understanding of what is in the box, and what we care about (i.e., the functions of the sheep and the box).

26.3 Summary

The story of *The Little Prince* teaches that responsibility is closely tied to relationships with others and inter-systems, requiring an understanding and appreciation of one's responsibilities. We hope this book delivers, as promised, a practical, inter-systems, systems engineering approach to addressing critical infrastructure in the cyber world. We hope it is useful and shared. Indeed, "A rising tide raises all boats." The final information shared is an example questionnaire on how to start the systems engineering journey of stakeholder alignment to overcome cyber vulnerabilities. "Cyber" does not have to be a misunderstood term. When we understand the term "cyber" in its contextual meaning and the cognitive shortcuts made about the word's meaning, we can extrapolate the concept as it is applied in various instances and gain a better understanding of the term in relation to the world around us. It takes all disciplines for an organization to work together, each from its own "cyber" perspective, to achieve an organizational vision. It is also essential for stakeholders to have a shared understanding to implement cyber policies, regulations, and guidelines effectively. Systems engineering practice can be considered the technical leadership that guides a group in cyberspace or towards a common goal, addressing cyberspace events and outcomes, such as mitigating or overcoming cyber vulnerabilities.

We hope this book helps you to use key systems engineering principles and practices to build a better team for designing and operating critical infrastructure in a cyber world. In the novel, *The Hitchhiker's Guide to the Galaxy*, dolphins use the phrase "So long and thanks for all the fish" as they leave earth just before it is destroyed. Bruce Schneier warns us in his book, "Click Here to Kill Everybody: Security and Survival in a Hyper-connected World," that computing, which connects us all, can pose a real threat (Schneier, 2018). As described in earlier chapters, cyber can introduce unintentional vulnerabilities to the system. However, the objective should not be to stop a few malware threats. Our goal should be to design systems that are less susceptible to attack in the first place (Ibid.). Incorporating secure software development into software engineering activities is a vital and continuous process. The entire team must embrace core cybersecurity elements. Technologies such as artificial intelligence (AI), machine learning (ML), and software-defined networking (SDN) are valuable software approaches to focus on as key enablers of systems engineering for critical infrastructure in a cyber world.

We prefer to believe in cyber as a medium rather than a domain. The medium is the commonality of all the creation and consumption across the system and systems of systems that have an impact. "Cyber," in its origins, is the study of control systems and communications and the interface between humans and machines. The key to systems engineering for critical infrastructure in a cyber world is to steer and steward an organization and individuals through the medium.

The thought we prefer to leave you with is:

> "This is your hockey moment. You're meant to be here." 1980 Winter Olympics US National Ice Hockey Team Coach Herb Brooks.

Best of luck to you! We're definitely rooting for you to succeed.

References

Adams, D. 1979. *The Hitchhiker's Guide to the Galaxy,* New York, Harmony Books.

Halford, G. 2005. *How much information is too much information?* [Online]. Available: https://phys.org/news/2005-02-how-much-information-is-too.html#:~:text=Emeritus%20Professor%20Graeme%20Halford%20and,between%20more%20than%20four%20variables. [Accessed].

Orzel, C. 2009. How Much Information Is Too Much Information? *Communications of the ACM,* 443.

Quinn, M. J. 2009. *Ethics for the Information Age.*

Saint-Exupéry, A. D. 1943. *The Little Prince,* New York, Reynal & Hitchcock.

Schneier, B. 2018. *Click Here to Kill Everybody: Security and Survival in a Hyper-Connected World,* New York, N.Y., W. W. Norton & Company, Inc.

Toffler, A. 1970. *Future Shock.*

Townsend, A. M. & Bennett, J. T. 2001. Electronic Empire. *Communications of the Association for Computing Machinery (ACM),* 44.

Appendix A

Questionnaire Resource

"A rising tide lifts all boats."
– John F. Kennedy (1917–1963),
President of the United States from 1961 to 1963

Abstract

This book offers a practical, inter-systems, systems engineering, and stakeholder alignment approach that system owners, stakeholders, and workforces can use to overcome cyber vulnerabilities arising within control systems in critical infrastructure. This book shares ideas and references to best practices, using composed solutions and approaches that can be immediately helpful to teams in successfully managing their work. If you have ever wondered where to start to help unlock a team's full potential to manage and operate systems, interface critical infrastructure components, and sensors with other technologies and systems, here are some good questions to ask. These are some of the questions that were used as part of research conducted at Colorado State University from 2020 to 2021. Colorado State University (CSU) Institutional Review Board (IRB) Protocol Number: 20-10209H (Scalco and Simske, 2019). Research Director Steve Simske, steve.simske@colostate.edu, Aleksandra Scalco, aleksandra.scalco22@alumni.colostate.edu, or RICRO_IRB@mail.colostate.edu.

1. I have cyber awareness training.

2. Cyber awareness training is required for my position.

3. Cyber awareness training is fully funded and available for my position.

4. I am familiar with the Purdue enterprise reference architecture, also known as the "Purdue model" (Figure A1. Purdue enterprise reference architecture (Control, 2020)).

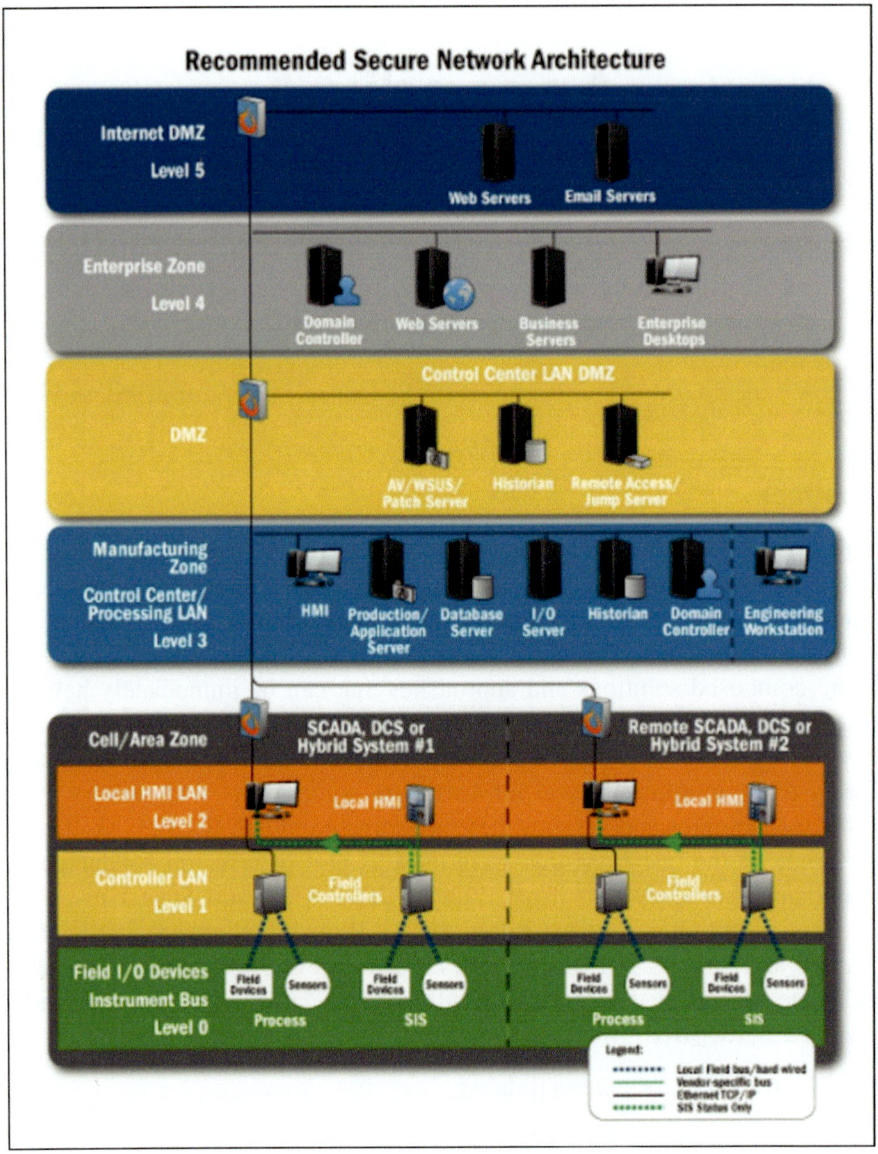

Figure A1 Purdue enterprise reference architecture (Control, 2020).

5. My work responsibilities include the enterprise security zone (Levels 4–5) of the Purdue model (Figure A1. Purdue enterprise reference architecture (Control, 2020)).
6. My work responsibilities include the industrial zone (Level 3) of the Purdue model (Figure A1. Purdue enterprise reference architecture (Control, 2020)).
7. My work responsibilities include the cell area (Levels 0–2) of the Purdue model (Figure A1. Purdue enterprise reference architecture (PERA), (Control, 2020)).
8. I recall when our latest risk assessment in our organization regarding our cyber-physical systems (CPS) was conducted.
9. I know how often system backups are tested.
10. Our organization uses threat detection capabilities.
11. Threat signatures are updated in our organization.
12. We maintain system backups.
13. System backup storage is secure in our organization.
14. Our organization has an employee security vetting process in place.
15. System backup storage is maintained securely.
16. The acquisition process has a cybersecurity element.
17. Our organization implements industry best practices for industrial control system (ICS) and supervisory control and data acquisition (SCADA) capabilities.
18. I have access to the industrial control system (ICS) and supervisory control and data acquisition (SCADA) capabilities.
19. Another outside organization has access to the industrial control system (ICS) and supervisory control and data acquisition (SCADA) capabilities of our organization.
20. I have decision authority for which risks are accepted as related to cyber-physical systems (CPS).
21. I have decision authority for implementing risk mitigations related to cyber-physical systems (CPS).

22. In our organization's system network, the architecture is partitioned or segmented to secure system backups from compromise as a security measure.
23. I have access to the physical network topology.
24. I know who has the physical network topology.
25. I have access to the logical network topology.
26. I know who has the logical network topology.
27. The critical assets related to control systems in our organization have been identified.
28. The critical assets of our organization are well-defined.
29. I know the infrastructure supporting these critical assets.
30. I am aware of the single points of failure.
31. I am aware of the infrastructure that supports single points of failure.
32. Defensive countermeasures are in place to protect operations.
33. Defensive countermeasures are in place to detect and respond to threat activity.
34. In our organization, timestamp data is configured accurately, synchronized, and managed across the system.
35. In our organization, timestamp data is used to reference persistent time-based trends.
36. Control system information publicly available is managed for security purposes (e.g., public-facing websites and job postings).
37. Long-term system monitoring capability is practiced.
38. Standard operations and activities are characterized.
39. Vendors remotely support the network.
40. Vendors directly connect their equipment to the network.
41. Vendors can directly update their software over the network.
42. I am responsible for making changes to process controls.
43. I am responsible for industrial control systems (ICS), including supervisory control systems used to control geographically dispersed assets.

44. I am responsible for an information system used to control industrial processes, including manufacturing, product handling, production, and distribution.
45. Industrial control systems, including supervisory control and data acquisition (SCADA) systems, are used to control geographically dispersed assets.
46. I am responsible for an information system used to control distributed control systems (DCSs) and smaller control systems utilizing programmable logic controllers (PLCs) to manage localized processes.
47. Cyber incident response policies, procedures, and logistics are well-documented within our organization.
48. Our incident response plan includes a cybersecurity element.
49. Our defensive countermeasures capture useful information.
50. Upstream and downstream impacts are known.
51. Upstream and downstream points of contact (POC) are known.
52. A list of useful incident response services is available.
53. Our organization has hazardous operations.
54. Our organizations are aware of the US government's incident response support and resources.
55. Our organization has contracted with a vendor for technical support.
56. Our vendor-supplied support contract includes a cybersecurity element.
57. Our vendor-supplied support contract includes an incident response element.
58. Our organization has cleared employees authorized to access classified information.
59. Our organization has a formalized relationship with government agencies.
60. Our organization is aware of the available Department of Homeland Security (DHS) support resources.
61. Our organization has used the available DHS support services.
62. Our organization is aware of the available Federal Bureau of Investigation (FBI) support services.

63. Our organization has used the available FBI support services.
64. Our organization is aware of the Department of Energy (DOE) Cybersecurity Risk Information Sharing Program (CRISP).
65. Our organization has used DOE CRISP.
66. I maintain configuration management of a control system(s).
67. I maintain awareness of vulnerabilities to legacy or older systems (due to age).
68. I ensure workforce training and procedures for the operation of the control system
69. equipment.
70. I maintain network awareness.
71. I provide institutional knowledge of the system's purpose and architecture.
72. I provide for the identification of new vulnerabilities.
73. I am responsible for maintaining the risk management framework (RMF) authorization to operate (ATO) for control systems.
74. I establish procedures to minimize the exploitation of vulnerabilities.
75. I ensure external vendors and contractors follow cyber hygiene requirements.
76. I can identify if an individual device is being tampered with within a complex system.
77. I ensure that monitoring policies allow for the identification of anomalous behaviors during workforce downtimes.
78. I provide coordination of information from cyber systems.
79. I provide policies and procedures for responding to normal conditions promptly to minimize false alarms.
80. I provide continuous assessment of normal and off-normal operations.
81. I provide support for the use of known vulnerable systems (when vulnerabilities cannot be avoided).
82. I integrate systems to meet the requirements specification(s).

83. I maintain awareness and active, valued communication between the workforce groups about ongoing vulnerabilities and newly discovered threats.
84. I provide for the identification of vulnerabilities as systems age or new systems are integrated into the existing infrastructure.
85. I identify false positives from vulnerability scans and security event logs.
86. I ensure that maintenance procedures or workarounds do not void the anomaly detection in control systems.
87. I ensure that regulations do not engender cybersecurity gaps in control systems.
88. I ensure conflicting or competing regulations do not create vulnerabilities in control systems.
89. I ensure that cybersecurity in design does not impact the ability to perform the work functions required by control system requirements.
90. I am able to identify anomalies in usage and trace them to cyber activity.
91. I am able to identify control system interactions.
92. I am able to identify cyber connections to critical physical systems.
93. I am able to identify the physical impacts of minor systems on critical cybersystems.
94. I can provide updates on vulnerabilities and identify changes in them.
95. I am familiar with countermeasures designed to address identified security risks.
96. In my current job function, I utilize data collected from various cyber defense tools (e.g., intrusion detection system (IDS) alerts, firewalls, and network traffic logs) to analyze events that occur within their environments and mitigate threats.
97. In my current job function, I design, develop, test, and evaluate information system security throughout the systems development life cycle.
98. In my current job function, I am responsible for analyzing, developing, testing, and maintaining system security integration, operations, and maintenance.

99. In my current job function, I develop processes to identify remedial technologies and cyber systems related to control systems.

100. In my current job function, I implement procedures for detecting, reporting, and responding to cybersecurity incidents related to control systems.

101. In my current job function, I perform assessments of systems and networks within the network environment or enclave, identifying where these systems and networks deviate from acceptable configurations, enclave policy, or local policy related to control systems.

102. In my current job function, I translate operational requirements into protection needs (e.g., security controls) related to control systems.

103. In my current job function, I ensure that the stakeholder security requirements necessary to protect the organization's mission and business processes are adequately addressed in all aspects of enterprise architecture, including reference models, segment and solution architectures, and the resulting systems supporting those missions and business processes related to control systems.

104. In my current job function, I test, implement, deploy, maintain, and administer the infrastructure related to control systems.

105. In my current job function, I analyze data from one or multiple sources to prepare the environment, respond to requests for information, and submit intelligence collection and production requirements in support of planning and operations related to control systems.

106. In my current job function, I conduct strategic and operational effectiveness assessments as required for cyber events related to control systems.

107. In my current job function, I determine whether systems performed as expected and provide input to the determination of operational effectiveness related to control systems.

108. In my current job function, I conduct strategic and operational effectiveness assessments as required for cyber events related to control systems.

109. In my current job function, I perform target system analysis and build and maintain electronic target folders that include inputs from environmental preparation and internal or external intelligence sources related to control systems.

110. In my current job function, I develop cyber indicators to maintain awareness of the status of a highly dynamic control system operating environment.

111. In my current job function, I collect, process, analyze, and disseminate cyber threat and warning assessments related to control systems.

112. In my current role, I develop, plan, coordinate, and evaluate cyber training and education courses, methods, and techniques based on instructional needs related to control systems.

113. In my current role, I develop detailed intelligence plans to meet the requirements of cyber operations.

114. In my current job function, I collaborate with cyber operations planners to identify, validate, and layout requirements for collection and analysis.

115. In my current role, I participate in the targeting selection, validation, synchronization, and execution of cyber actions.

116. In my current job function, I synchronize intelligence activities to support organizational objectives in cyberspace.

117. In my current job function, I advise senior management (e.g., CIO) on cost/benefit analysis of information security programs, policies, processes, systems, and elements.

118. In my current job function, I conduct independent comprehensive assessments of the management, operational, and technical security controls and control enhancements employed within or inherited by an information technology (IT) system to determine the overall effectiveness of the controls (as defined in NIST SP 800-37).

119. In my current role, I consult with customers to evaluate their functional requirements and translate these requirements into technical solutions.

120. In my current job function, I conduct software and systems engineering, as well as research on software systems, to develop new capabilities, ensuring that cybersecurity is fully integrated with control systems.

121. In my current job function, I exercise decision-making authority and establish vision and direction for an organization's cyber and cyber-related resources and operations related to control systems.

122. I coordinate with partner target activities and intelligence organizations and present candidate targets for vetting and validation.
123. Red teams communicate vulnerabilities discovered to enable patching or remediation.
124. Red teams meet current mission requests made by our organization.
125. Red team recommendations are tracked to resolution (e.g., patch or remediation) as part of the risk management process.
126. I present a formal debrief to senior management and other key staff members on the cyber Red team's findings.
127. In my current job function, timestamp data is used to reference persistent time-based trends for IT systems (e.g., enterprise network systems).
128. In my current job function, timestamp data is used to reference persistent time-based trends for operational technology (OT) systems, such as cyber-physical systems and control systems.
129. In our organization, time services are managed and provisioned separately for IT and OT systems.
130. Timestamps are a useful tool for identifying trends in system data.
131. A timestamp is a valuable tool for predicting future values of control system data.
132. Traditional statistical forecasting strategies (e.g., dynamic regression) are used in our organization as a baseline for the prediction of network performance.
133. Penetration testing is used in our organization as a method for determining the risk to the network.
134. I perform penetration testing as a method for determining the risk to the network.
135. Network traffic is externally encrypted using a network-based encryption appliance on our organization's network systems.
136. Sensitive network connections between traffic sources and points of encryption are monitored on our organization's network systems.

137. Aggregate subsets of data are logged within each zone and then aggregated to a central aggregation facility for centralized log analysis and reporting on our organization's network systems.

138. Zone perimeters are configured to allow the movement of log files in a one-way direction of transmission, thereby avoiding potential compromise of our organization's network systems.

139. Physical separation using a data diode or unidirectional gateway is used for monitoring across the critical zone boundaries to ensure that all log transmissions occur in one direction within our organization's network systems.

140. Security analytics, such as security information and event management systems (SIEMS), are used for anomaly detection of our organization's network systems.

141. Control system vendors have specific limits on the number of network applications that can be stacked in a given segment.

142. Modbus is used to communicate between sets of known devices, utilizing the expected function codes.

143. Intrusion prevention systems (IPS) deployed on control systems to actively block suspect traffic are tested to ensure that a given signature will not block a legitimate control command.

144. Separate standards regarding the use of programmable logic controls (PLCs) and field devices are established for critical safety operations within control systems.

145. Our organization has well-established incident response plans for control system operations.

146. Control system operations and incident response plans are practiced in our organization.

147. Software-defined networking (SDN) is implemented in our organization.

148. Software-defined networking (SDN) solutions are being evaluated for use within our organization.

References

Control 2019. Purdue Enterprise Reference Architecture (PERA). Control.

Scalco, A. & Simske, S. J. 2019. Cyber-physical Systems/Control System (CPS/CS) Workforce Questionnaire. *Protocol Number 20-102009H.* Colorado State University (CSU) Institutional Review Board (IRB).

Index

C

S

V

W

Z

About the Authors

Dr. Aleksandra Scalco is currently Professor of Systems Engineering at the Department of Defense (DOD), Defense Acquisition University (DAU), USA (since 2024). Prior to DAU, Aleksandra held various government systems engineering and information systems security designer (ISSD) positions and was designated as a cybersecurity subject matter expert for control systems. She draws on more than 25 years of experience in systems engineering to provide insight into cybersecurity for control systems. She has helped teams design, develop, and field solutions to enhance cybersecurity for high-impact, national-level initiatives, employing systems engineering principles, along with knowledge of the "problem space" and available products and practices (the "solution space"). She is the author of numerous publications and has authored curriculum and instructional design content on cybersecurity for critical infrastructure. Dr. Scalco is a Senior Member of the Institute of Electrical and Electronics Engineers (IEEE). She is a member of the seventh cohort of the International Council on Systems Engineering (INCOSE) Institute for Technical Leadership and holds a Certified Systems Engineering Professional (CSEP) designation. She is an Expert Certified in Information Technology Infrastructure Library (ITIL) IT Service Management. Aleksandra was a Colorado State University (CSU) Systems Engineering Outstanding Dissertation Awardee, 2021–2022. She is happily married (Rich).

Dr. Steven Simske is currently (since 2018) Professor of Systems Engineering at Colorado State University (CSU). Prior to CSU, Steve held various positions at HP, HP Inc, and HP Labs, including the role of technical Fellow and Vice President/Director from 2011–2018. Before HP, Steve worked in a NASA CCDS (Center for the Commercial Development of Space) for 20 years (the last 13 years part-time) as a Ph.D. candidate, Post Doc and Research

Professor, flying 18 experiments on 13 Space Shuttle missions (including ISS and KC-135 elements) as payload manager/designer, along with various sounding rockets. His degrees and PostDoc are in Biomedical, Electrical, and Aerospace Engineering. He has more than 240 US Patents and has helped create exercise-responsive pacemakers, impedance tomography systems, several plant and animal modules for spaceflight, digital scanners, digital senders, digital cameras, document digitization services, wireless wearable health monitors, forensic imaging devices, mobile authentication services, and 3D printing materials. Steve has written seven books and four more are in process. Steve is a Fellow of IEEE, NAI, IST, and CSU FIIE, and a winner of the 2022 CSU Best Teacher Award. Steve is happily married (Tess) and has two wonderful sons (Kieran, Dallen).

For Product Safety Concerns and Information please contact our EU representative GPSR@taylorandfrancis.com Taylor & Francis Verlag GmbH, Kaufingerstraße 24, 80331 München, Germany